# Fundamentals of Cybernetics

# Fundamentals of Cybernetics

A. Ya. LERNER

*Translated from Russian by* E. GROS

*Translation edited by* F. H. GEORGE

A PLENUM/ROSETTA EDITION

Library of Congress Cataloging in Publication Data

Lerner, Aleksandr IAkovlevich.
Fundamentals of cybernetics.

Translation of Nachala kibernetiki.
"A Plenum/Rosetta edition."
Bibliography: p.
Includes index.
1. Cybernetics. I. Title.
Q310.L3913 1975 001.53 75-30901
ISBN 0-306-20018-X

Reprinted by arrangement with Scientific Information Consultants Ltd.

First paperback printing 1975



A Plenum/Rosetta Edition
Published by Plenum Publishing Corporation
227 West 17th Street, New York, N.Y. 10011

United Kingdom edition published by Plenum Press, London
A Division of Plenum Publishing Company, Ltd.
Davis House (4th Floor), 8 Scrubs Lane, Harlesden, London, NW10 6SE, England

# Contents

# Foreword

by the Editor of the English translation

In editing this translation I have tried to assure a clear English version of Professor Lerner's excellent Russian text, in rendering which the Translator deliberately used everyday English so far as practicable. The aim was to retain the fullness of the Author's meaning while preserving every technical detail.

The book itself, which is called *Fundamentals of Cybernetics*, is a treatment which has the breadth it claims. In saying this one has to add that while it successfully covers the bulk of cybernetic material, it does so, inevitably, from a particular point of view. The point of view adopted is that of the engineer and the applied mathematician. The emphasis is very much on control theory, and although the book carefully eschews detailed mathematical analysis, a certain number of equations are included which amplify the descriptions given. These equations which are the mathematical descriptions of the control functions involved can generally be omitted without vitiating the reader's understanding of the text.

One of the most interesting features of this book is the approach to theory of games, communication theory, behavioural and biological systems, all through the eyes of engineering cybernetics.

This is a book which provides a good insight into the contemporary Russian approach to cybernetics, and one that should prove of interest and value to all Western readers.

F. H. GEORGE

# Preface to the English edition

Is it possible to build a machine which will be cleverer than man?

Do we live at the dawn of the 'era of robots'?

Is the machine a danger to people who are fighting for their existence?

Such and similar questions worry many people whose field of activity is far removed from cybernetics. These questions are hotly discussed, and many popular and semi-popular books and articles have been written about them.

On the other hand, more topical although less sensational problems as, for instance, the use of computing techniques and data processing systems for controlling production, for commercial and banking uses, in medical diagnostics, etc., seem to attract less interest.

In order to be able to examine knowledgeably vital contemporary and long-term problems which are associated with the emergence and development of cybernetics, it is only necessary to know the concrete contents of those ideas and methods which form the basis of this remarkable science. To impart this information to the reader was the task which the author of this book has set out to accomplish.

The author is very glad that readers in English speaking countries will now be able to acquaint themselves with the contents of this book. If the book will give even a little help towards developing science and utilising its achievements in the interest of progress, the author will consider that his work has contributed to the efforts of other colleagues who devote their lives to this aim.

A. Ya. LERNER

# Preface to the Russian edition

The need for a book which presents the fundamental concepts of cybernetics with a sufficiently strict scientific treatment, yet in a form which is acceptable to people who have no deep specialist knowledge of mathematics, has been obvious for a long time. However, most people have recognized the difficulty of writing such a book.

How can one single subject encompass such problems as modelling cybernetics on one hand and coding signals on the other? How should the theory be presented without the use of differential equations or other mathematical methods? Is it possible to select examples that reveal sufficiently clearly internal similarity of the 'behavioural' features of the different economic, biological and engineering systems? These systems apparently have nothing in common, but must be shown to be similar to each other, as they in fact are.

The problem of presenting the main ideas and methods of cybernetics to a wide circle of readers is considered sufficiently important to justify this attempt.

Obviously the author has tried not to omit important or interesting sections of cybernetics. But this was not the main task. The most important consideration was to find methods of presentation to make the book suitable for teachers and engineers, doctors, students and managers. The author wished particularly to assist people with inadequate mathematical knowledge and without the necessary time to study the works of Wiener, Shannon, Andronov, Kolmogorov and others, but who, in spite of this, wanted to study the methods of solving cybernetic problems. These problems were to be studied more deeply than can be done, for instance, from reading the excellent work of Stanislaus Lem.

Another secret aim of the author should also be mentioned: he wanted to dampen enthusiasts who see cybernetics as a new magic or a panacea. On the other hand, he wished to reduce the pessimism of sceptics who do not anticipate anything useful from cybernetics. Extreme views about the evolution of cybernetics are usually due to ignorance of the real content of this science.

Of course, the contents of the book and the method of presentation of the material may become the subject of sharp criticism and may give many pleasant minutes to friends and colleagues of the author, who will find in it a sufficient cause for discussion.

Thanks to the participation of the many people involved in the production, the book has been completed fairly quickly. Some of the people helped to collect examples and problems (L. Mikulich, L. Dronfort, V. Epshteyn, V. Burkov, *et al.*), others have read the text and made critical comments (Ya. Khurgin, A. Butkovskiy, B. Biryukov, B. Kogan), while others helped with shaping the text (E. Mezenov and I. Doronkin). The reviewer, A. Kobrinskiy, and the editor, L. Levitin, also worked very hard on the manuscript. L. Levitin and Yu. Shmukler not only improved considerably the presentation of the material, but also filled a number of gaps in the contents, especially in Chapters 18 and 19. To all these friends the author is deeply grateful and takes full responsibility for all the inadequacies of the book and errors which apparently occur, in spite of all the help he received from so many quarters.

A. Ya. LERNER

# 1 Introduction

The development of science consists not only of deepening and widening the already established scientific disciplines but also depends on the emergence of new ones. The emergence and development of new sciences is influenced primarily by two factors: isolation and generalisation.

Isolation of scientific disciplines is due to the discovery of new objects of investigation and the emergence of specific scientific trends. This leads to the study of a relatively narrow class of objects which are characterised by their specific approach to both the formulation and the solution of problems. Examples of this type of specific scientific disciplines include, for instance, chemistry of high molecular compounds and the theory of electrical machines, which are both devoted to the study of a relatively narrow field. In addition there are the more general scientific disciplines, whose characteristics are that they are created for the purpose of studying such natural phenomena as occur in a very wide class of objects. Disciplines of this type are, for instance, the theory of dimensions and the theory of similarity, the theory of dynamic systems and thermodynamics.

The very general, as opposed to the very specific, sciences tend by their nature to be more theoretical and depend much more on the language, mathematical or otherwise, used to describe them.

Cybernetics belongs to this same category of generalising disciplines. The founder of cybernetics, Norbert Wiener, defined it as a science of control and communication in mechanisms, organisms and society. At present cybernetics represents the general theory of control which can be applied to any system. Here, by 'system' we mean a group of elements of whatever kind, considered as an interconnected whole.

Strictly speaking, the elements unified in the world around us form interconnected wholes and should be considered as systems. However, for solving many theoretical and practical problems it is useful to consider systems made up of smaller groups of elements. So we consider a growing tree or business, as opposed to a whole city or a country. The more the elements are interconnected inside the system the greater will be the isolation of this group of elements from the outside world, and thus the greater will be the justification to consider it as a 'separate system'.

In studying the control processes in systems, by any type of cybernetic methods, man aims to recognise the objective relationships characteristic of control process and then to use these relationships for improving the natural and creating artificial control systems for his biological and social aims. Talking of control, it is necessary to consider that any control will follow from information on the selection of the control actions, even the control actions themselves are based on information contained in the control commands.

The source of any information is observation accompanied by active experiment. Therefore, control is always associated with the use of observations, the use of information both on the controlled system and on the external medium with which it interacts, as well as on the results of the control actions. The exchange of information between the system and the medium, and between the internal parts of the system, is made possible by means of various types of connections through which information circulates. The presence of such connections is a characteristic feature of any cybernetic system.

Feed-back – a channel along which data on the results of control are fed back into the system – is particularly important in cybernetic systems. It is due to feedback that cybernetic systems are, in principle, capable of going beyond the limits of actions predetermined by the designer. It is this feature, above all others, which underlines the enormous potentialities of cybernetic systems.

## 1.1. The sources of cybernetics

Cybernetics is a science which began to develop only after the second world war. Its development, however, has been so rapid that it already exerts a great influence on the methods of investigation and the solution of practical problems in the most varied branches of science and engineering: in biology and medicine, in communications and automation, in computing and economics. Cybernetics is based on the idea that it is possible to evolve a *general* approach to the investigation of control processes in *various* types of system. The importance of this idea is that, in addition to general methodological considerations, it offers a powerful tool for quantitative description of processes relating to the solution of complex problems, which is based in turn on methods of information theory, the theory of dynamic systems, the theory of algorithms and the theory of probability.

The birth of cybernetics is generally associated with the date of publication (1948) by Norbert Wiener of his remarkable book *Cybernetics, or Control and Communication in the Animal and the Machine.* In this work, the outstanding American mathematician gives a clear outline of the means of developing a general control theory, and laid the foundations for the methods of considering problems of control and communication for various systems from a single unified point of view.

These problems attracted the attention of Wiener, Rosenblueth, and other scientists who contributed to the birth and development of cybernetics, not only by their desire to understand the trend of development and the methods of the science, or by their aspiration to generalise upon the achievements of various branches of science (this also played a role), but by a direct stimulus for intensive work on control problems from the most generalised points of view. All this was also meant to provide the solution to concrete practical problems, as for example in the production and use of computers and in particular in the use of computing devices for directing the fire of anti-aircraft guns, the separation of a useful signal from the accompanying noise, the design of machines for reading aloud, some problems of neurophysiology, etc.

Without detracting from the work of Norbert Wiener and his colleagues, it must be pointed out that a number of scientific trends which form the basic concept of cybernetics has been worked out many years before and some of them even centuries ago. As long as 100 years ago (starting from the work of Maxwell and Vyshnegradskiy) the theories of control and feedback systems were developed. Over 30 years have passed since the publication of work on the application of logical algebra to the investigation of switching circuits (the work of Shestakov and Gavrilov in the USSR and of Nakashimo in Japan). The idea of designing digital computers was dealt with by Pascal and Leibniz in the 17th century and in a more sophisticated form by C. Babbage in the 19th century. However, it was the work of Wiener which produced a 'chain reaction' in the formation of a generalised theory of control.

There was rapid development and cross fertilisation in the theories of information, switching circuits, automatic control and neural networks. New engineering tools appeared in the form of analog and digital computers; it then became possible to carry out cybernetic experiments which were based on modelling of control processes by means of computers.

Despite the fact that many classical scientists regarded this new science with suspicion, cybernetics continued its advance, proving the right to its existence, not only by its theoretical results but also by its enormous contribution to the solution of many complex practical problems. The development of large digital computers, optimal and self-adaptive control systems, effective methods of operations control and many other important scientific and practical results, were due directly to the development of work in the field of cybernetics: therefore this science has justified itself.

## 1.2. The aims and tasks of cybernetics

One of the basic ideas introduced into our 'Weltanschauung' by cybernetics is the new outlook it engenders on the components of the world around us. The classical concept of the world consisting of matter and energy has had to give way to the concept that the world consists of three components; energy, matter and information, since without information, organised systems are unthinkable and the living organisms which can be observed in nature and in control systems produced by man, represent organised systems. Moreover, these systems are not only organised but they remain in this state with the progress of time, and do so without losing their organisation as would be expected in accordance with the second law of thermodynamics.

The only possible mechanistic explanation of the fact that the organised state is retained is the continuous drawing of information from the outside world. This information refers to the phenomena occurring in that outside world and the processes occurring in the systems themselves.

One of the basic features of cybernetics is that it does not only consider control systems in their static state but also during movement and development. Consideration of systems in the process of change fundamentally alters the approach to their study. In a number of cases such a dynamic approach also reveals relationships and facts which otherwise would remain undiscovered. Such a functional property of systems as *stability*, which is of decisive importance for evaluating the serviceability of many systems and even for elucidating the possibilities of their existence over a long period, would be impossible without considering the dynamics of their internal organisation.

Cybernetics does not consider isolated systems, but groups of systems which generally speaking encompass the whole universe. This science should and does consider the multitude of interconnections which will necessarily occur between individual parts of complex systems, and it attempts to determine their properties, their behaviour, their development, destruction, and as well, it studies their reproduction.

The distinguishing feature of the cybernetic approach is the relative nature of the point of view, in the sense that one and the same group of elements will sometimes be thought of as a system, while on other occasions it will be thought of only as a part of a system, or subsystem. Thus for instance a pneumatic drill can be considered as a dynamic system in itself, but we can also consider as a system the man who performs the work by means of the pneumatic drill. This group in turn is part of a system comprising the organisation responsible for the work which this man performs, etc. The properties and features of any object cannot be correctly evaluated and taken into consideration without considering the multitude of connections and interactions which may form between individual objects and the medium surrounding them. Consideration of the influence of the medium is characteristic of the cybernetic approach to the investigation of phenomena occuring in controlled systems.

However detailed and exacting the study of the behaviour of the system may be, we will never manage to take into consideration all the factors which influence its behaviour, either directly or indirectly. Therefore it is necessary to consider the inevitability of the existence of some random factors due to causal processes which have not been taken into consideration. Cybernetics makes wide use of statistical methods for investigating the behaviour of systems subjected to random stimulation. Statistical methods permit forecasting – although only from the probability aspect – strictly and accurately the 'average behaviour' of complex systems.

The ideas and methods developed in cybernetics are directed to achieving the following aims:

(*a*) To establish important facts, which are general for all or for some classes of control systems. As in any theory, factual data are vital and serve as a basis for advancing hypotheses, construction of theories and the establishing of rules.

(*b*) To reveal limitations characteristic of controlled systems and establish their origin, i.e. thus also to establish those boundaries within whose limits the designer is free to select the control equipment which is capable of changing the controlling action, and within which the controlled system can change its state.

(*c*) To find general laws to which the systems comply. Based on factual data, by proposing the respective axioms, evolving proofs of concepts which are based on the accepted axioms, cybernetics, as in the case of any other accurate science, can and should gradually create a valid set of theoretical concepts, laws and principles which will form the central core of the science.

(*d*) To indicate ways of utilising facts and relationships which form the theory for practical activity of man. This practical trend of cybernetics will obviously be no less important than its theoretical development. It is obvious that it would be senseless to study the behaviour of systems, to establish facts and relationships if these could not be utilised for practical purposes. However, theory itself will not provide a direct solution of many applied problems. To solve practical problems it is necessary to build a bridge between the theoretical concepts and the applied methods used for the solution of the problems. In doing this it is necessary to take into consideration the specific features of certain classes of control systems. Therefore application of general methods of cybernetics for solving practical problems is studied in the applied sciences of *engineering cybernetics, economic cybernetics, biological cybernetics*, all of which in turn have originated from cybernetics itself.

## 1.3 Cybernetic systems

Cybernetics, being the science of control, does not study all systems generally, but only *control systems*. However, the range of application of cybernetics extends to a great variety of systems: engineering, biological and economic systems in which control exists.

One of the characteristic features of a control system (and the system controlled) is its ability to change its movement, to pass into various states under the effect of control actions. For instance, a motor car can assume various positions in space, it can move in various directions and with various velocities, depending on how it is driven. A military unit will carry out a specific manoeuvre at a given command which differs from manoeuvres which the unit is capable of carrying out if different commands are given. The temperature in a refrigerator can be lowered or raised depending on whether the cooling unit is switched on or switched off.

There will always exist a certain number of movements and if we talk of a control system, then a preferred movement is chosen. *If there is no choice, then there can be no control.*

Since all processes in the world are interconnected and influence each other, if any controlled object is chosen, then we have to take into consideration the influence of the surrounding medium on this object, and the influence of the object on the surrounding medium. Therefore a study of the behaviour of any controlled system has to be carried out taking into consideration its ties with the medium.

Every cybernetic system is therefore characterised by the properties which make up the system and the ties which reflect the interconnection between the system and the medium. A concrete cybernetic system will consist of concrete objects such as machines, natural resources, people, etc. Its relations with the surrounding medium are expressed in the form of certain physical-chemical quantities (forces, energy or matter flows, etc.).

Digressing from the concrete features of individual cybernetic systems and singling out the general relationships which describe the changes in state in the case of various control actions, we arrive at the concept of an *abstract cybernetic system*. Its distinguishing feature is that its components are not described in terms of the names of the objects but are represented as an abstract grouping of elements which are defined by certain properties which are common for a wide class of objects. The relations between the abstract cybernetic system and the medium are also defined in the form of *quantitative* characteristics irrespective of the qualitative nature of the concrete couplings.

The switching from the investigation of concrete systems to an abstract cybernetic system is of the same nature as the transition from studying operations with concrete numbers, as in arithmetic, to operations with abstract numbers, as in algebra. For brevity we will in the following omit the word 'cybernetic' from the term 'cybernetic system'.

Fig. 1.1 shows diagrammatically a system in the form of a part of space in which all its elements are concentrated, and the connections of the system with the ambient medium. The arrows on the lines, representing couplings, show the direction in which the actions are transmitted. $X$ denotes the actions of the system on the medium, $Y$ the actions of the medium on the system under consideration. These connections may express effects lumped into certain points in the system, for instance, in the form of a force applied to an element of the system; the $Y$ may also be distributed, acting on a surface or on each point of the entire system or on some of its parts. For instance, the effects of the temperature or pressure on the surface of some parts of the system, the effects

Fig. 1.1. *System and medium: Y-Input, X-Output.*

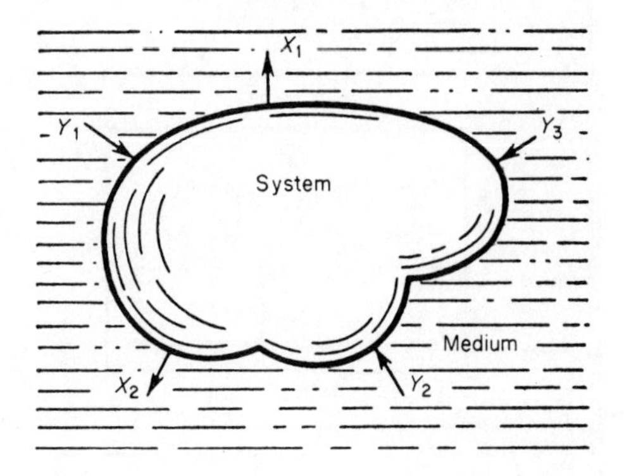


of the gravitational or magnetic field, etc., can be considered as distributed effects.

With cybernetic systems, we consider control and controlled systems and these should always contain a mechanism which realises the control function. In some cases this function is accomplished incidentally by elements provided for other purposes, but frequently this controlling mechanism is localised in elements which have been specially designed for control. In the latter case, a controlled system can be schematically represented as an assembly of the controlling (a) and the controlled (b) part of the system, as is shown in Fig. 1.2. Here the arrows indicate the paths of the actions exchanged by the parts of the system.

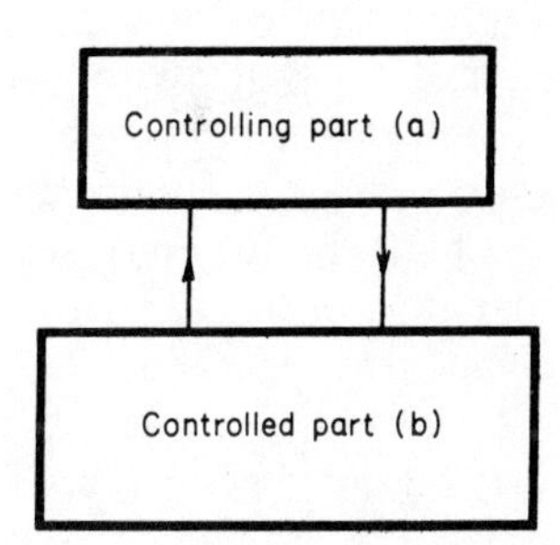


Fig. 1.2. *Simplest structure of a cubernetic system*

It is pointed out that such very simple control or controlled systems never in fact exist in isolation. They interact with the surrounding medium and with each other, they may make up more complex control systems, form part of a control, or of a controlling part of a complex system, or form a hierarchy of controlled and control systems as shown in Fig. 1.3.

'Cybernetic systems' have so far not achieved such an accuracy and definiteness as did the 'systems of equations' in mathematics, or 'systems of material bodies' in mechanics. The applicability of this concept to a specific system depends not only on the system itself, but also on the point of view and the aims of the research worker who investigates this system. Therefore one and the same system should not always be considered as a single system in

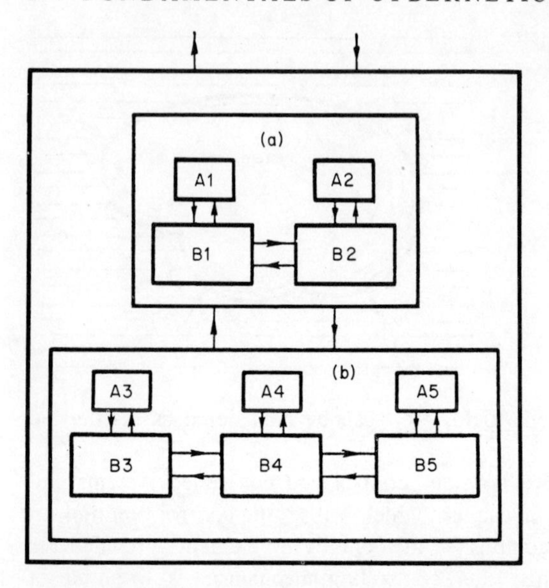


Fig. 1.3. *Hierarchy of cybernetic systems*

cybernetics. Thus, for example, an aircraft can be considered as a controlled, and consequently a cybernetic system, during the design of the auto pilot, or when solving the problem of selection of the optimum flight program. However, when solving other problems the same aircraft can be considered as a body characterised by a certain resistance to the air flowing past it or as a design with a certain rigidity.

In this way the term 'cybernetic system' defines not so much a definite class of systems but more an approach to its investigation; an approach based on the study of the properties and features of the system as a *controlled one*.

Obviously not every system will be controllable. The necessary condition is that it is *organised*, i.e. the presence of a specific structure expressed by its component elements and the connections between them. The concept of organisation is very difficult to define accurately, but from intuition it is clear that organised systems are far from being in 'thermodynamic equilibrium'. For instance, a gas consisting of molecules moving at random can be considered as having a zero degree of organisation, whilst organisms capable of maintaining their existence and reproducing themselves have a high degree of organisation.

Obviously not all the organised systems are cybernetic ones although all cybernetic systems have a certain degree of organisation.

## 1.4. Exposition

This book will touch on a wide range of problems which, in the view of the author, form the main content of cybernetics. Due to the breadth and the variety of these problems, it is obviously not possible to present the material in a book of a limited size as completely as some of these important problems deserve to be presented. The aim is to make this book accessible to a wide range of readers who do not have a great knowledge of mathematics and this has forced a limit on the number of problems which can be explained, where we must do so without using the apparatus of differential equations, statistical dynamics, etc. As a result, a number of interesting problems could not be dealt with at all, and some are considered in a simplified form. Nevertheless, an attempt is made to present not only ideas which are developed in basic works on cybernetics, but also to communicate, as far as possible, information and methods characterising the apparatus used for the quantitative analysis of phenomena in cybernetic systems.

The concepts described in the book are presented in the order of increasing complexity. The exposition begins with the fundamental concepts of the kinematics of control and controlled systems, where an idea is given on the form of movement of systems and methods of describing these. Later on one of the fundamental concepts of cybernetics, the concept of models, is introduced, which determines the approach to an investigation of cybernetic systems of any type.

The dynamics of the movement of systems is considered on the basis of the geometrical interpretation of movements – the *phase space method*. Introduction of the information concept, one of the fundamental concepts of cybernetics, permits the use of quantitative evaluations for communication contained in the signals, and for control purposes.

After the first five chapters, which are essentially introductory, consideration of the more serious control problems begins. Here the essence of the fundamental method of control, based on the use of feedback is explained and the fundamental ideas of optimal control are presented. In the subsequent chapters, the fundamental concepts of discrete automata, including such universal automata as the digital computer, are described.

In the chapters on 'adaptation', 'games', and 'learning', the basic ideas and methods of constructing artificial systems, which simulate the intellectual functions of the brain, are described.

Chapters 14 and 15 are devoted to problems of controlling large systems. The last three chapters deal with the biological, psychological and social aspects of cybernetics. Here the idea of formal neural networks is presented, the problems of man-machine symbiosis is considered and the future prospects of development of cybernetics are outlined.

The book is written on the assumption that the reader will read it in the proper sequence. Therefore in the subsequent chapters, the author based his exposition on material presented in the previous chapters. In places where a term could not be considered to be generally known, it is defined. However, this definition is not always given where the term is mentioned for the first time. Therefore, if the reader experiences difficulty in understanding the meaning of the terms, he should refer to the subject index where the page on which the definition is given is marked.

To facilitate an understanding of the subject, each chapter contains problems and exercises. It is recommended that the reader works through the exercises and problems not only because it will help him to understand better the information presented in the basic text, but also because he will recognise some new facts or he will himself obtain additional results in answer to the question.

Some of the readers may be interested in studying more deeply the individual problems dealt with in this book. To guide them through the extensive literature on cybernetics and allied disciplines, each chapter contains a list of recommended works which contain more detailed information on the problems dealt with in that particular chapter.

# 2 Movement

The term 'motion' in mechanics is used in a narrow sense and means the *change* of the position of any object in space with the progress of time. In cybernetics, 'movement' has a more general meaning – *any* change in an object with the progress of time.*

Motion is for instance a change in the temperature of a body, a change in the charge of a condenser, a change in the volume or pressure of a gas, a change in one's bank balance, a change in raw material reserves, and finally such processes as life and thought also can and must be capable of showing change.

Since the movements of a great variety of objects are subjected to many generally valid laws, particularly from the point of view of controlling the processes involved, it may be useful to consider the laws of movement not of real systems (of which there are many) but of abstract cybernetic systems, which were described in section 1.3. The methods of describing movement, used in this chapter, will be essential to the understanding of the material given in all the subsequent chapters of the book.

## 2.1. The state of a system, the state space

The state of any system can be described with a certain degree of accuracy, by a group of values which determine its behaviour. By means of these values the states of individual systems can be compared and the differences between them judged. They also enable comparison of the states of one and the same system at different times in order to elucidate its degree of change, which we may, if we choose, think of as *movement*.

There are many different ways of describing the state of a system. For

* This essentially is the traditional point of view of dialectics. The well-known Hegel formula: "Movement is change generally".

instance, it is possible to list the range of values of the quantities $X_1, X_2, \ldots,$ $X_n$, which determine the state of a system at a given instant of time and then list their values for different instants of time. In this case the sequence of the states of the system can be represented as a Table. For instance, the state of a patient suffering from acute nephritis can be characterised by the data given in Table 2.1.

Table 2.1 *Sequence of the states of a patient suffering from acute nephritis*

| | | | *Arterial blood* | | |
|---|---|---|---|---|---|
| *Date* | *Time* | *Temperature* °C | *pressure maximal* | *mm Hg minimal* | *Residual nitrogen content of the blood mg %* |
| 7·2 | 9 a.m. | 37·8 | 190 | 120 | 103 |
| 8·2 | 9 a.m. | 37·6 | 180 | 120 | 95 |
| 9·2 | 9 a.m. | 37·4 | 170 | 95 | 89 |
| 10·2 | 9 a.m. | 37·0 | 165 | 90 | 90 |
| 11·2 | 9 a.m. | 37·1 | 175 | 100 | 102 |
| 12·2 | 9 a.m. | 37·6 | 180 | 100 | 100 |
| 13·2 | 9 a.m. | 37·5 | 170 | 90 | 91 |
| 14·2 | 9 a.m. | 37·1 | 160 | 90 | 84 |
| 15·2 | 9 a.m. | 37·1 | 150 | 90 | 60 |
| 16·2 | 9 a.m. | 36·7 | 145 | 85 | 49 |

The state of a system can also be described graphically by representing the values of each of a sequence of quantities $X_1, X_2, \ldots, X_n$, by a point on straight line, whose position on a given scale corresponds to the quantity $X_i$ ($i = 1, 2, \ldots, m, n$). If the state of the system changes with the progress of time, the movement can be represented by a family of graphs: $X_1(t), X_2(t), \ldots, X_n(t)$, where $t$ is the time, referred to a certain instant, which has been arbitrarily chosen as the initial reference point for time. Fig. 2.1 shows a family of graphs which depict the movement of a ship. Here $X_1$ is the source of the ship, $X_2$ its speed, $X_3$ and $X_4$ the longitude and latitude of its location.

Usually another method of representing the state and movement of a system is used, this being based on the concept of the *state space of the system*. This is more convenient for a given purpose.

The method depicting the value of any variable quantity by a point on a straight line has already been used above, and this represents a space with one dimension, or, as we usually say, a one-dimensional space. If a point has to be represented as a set of two quantities ($X_1$ and $X_2$), then the one-dimensional space will be inadequate and in the given case a two-dimensional space must be used. Let us take a plane with a system of rectangular coordinates on it, as shown in Fig. 2.2. Here the point a depicts the state of the system characterised by the values: $X_1 = X_{1a}$, $X_2 = X_{2a}$, and the point b corresponds to $X_1 = X_{1b}$, $X_2 = X_{2b}$.

If the state of the system is determined by the values of three quantities, it

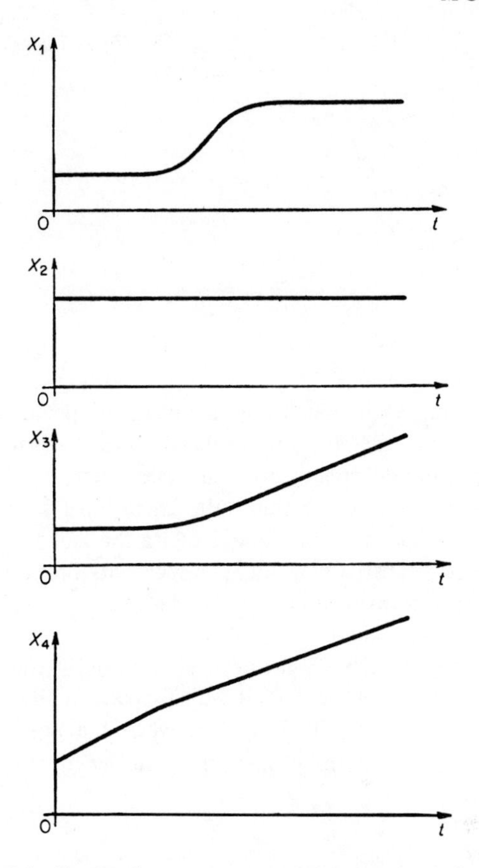


Fig. 2.1. *Graphical representation of the movement of a ship.*

will obviously have to be depicted by a point in the three-dimensional space, as is shown in Fig. 2.3.

We will designate the number of values which determine the state of the system under consideration by $n$. For $n = 1$, $n = 2$ and $n = 3$, the state of the system can be clearly represented in graphical form, but when the number of dimensions is greater than 3 this graphical representation cannot easily be used.

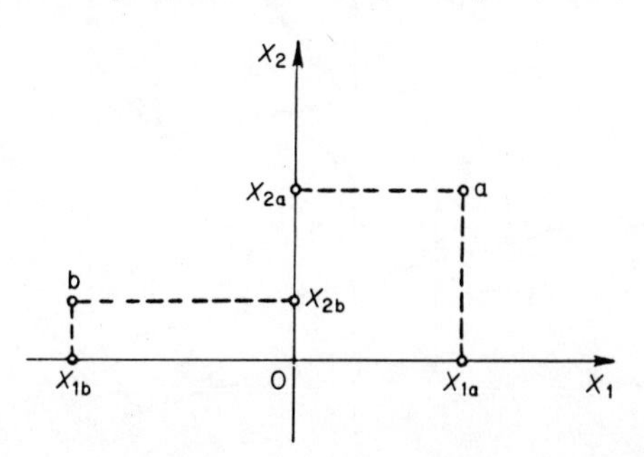


Fig. 2.2. *Two-dimensional space state of a system.*

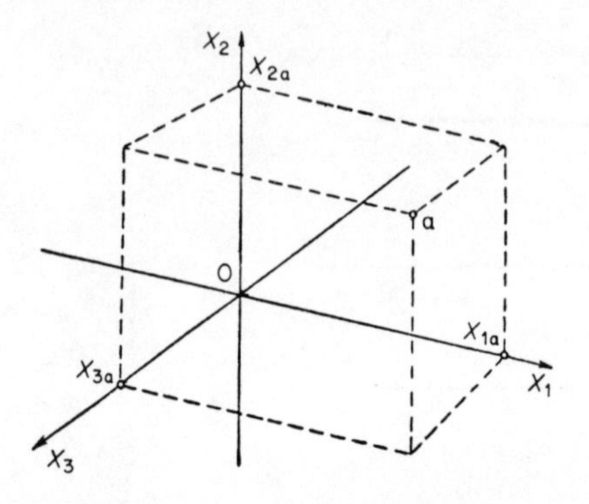


Fig. 2.3. *Three-dimensional space state of a system.*

In spite of this, important conclusions about the properties of the system can be drawn from reasoning based on using the state of the system in the form of points in the appropriate space, even in cases when $n > 3$. This obviously necessitates using the concept of the multi-dimensional space (or, as it is also called, *hyperspace*), which may be done. Despite the fact that the notion of an $n$-dimensional space is abstract, in many respects its properties are similar to those to which we are accustomed in the familiar one-, two- or three-dimensional spaces.

In particular, one of the basic geometrical concepts – the distance between two points – can be used in an $n$-dimensional space in the same way as in a three-dimensional space. The distance $d$ between the points $a$ and $b$ in the three-dimensional space is nothing more than the length of the diagonal of a

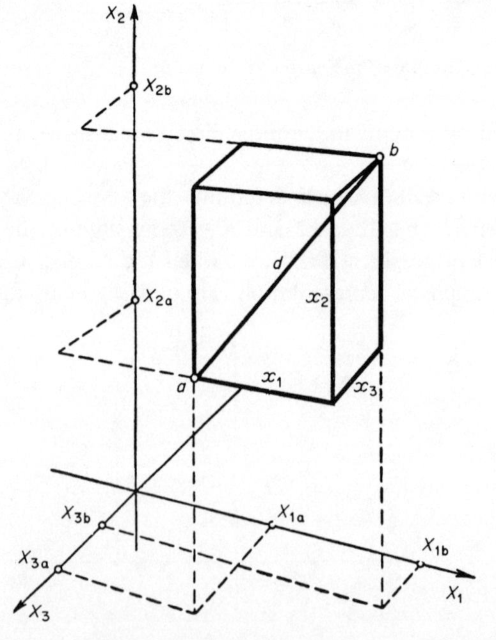


Fig. 2.4. *Distance between points in Euclidean space.*

parallelipiped, shown in Fig. 2.4, the apices of which contain these points, and the edges are parallel to the coordinate axes.

As is familiar from solid geometry, the length $d$ of the diagonal of a rectangular parallelipiped with the edges

$$x_1 = X_{1b} - X_{1a}, \quad x_2 = X_{2b} - X_{2a}, \quad x_3 = X_{3b} - X_{3a}$$

can be determined from the formula

$$d = \sqrt{x_1^2 + x_2^2 + x_3^2}. \tag{2.1}$$

In the same way, for an $n$-dimensional space, the distance between the point $a$ with coordinates $(X_{1a}, X_{2a}, \ldots, X_{na})$ and the point $b$ with coordinates $(X_{1b}, X_{2b}, \ldots, X_{nb})$ can be determined as the value

$$d = \sqrt{x_1^2 + x_2^2 + \ldots + x_n^2}, \tag{2.2}$$

where $x_1 = X_{1b} - X_{1a}$, $x_2 = X_{2a} - X_{2a}$, ..., $x_n = X_{nb} - X_{na}$. The values $x_1$, ..., $x_n$ are equal to the length of the sides of an $n$-dimensional rectangular parallelipiped, similar to the three-dimensional parallelipiped shown in Fig. 2.4.

The space in which the distance is determined by Eq. 2.2 is called Euclidean space. In this space all the theorems of classical Euclidean geometry, in particular the Pythagorean theorem, are valid.

The space in which each state of the system is represented by a single point will be referred to as the *state space of the system.* The number of dimensions of the space states equals the number of the independent quantities which determine the state of the system. These independent variables are frequently referred to as the *degree of freedom* of the system. Each state of the system is characterised by a set of definite values of the variables $X_1, X_2, \ldots, X_n$. In state space this corresponds to a point with the same values of the coordinates $X_1, X_2, \ldots, X_n$. This point is called the *representative point* (it 'represents' the given state of the system), and the variables $X_1, X_2, \ldots, X_n$ are called the *coordinates of the system.*

In real systems not all the coordinates can change without limit (for instance $-\infty < X_i < \infty$). Most of the coordinates can only assume values within a limited interval, i.e. an interval which satisfies the condition

$$X_i' \leqslant X_i \leqslant X_i''$$

where $X_j'$ and $X_i''$ are the boundaries of the interval of the possible values of the coordinates $X_i$. The area of the state space in which the representative point can lie is called the *area of permissible states.* In what follows, when speaking about the state space, we will only have this permissible area in mind.

Even within the limits of the permissible states, an arbitrary point will not always represent the possible state of the system. Only *continuous state spaces* corresponding to a system whose coordinates may assume *any value* (within the permitted limits) will have such a property. However, there are systems, referred to as discrete, in which the coordinates can assume only a finite number of fixed values. The state space of such systems is discrete. In this case the representative point can only occupy a finite number $S$ of positions

$$S = s_1 \cdot s_2 \cdot \ldots \cdot s_n,$$

where $s_i$ is the number of discrete states of the $i^{\text{th}}$ coordinate.

If the systems are changing then the coordinates change with time. In this process the representative point changes its position in state space, describing, as a result, some trajectory.

As an example we consider the process in a piston compressor cylinder, as shown in Fig. 2.5. The state of the system can be described by two values: the distance of the piston from the rear wall of the cylinder $l$ and the air pressure $p$ in the cylinder. Consequently the state space is two-dimensional, and the representative point has the coordinates $X_1 = l$ and $X_2 = p$. The operating cycle begins with the movement of the piston 3 to the right. The air under the

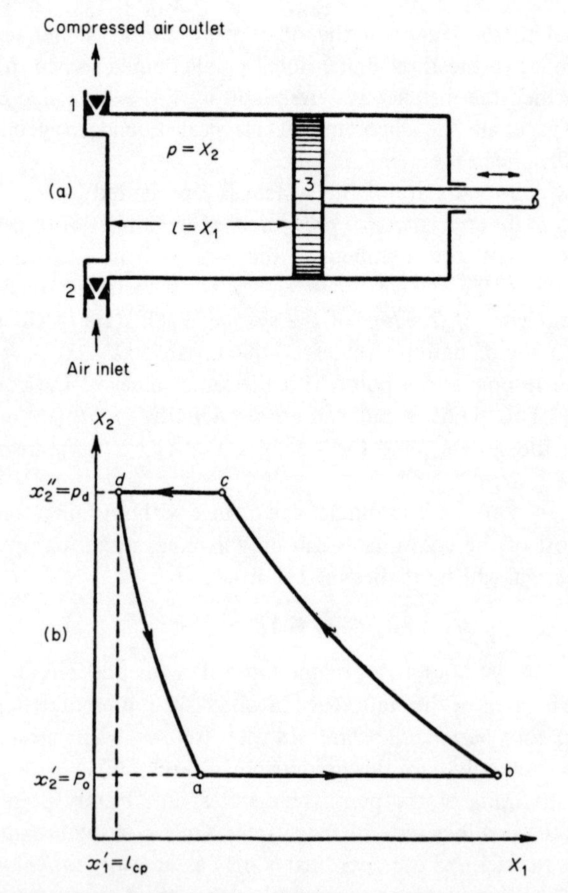


Fig. 2.5. *Example of the description of change in the state space; (a) Sketch of a piston compressor; (b) Operating cycle.*

pressure $p_d$ (point d) which remains in the 'dead space' $l_d$ expands to the intake pressure $p_0$ (point $a$). Fresh air then enters the cylinder through the inlet valve 1. On the indicated diagram this is represented by the line $a - b$.

When the piston moves to the left the air is compressed to the pressure $p_d$ (point $c$), is forced by the piston through the outlet valve 2 (line $c - d$) to the end of the compression stroke and then ejected. Usually the piston does not travel right up to the rear wall of the cylinder but remains at distance $l_d$, a certain volume of compressed air remains in the cylinder and this is the dead volume.

This terminates the working cycle, and all the processes then repeat themselves in the same sequence. The representative point describes on the plane $X_1$, $X_2$ a closed trajectory in the direction indicated by the arrows. In the example under consideration the representative point always returns to one and the same position, which means that the same states repeat themselves many times in the system.

## 2.2. Input and output values

The motion of a system – changes in its state – may occur under the influence of external factors as well as a result of processes occurring in the system itself.

Strictly speaking, each system is influenced by an infinite number of external forces, but not all of these are equally important. It is obvious that the force of attraction of the moon has no great effect on the movement of the motor car relative to the Earth, although in principle such an influence exists. Only those external influences which have an important influence on its state of the system must be considered, with respect to a particular problem. These external forces are called input values or input signals (or sometimes input variables) of the system, and the elements of the system to which these input signals are applied are called system *inputs*.

For instance, the movement of an aircraft is greatly influenced by such factors as the force and direction of the wind, the density of the atmosphere, the position of the rudders, elevators and ailerons, and the engine thrust. All these factors are considered as the input variables of an aircraft.

It is frequently advisable to consider as the input quantities of a system not the coordinates $X$, which determine its state, but some other quantities $Z$ which are determined uniquely by the coordinates of this system. Each of $l$ input values $Z_i$ is associated with the coordinates of the system by the functional dependence

$$Z_i = \phi_i(X) \quad (i = 1, 2, \ldots, l). \tag{2.3}$$

The controlled system in this case can be represented in the form of a part $S$, which transforms the input actions $Y$ to the coordinates $X$, and a set of functional converters $\phi$ which transform the coordinate systems into output values, Fig. 2.6.

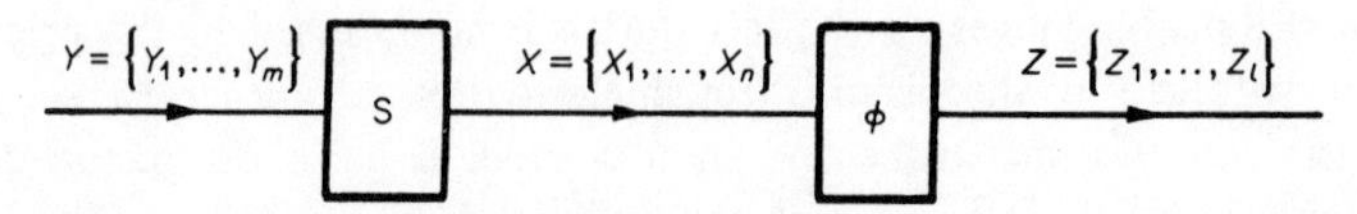


Fig. 2.6. *Sketch showing the transformation of input into output values:* Y – *input values,* X – *coordinates of the controlled system,* Z – *output values,* S – *input value – coordinate system converter,* $\phi$ – *coordinate output value converter.*

The need to introduce into the investigation output values which do not directly enter into the set of coordinates that determine the state of the system, arises when the control problem is not to bring the system into a given state but to achieve aims which are functionally associated with the state of the controlled system.

The aim of controlling the manufacture of synthetic fibres, for example, is to obtain a fibre of the required strength $Z_1$ and elasticity $Z_2$. These quantities depend functionally on the coordinates of the process: the temperature of the mass ($X_1$), the admixture content ($X_2, X_3, \ldots$) in the basic raw material, etc. It is obvious that in such cases it is necessary to distinguish the input quantities from the coordinates which characterise the state of the system.

In the solution of control problems it is important to distinguish two types of input signals: those which are part of the *control* and those which are *disturbances.* The control signals include quantities which can be regulated during the control of the system and changed so as to realise a movement which is preferable compared to the other possible movements of the system. In the example of the aircraft mentioned earlier the control actions will be those produced by the rudder, ailerons and elevators, and the engine thrust, which the pilot will manipulate as he thinks fit.

The disturbances include other important effects on the system, for instance the influence of the wind and the density of the atmosphere on the movement of the aircraft. Where it is necessary to distinguish the types of input the control effects are designated by the symbols $Y_1, Y_2, \ldots, Y_r$, and disturbances by the symbols $M_1, M_2, \ldots, M_s$.

The effects of the system on the surrounding medium are characterised by its *output signals*. The set of outputs and their changes determine the behaviour of the system; it is these which permit the outside observer to estimate whether the movements of the system correspond to the aims of the control.

In the example of controlling the movement of an aircraft, the output values are its course and speed, since these values determine the direction and speed of movement of the load, and the aim of control in the given case is to transport the load to a given place within a given time. The input actions on the body of an animal are in particular the actions which are perceived by its sense organs, the output values are the movements of its organs.

Usually, changes in input will produce changes in output. However, the changes in the output are not always immediate; in some cases they may lag behind, but they will never be in advance of the changes in the input because the output are the consequences of the input whilst the latter are the cause of movement of the system. Fig. 2.7 shows a system S and its input and output values.

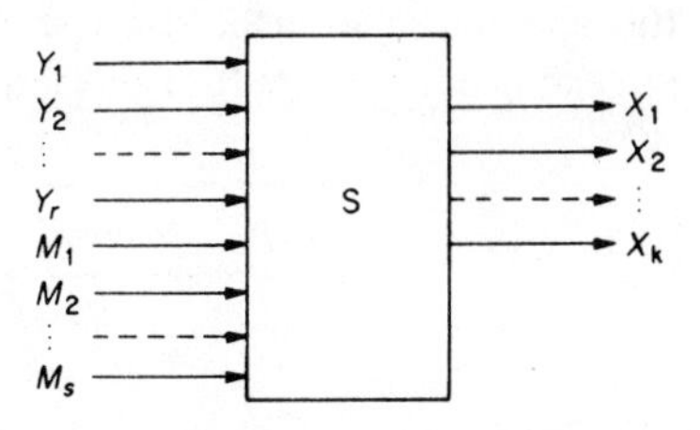


Fig. 2.7. *Input and output values.*

It is pointed out that the disturbances which influence the motion of a system may not only be external, but can also originate inside the system, for instance as a result of changes in the properties of its elements after prolonged operation and generally as a result of perturbations in the normal functioning of the elements of the system.

In some cases cybernetic systems can conveniently be considered as being broken up into parts which interact with each other. In this case the output of one part may at the same time be the input to another part of the system, for instance as is shown in Fig. 2.8.

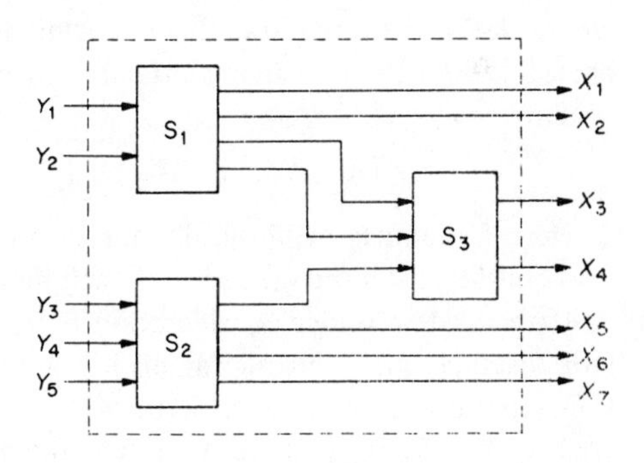


Fig. 2.8. *Interaction parts of a cybernetic system.*

## 2.3. Transformations

The motion of a system can be considered as a sequence of transformations of its states. It can be assumed that the transition of the system from the state $a_1$ at the instant $t_1$ to the state $a_2$ at the instant $t_2$ is the result of transformation of $a_1$, $t_1$ into $a_2$, $t_2$. The change of output of any system or element under the effect of changes in input can also be considered as their transformation.

We say that *transformation* of one object into another is realised by the action of the *operator* on the object. The object subjected to transformation is called the *operand*, and the result of the transformation is the transform (the operator 'maps' one object onto another).

Using the terms which have now been introduced, we can describe any transformation as follows: the operator, being applied to the operations, changes the latter into the transform. Obviously, under the subsequent action of the operator the transform obtained in the previous operation must be considered as the operand. The successive transition of the system into the states $a_0$, $a_1$, $a_2$, . . . can occur as a result of the action of the operator $P$ according to the scheme:

| *Step* | *Operand* | *Transformation* | *Transform* |
|---|---|---|---|
| 1 | $a_0$ | $Pa_0$ | $a_1$ |
| 2 | $a_1$ | $Pa_1$ | $a_2$ |
| 3 | $a_2$ | $Pa_2$ | $a_3$ |
| . . . | . . . | . . . | |

Denoting the $n$-fold action of the operator $P$ by $P^n$, we obtain

$$a_n = P^n a_0$$

If the output coordinate of the system $X$ is considered as the result of transformation of the input value $Y$, then the relation between $X$ and $Y$ can be written in the form

$$X = KY$$

where $K$ is the operator characterising the properties of the system under consideration. If the system contains $n$ output and $m$ input coordinates, then we obtain

$$\{X_1, X_2, \ldots, X_n\} = \{K\} \{Y_1, Y_2, \ldots, Y_m\}.$$

Here $K$ denotes symbolically the entire set of transformations, which take into consideration the influence of each input on each output.

If the system under consideration represents an inertia-free linear transducer (for instance, an electronic amplifier, a mechanical transducer or a photocell), then the operator $K$ will perform the role of a transformation coefficient transmission factor and will represent the number $k$ by which the input must be multiplied to obtain the output of the transducer.

$$X = kY.$$

For a nonlinear inertia-free converter the output value is some function of the input value and the operator will be denoted by the symbol $F$, giving a nonlinear transformation

$$X = F(Y).$$

Consider, as an example, equipment for measuring the flow rate of a liquid, Fig. 2.9, where the input value $Y$ is the flow rate $v$ of the liquid, and the output value $X$ is the pressure differential $\Delta p$ across the hydraulic resistance. The relation between the input and the output values in this case will be

$$X = cY^2,$$

where $c$ is a constant. Consequently the transformation operator in the given case will be the square multiplied by the constant.

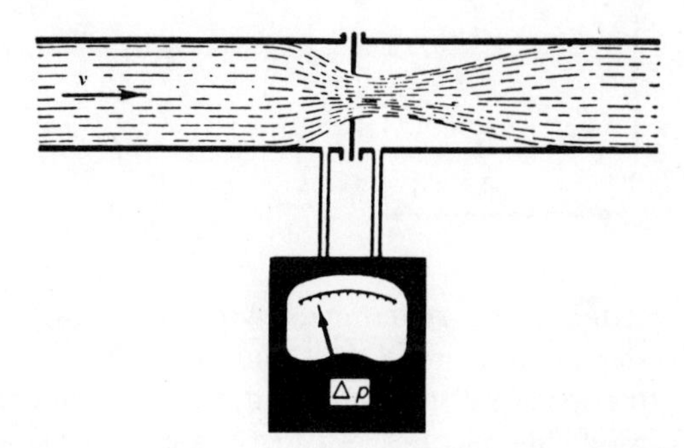


Fig. 2.9. *Equipment for measuring the flow rate of a liquid.*

If the output coordinates are not established instantaneously at values corresponding to the input values, then the operator will be more complicated and cannot be expressed by algebraic actions on the operand alone.

**Exercises**

1. On a telephone, a six-digit number is selected. Determine the dimension of the state space of the system. To what do the coordinates of the system correspond? How will the representative point behave during the successive change of the last digit into the telephone number from 1 to 0 (1, 2, . . .9, 0)?

*Solution.* The state space will have the dimension $n = 6$. The coordinates of the system correspond to the digits in the number (assuming that the entire number consists of digits). The representative point will move along the last coordinate.

2. A factory specialises in the assembly of motor cars from finished components. What are the input and output values for the given system? What are the disturbing inputs in this case?

*Solution.* The main input actions are the finished components delivered to the factory, and the production plan of the motor cars delivered by the planning organisations, the output value – the number of motor cars produced. Disturbances occur if the supply of finished components is interrupted or if the rhythm of the work in the factory is interrupted.

**3.** Some systems can be in four different states, and it is known that the trajectory which describes the representative point is closed, Fig. 2.10. Two operators may act on the system: $P$, which changes the system from a given state into the next state clockwise, and $Q$, which transforms the system from the given

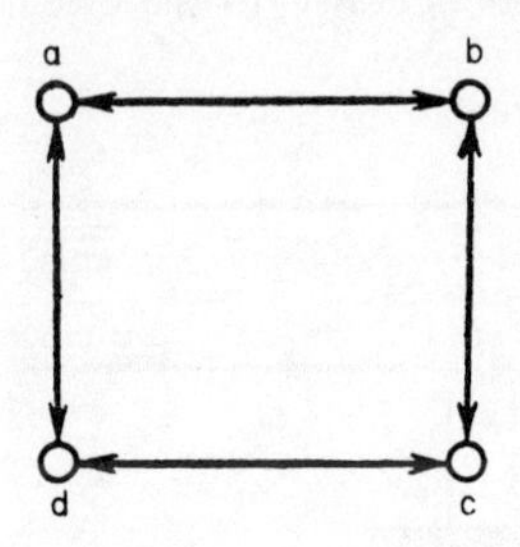


Fig. 2.10. *Sketch to illustrate Example No. 3.*

state into the state minus two, anticlockwise. Assume the given system has the states $a$, $b$, $c$ and $d$. It is known that the image $c$ was obtained by applying first the operator $P$ to the operand, and then the operator $Q^2$. Find the operand. What would the picture be if first the operator $Q$ and then $P^2$ acted on the operand thus found?

*Solution.* The operand is the state $b$. The successive application to the operand $b$ of the operators $Q$ and then $P^2$ would yield the image $b$.

# 3 Models

For a long time similarities in the characteristics of various objects have been the basis of the scientific approach to the study of a great variety of phenomena in nature. Essentially, the concept of a model which has characteristics similar to the studied phenomena has been introduced in either an explicit or an implicit form in all the various branches of science. However, in no field has the concept of modelling been more explicit and thorough than in cybernetics, where it figures in the most general form. This is a basic concept which is decisive from the point of view of the technique of studying the behaviour of cybernetic systems.

The problems surrounding the construction and utilisation of models are formulated and solved in various ways. We may be interested in a model which differs from the original in the scale of the geometrical dimensions, or in the duration of the processes, and the model may need certain convenient features from the point of view of experimental investigations.

Frequently it is convenient to study the properties of an object, using a model of a different physical nature, based on a formal functional similarity of equations which describe the motion of both the original and the model.

Of great importance in science is the concept of the simplified (or abstract) model, which also permits studying very complex objects and systems, retaining in the model only those characteristics of the original which are important in the particular context of the phenomena studied. The process of modelling is important in that it allows us to study the model rather than reality, and this means the problem becomes unmanageable. It also allows us to dwell only on what is relevant.

## 3.1. The original and the model

The concept of the model is based on the presence of some similarity between itself and some object. Here, the words 'similarity' and 'object' are understood in the widest sense. The similarity may be purely external, it may relate to the internal structure of objects which show no external similarities at all with the model, or to certain general features of the behaviour of objects which have nothing in common either as regards shape or structure. The concept of similarity is applied to a very wide class of material *objects*, including living and inanimate objects in nature, artificial objects produced by man, pictures, symbols, etc.

*If any similarity can be established between two objects then the original-model relation exists between these objects.*

This means that one of these objects can be considered as the original, the other as the model. For certain purposes anything can be the model of any original, the similarity need only be slight.

The similarity between the original and the model will be denoted by the sign $\sim$ so that if the object A is the model of the object B, then we write this in the form $A \sim B$. Thus we can also write $B \sim A$ and this will always be valid, since the similarity of objects is always mutual. The original-model relationship can occur not only between two but between any number of objects. For instance, for a set of objects $A \sim B \sim C \sim D$ any of these, for instance B, can be

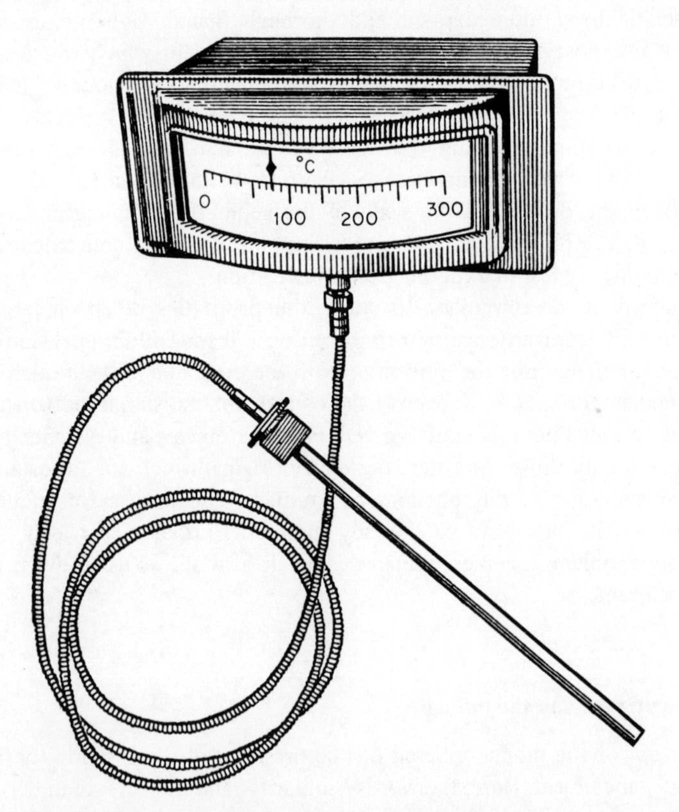


Fig. 3.1. *External appearance of an industrial pyrometer.*

considered as a model of the objects A, C and D or as the original for the models A, C and D.

External similarity – similarity in shape, say – will occur for such objects as a ship and some representation of it in the form of a sketch, a three-dimensional model or a set of drawings; between a metallic casting and its wooden model. There may also be a similarity in structure: the system for controlling an economy and its structural diagram or an urban water supply network and the urban electricity supply network. The most important similarity between systems leading to relationships between an original and a model from the cybernetic point of view is the similarity in *behaviour* which permits modelling its motion, or change. The basis of modelling behaviour is the fact that the same behaviour can be observed under certain conditions in systems which are quite different from each other in form, structure and in the physical nature of the operating processes.

As an example, let us consider a gas pyrometer (system A) and a thermoelectric pyrometer (system B). These instruments may be designed to be similar in shape and appear as shown in the diagram Fig. 3.1.

As both instruments are used for measuring temperature, cartridge 1 (Fig. 3.2) is submerged into the medium whose temperature is to be measured. The pointer of the metering instrument which is connected to the cartridge by tube 2 is deflected and will be moved to a position representing the scale division that corresponds to the temperature being measured. Both instruments have the same external appearance and will react equally to changes of the input values

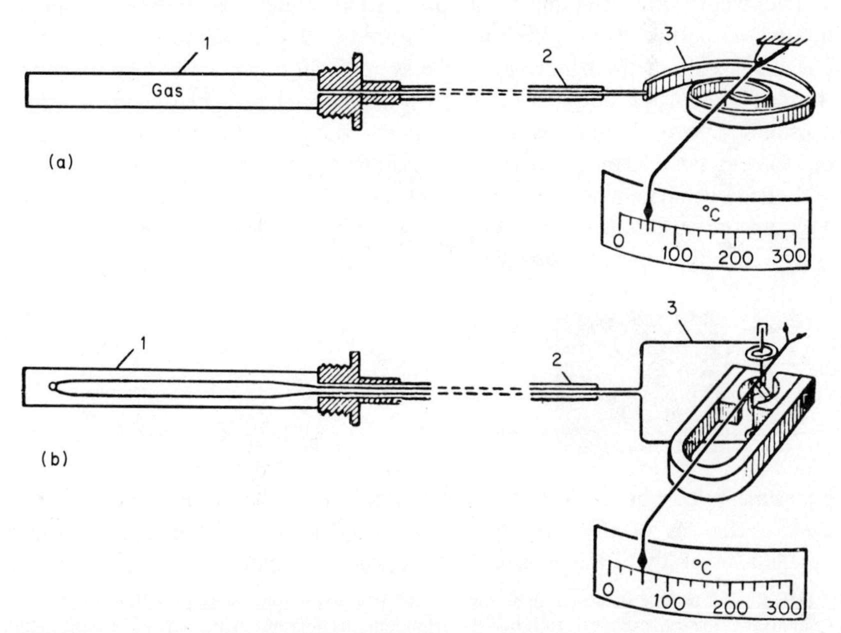


Fig. 3.2. *Industrial pyrometers: (a) pressure-operated; (b) thermoelectric.*

(temperature of the medium surrounding the cartridge). However, internally these instruments differ greatly. The cartridge A is filled with gas the pressure of which varies with the temperature. This pressure is measured by a pressure gauge connected with the cartridge through a tube, placed in the tube as shown in Fig. 3.2(a). A thermocouple is placed in the cartridge of the instrument B to generate an electromotive force, the magnitude of which depends on the temperature of its hot end. This electromotive force is measured by a millivoltmeter connected to the thermocouple by leads which are inside the tube, as shown in Fig. 3.2(b).

If the design elements of both instruments are chosen so that at an equal temperature of the medium the readings are the same, and if their inertia properties are also such that under equal conditions both instrument readings become established at the same speed, then from the point of view of reaction to input actuations the instruments A and B will become indistinguishable, at least for the specified limits of changes in the input values. Each of these instruments can then be considered as a model of the other, accurately reproducing the properties of the original which is of interest to us.

## 3.2. The 'Black Box'

For formulating and solving problems of modelling controlled systems, the concept of the 'black box' seems to be very fruitful. By a black box we mean a system where only the input and the output values are accessible, and the internal structure is not known. It is found that a number of important conclusions about the behaviour of the system can be drawn by observing only the reactions of the output to changes of input. Such an approach opens up the possibility of objective study of the system whose structure is either unknown, or too complicated for conclusions to be drawn about it from its behaviour.

Let the behaviour of a system be determined by its input values $Y_1, Y_2, \ldots, Y_m$ and the output values $X_1, X_2, \ldots, X_n$ (Fig. 3.3). If the behaviour of such a system is observed for a sufficiently long time and, if necessary, some active

Fig. 3.3. *The 'black box'*

experiments can be made,* it is possible to learn enough about the system to predict changes of the output for any given changes in the input. However detailed the study of the behaviour of the black box may be, it is not possible to

* By active experiment, we mean actions of the observer (experimenter) on the input of the system for the purpose of studying its behaviour, as distinct from a passive experiment consisting only in observing the behaviour of the system.

derive a justified conclusion on its internal structure because different systems may produce the same behaviour.

Systems which are characterised by the same sets of input and output values and which react similarly to external actuations, are called *isomorphous.* Obviously isomorphous systems are indistinguishable from each other to an observer who has access only to the input and output values. An example of isomorphism is the experimenter's inability to distinguish between the gas and the thermoelectric temperature gauge described in 3.1, whatever his experiments, so long, of course, as he does not 'open the black box.'

The conditions of isomorphism of the systems A and B can be expressed by the following system of equations:

if

$$Y_{1\mathrm{A}}(t) = Y_{1\mathrm{B}}(t), Y_{2\mathrm{A}}(t) = Y_{2\mathrm{B}}(t), \ldots, Y_{m\mathrm{A}}(t) = Y_{m\mathrm{B}}(t)$$

then (3.1)

$$X_{1\mathrm{A}}(t) = X_{1\mathrm{B}}(t), X_{2\mathrm{A}}(t) = X_{2\mathrm{B}}(t), \ldots, X_{n\mathrm{A}}(t) = X_{n\mathrm{B}}(t)$$

for any instant of time $t$.

Thus, study of the system by the black box method cannot lead to an unequivocal conclusion on its internal structure, because the behaviour of a given system considered as a black box does not differ from that of all other systems with which it is isomorphous. We must remember that for any concrete system an infinite number of concrete systems which are isomorphous with it can be chosen.

Obviously the original-model relation will exist between any isomorphous systems, in the sense that any one of a set of isomorphous systems can be considered either as the original or as the model of the others; the context of their use will determine the nature of the relationship. The conditions of isomorphism are not necessarily conditions of correspondence between the model and the original. The system A can be a model of the behaviour of the system B even if the correspondence is not as complete as would be required according to conditions (3.1).

The 'black box' concept is extensively used in science and engineering, but not always in the above explicit form. Essentially, a black box is any object judged from studying its external properties without investigating its fine structure and the properties of its finer components which make up the object studied. Thus, we make and use electric conductors without penetrating into the fine point of the mechanism of passage of electric current through metals; agricultural methods are developed for growing plants, based on studying their behaviour and not on studying their molecular structure.

The black box method is particularly important in the study of the behaviour of complex systems. Since cybernetic systems are complex it is natural that black box methods are fundamental to their study.

## 3.3. Simplified model

Among the coordinates of a system which determine its state, may be included those of greater or lesser importance from the point of view of a particular problem. If the unimportant coordinates are disregarded, then instead of the initial system A with the state space dimension $n$ we obtain a simpler system B with the state space dimension $n' > n$. For each given state of the system A there will be a corresponding state of the system B (because the given values of the inessential coordinates do not disturb determination of the state B from the important coordinates of the system A). However, to each definite state of the system B there will be no single corresponding state of the system A, since by giving the state of the system B we do not fix the values of the unimportant coordinates. Therefore all the states of the system A for any combination of the values of the inessential coordinates will correspond to one state of the system B, determined by the given values of the important coordinates.

Simplification of the initial system can also be achieved by grouping together a set of states into a single state. For example, instead of considering all the possible positions of the representative point in the state space, it is possible to simplify the problem by considering the point of only one of the areas into which the state space is subdivided. Such a substitution has the property of setting up an unequivocal correspondence between the states of the system A and the system B, and non uniqueness of the correspondence the other way round (between the system B and the system A).

It can be seen from Fig. 3.4 that by giving the coordinates of the point a in the system A, we define uniquely the square that determines the state of the system B (the square (2,2)). However, by giving the square in the system B, we do not determine uniquely the position of the point in the system A. This is because all states of the system A in the hatched area of its state space correspond to the given square (3.4) of system B.

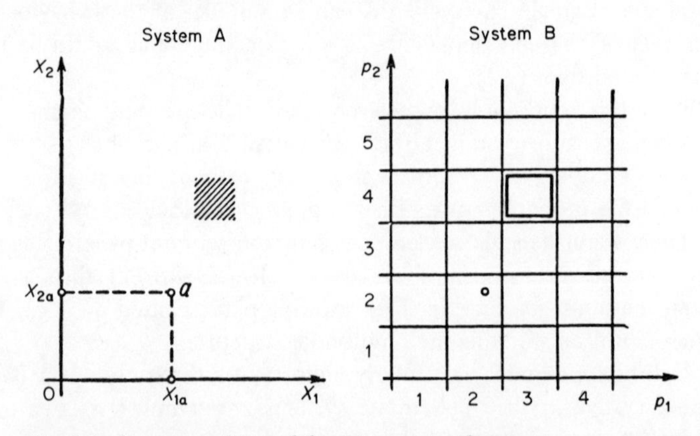


Fig. 3.4. *Homomorphism of the state spaces of systems.*

The system B obtained from the initial system A by simplification (by reducing the number of coordinates under consideration or by less accurate estimation of its variables), is called the *homomorphous* or the *simplified model* of the system A. The relations between the original A and its homomorphous model B do not have equal validity because A cannot be considered as the homomorphous (or homomorphic) model of B.

Consider as an example a machine which manufactures steel balls for bearings. For the machine operator, the important variables are: productivity, deviation of the diameter of the balls from the nominal diameter, the energy consumption, the pressure in the lubrication system and the flow of emulsion which cools the cutting tool. However, the dispatcher who controls the entire plant will be interested in a smaller number of variables on the machine. He will, for example, be interested in which of the following three states the machine is at any time:

1st state – machine not working.

2nd state – machine does work and the diameter of the balls is within the tolerance limits.

3rd state – machine working, but the diameter of the balls is outside the tolerance limits.

Substitution of the initial state space of the machine, which contains a multitude of values of all its coordinates, by a state consisting only of the three states enumerated, represents a transition from the original system to its homomorphous model, or homomorph.

### 3.4. Analogue systems

If B is to be a model of A, it is sufficient that at least one of the output values of the system B should correspond in some scale to any output value of the system A, provided that a certain correspondence between the conditions at the inputs of these systems is retained. These requirements to the model can be expressed by the condition

$$B \sim A$$

when

$$Y_{1A} = k_1 Y_{1B}(k_0 t),$$

$$Y_{2A}(t) = k_2 Y_{2B}(k_0 t),$$

At least one pair $X_{iA}$ and $X_{jB}$ can be found amongst the output coordinates, such that if for some instant of time $t = t_0$ there is a correspondence between the states of this system, then for any other instant of time

$$X_{iA}(t) = kX_{jB}(k_0 t), \qquad (3.2)$$

where $k, k_0, k_1, k_2, \ldots$ are scaling coefficients.

Systems which satisfy the condition (3.2) are called Analogue Systems. Analogue Systems include, for example, a pendulum and an electric oscillatory circuit whose output values ($X_A$ – deflection angle of the pendulum from the vertical and $X_B$ – voltage on the condenser) will perform damped sinusoidal oscillations after being removed from the position of equilibrium and then left undisturbed, see Fig. 3.5 ($X_{Ai}$, $X_{Bi}$ – initial values of the variables $X_A$ and $X_B$).

The existence of analogue systems is the result of a formal similarity between some features of the behaviour of homomorphous models of systems differing in nature and structure. This similarity arises only after extensive simplifications in the process of constructing homomorphic models of the initial systems. If an attempt is made to dispense with some simplifications, then the analogy may be lost. Thus, for instance, if the dry friction in the suspension of the pendulum or the radiation of electromagnetic waves from the oscillation circuit are taken into

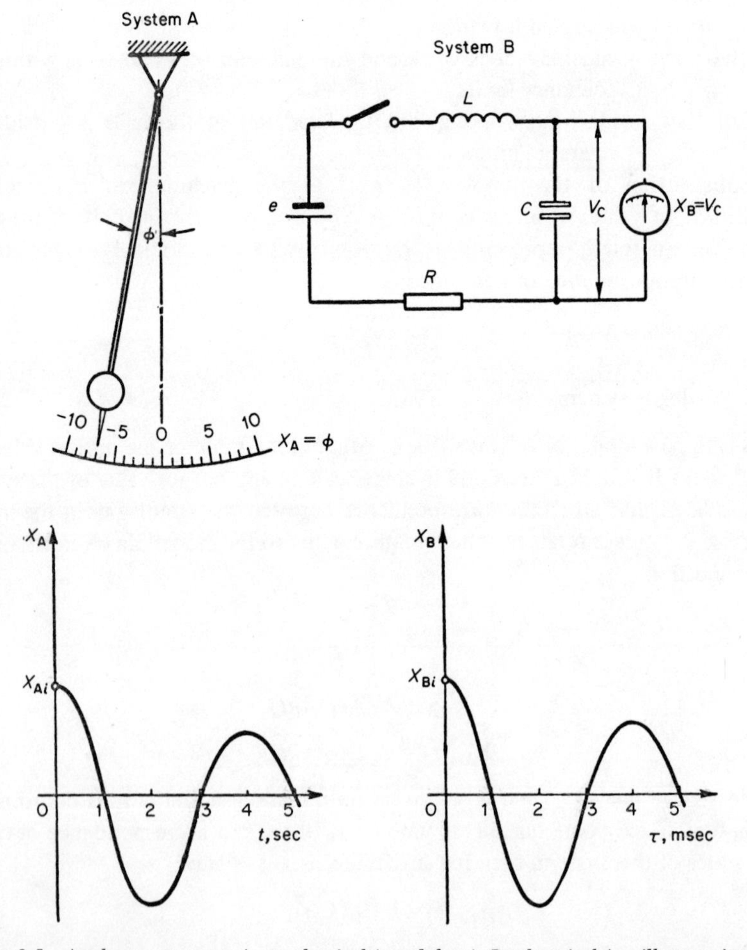


Fig. 3.5. *Analogue systems: A: mechanical (pendulum), B: electrical (oscillatory circuit).*

consideration, then the forms of movement of the coordinates $X_A$ and $X_B$ will not be analogous.

Analogous coordinates of the original A and its analogue B (or more accurately the coordinates of the homomorphic models A and B) are called *representative values.* The relations between the appropriate representative values are given by the scale coefficients $k_1, k_2, \ldots$. Each of these coefficients indicates how many units of the representative value in the model correspond to one unit of measurement of the appropriate quantity in the original. In the example under consideration the scaling coefficient which interrelates the voltage on the capacitor with the angle of deflection of the pendulum shows how many volts $V_c$ correspond to one degree of deflection $\phi$ on the pendulum. This coefficient has the dimension (volt/degree).

The difference in the time scales of processes in the original and the model is determined by the dimensionless coefficient $k_0$, which indicates how many times faster the processes are in the model than they are in the original.

Table 3.1 shows some system analogues used for simulating processes in controlled systems.

Table 3.1 *Analogue systems*

| Mechanical analogues | Pneumatic analogues | Electrical anologues |
|---|---|---|
| $X_1 = \Delta l$, $X_2 = F$ | $X_1 = \Delta l$, $X_2 = p$ | $X_1 = I$, $X_2 = \phi$ |
| $X_1 = F$, $X_2 = v$ | $X_1 = F$, $X_2 = v$, $r$ | $X_1 = I$, $R = \text{const}$, $X_2 = V$ |
| $X_1 = F$, $m$, $X_2 = v$ | — | $X_1 = V$, $L$, $X_2 = I$ |
| $X_1 = F_1$, $X_2 = F_2$ | $X_2 = F_2$, $X_1 = F_1$ | $L_1$, $L_2$, $X_1 = V_1$, $X_2 = V_2$ |
| $l = \text{const}$, $m = \text{const}$, $X_1 = \tilde{F}$, $X_2 = A$ | — | $L = \text{const}$, $X_1 = \tilde{V}$, $C$, $X_2 = I_a$ |

Fig. 3.6. *Universal analogue modelling device. (Photograph by courtesy of Electronic Associates Ltd.)*

During the last decade the use of a universal analogue modelling device (Fig. 3.6) has been considerably expanded for scientific investigations and engineering calculations with respect to processes in complex control systems. Such equipment is made up of assemblies of electronic amplifiers, capacitors and resistances. By setting up appropriate connections such elements yield analogue models for a wide class of systems, which permit a comparison of the behaviour of these systems under various conditions, by observation of the behaviour of the models.

### 3.5. Mathematical models

A mathematical model of a system is its description in any formal language which must allow conclusions about certain features of the behaviour of the system to be derived by applying formal procedures to the description.

Mathematical description cannot be all-embracing or completely accurate, so that mathematical models do not describe real systems but rather their simplified (homomorphic) models.

There are various types of mathematical model, some of which may represent the basic characteristics of a system, given, say, by functional relationships or graphs. Equations can also describe the movement of a system, while tables or graphs show transitions of the system from one set of states into others, etc.

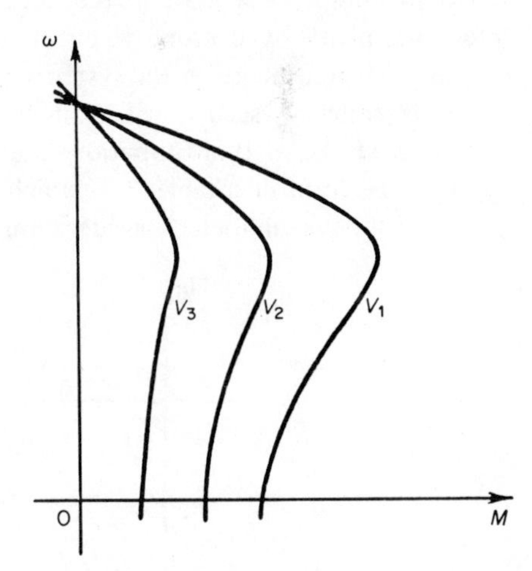


Fig. 3.7. *Family of characteristics of an induction motor.*

Fig. 3.7 shows an example of a mathematical model of an induction motor, given as the family of characteristics which interrelate the torque $M$ of the motor and the angular velocity $\omega$ for various values of applied voltage $V_1$, $V_2$, . . . . Using this model, it is possible to predict, for instance, how the speed (rpm) will change at various loads and various voltages in the supply mains.

The Newtonian expression

$$F = \frac{m_1 m_2}{r_2},$$

where $m_1$ and $m_2$ are masses of the material points at a distance $r$ from each other, and $F$ is the force of interaction between them, can be considered as a mathematical model of a system of two material points, because the expression allows conclusions to be drawn on their interaction under different conditions.

The behaviour of a system which passes successively into different states can be given in the form of a diagram or a table of transitions.* Thus the mathematical model of the life cycle of a plant (in a very simplified form) can be represented in the form of a graph as shown in Fig. 3.8. Here the nodes represent

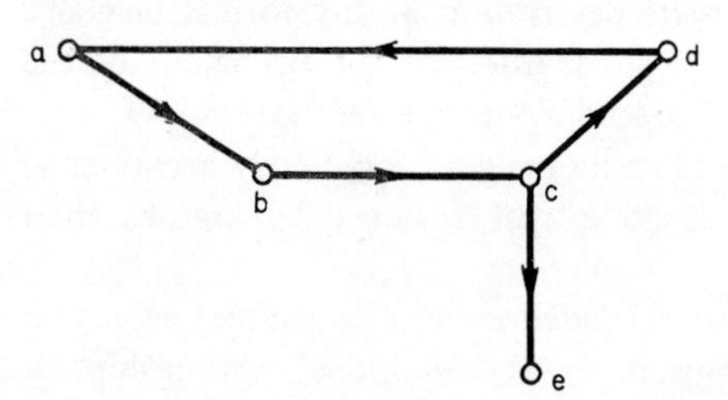


Fig. 3.8. *Diagrammatic representation of the life cycle of annual plant, in the form of a graph.*

the states of the system and the arrows represent transitions from one set of states into others. The state of the 'seed' is denoted by a, the 'plant' by b, the 'blooming plant' by c, from which the system can change either into the state e – 'unpollinated plant' or the system d – 'pollinated plant', as a result of which the initial state a – 'seed' – will again be obtained.

The sequence of transformations illustrated by the graph Fig. 3.8 can also be given in the form of a table, 3.2, which will serve in addition as a mathematical model of the system under consideration.

Table 3.2

| | *a* | *b* | *c* | *d* | *e* |
|---|---|---|---|---|---|
| *a* | | | | ← | |
| *b* | ← | | | | |
| *c* | | ← | | | |
| *d* | | | ← | | |
| *e* | | | ← | | ← |

* The diagram of transitions represents the so-called *graph*. The definition of a graph and the concepts relating to it are given in section 15.1.

It can be seen that from the state c the system does not change into one, but rather into two states d and e. This means that the transition depends on some particular causal factor; this, in the given case, depends on whether the flower is pollinated or not. It is obviously difficult to take into consideration all the factors leading to pollination, and to predict the fact of pollination for each individual plant. However, for a large number of plants, statistical data enable prediction of the average frequency of pollination. Then the ratio of pollinated plants to a total sufficiently large number of plants will yield the value $p$ characterising the probability of pollination of each plant. The value $1 - p$ will then be the probability of non-pollination of each plant.

For systems whose behaviour depends on random factors it is necessary to indicate not only the state into which the system passes but also the probability of this transition.

For the example under consideration the statistical model of the system is shown in Table 3.3.

Table 3.3

| | *a* | *b* | *c* | *d* | *e* |
|---|---|---|---|---|---|
| *a* | | | | *1* | |
| *b* | *1* | | | | |
| *c* | | *1* | | | |
| *d* | | | *p* | | |
| *e* | | | *1-p* | | *1* |

**Exercises**

1. Which of the objects listed below can be considered as the original and which the model, and in what sense: (a) book, (b) solar system, (c) oscillation circuit, (d) atom, (e) the pendulum (balancing wheel) of a clock, (f) a gramophone record, (g) the memory of a computer.

*Solution.* The book, gramophone record, memory of a computer are models. The solar system, an oscillation circuit, an atom, the pendulum (balancing wheel) of a clock are originals.

2. Are the photographic negative and the print from it isomorphic?

*Solution.* Yes.

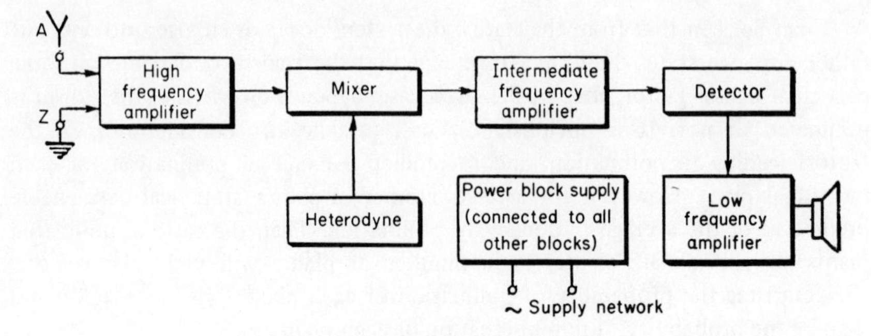


Fig. 3.9. *To Example 4.*

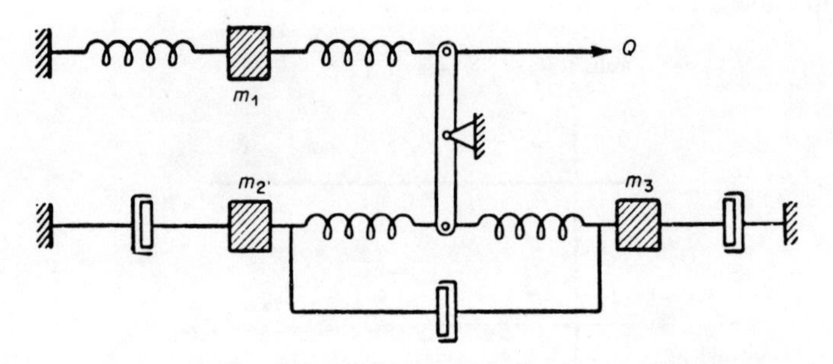


Fig. 3.10. *To Example 5.*

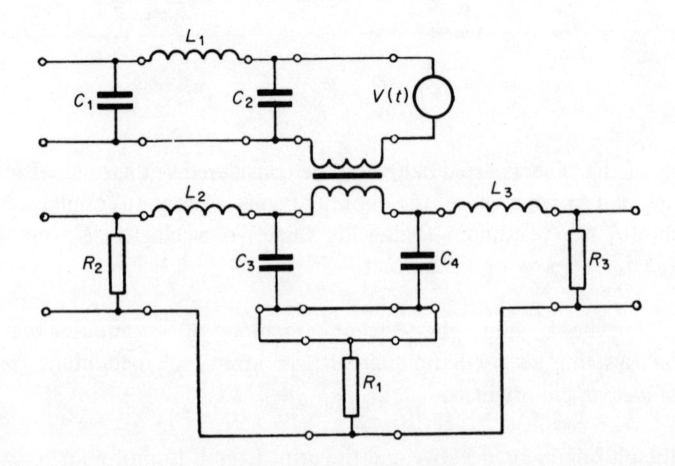


Fig. 3.11. *To Example 5.*

3. Try to name some systems which are isomorphic to the electromagnetic voltmeter.

*Solution.* The following systems: a thermal voltmeter, a magnetoelectric voltmeter, an electrostatic voltmeter, an electrodynamic voltmeter.

4. A typical industrially produced radio receiver contains several tubes, sets of resistances, condensers and circuits. Usually the basic circuit is reproduced on two pages of a book. Could the space occupied by the circuit be reduced considerably by constructing its homorphous model? If this is possible, how? (Consider the case of a superheterodyne receiver). What could serve as an isomorphic model of such a receiver?

*Solution.* We can use as a homomorphic model the block schematics shown in Fig. 3.9. As the isomorphous model any other circuit of a superheterodyne receiver can be used.

5. On the basis of Table 3.1, construct an electric analogue system for the following mechanical systems (Fig. 3.10).

*Solution.* The electric analogue circuit is given in Fig. 3.11.

6. Let the work of two systems A and B be given in the form of the following two transition tables which are their mathematical models:

A

| ↓ | $a$ | $b$ | $c$ | $d$ |
|---|---|---|---|---|
| $a$ | $b$ | $a$ | $b$ | $c$ |
| $y$ | $a$ | $b$ | $c$ | $b$ |
| $z$ | $a$ | $b$ | $b$ | $d$ |

B

| ↓ | $i$ | $k$ |
|---|---|---|
| $\alpha$ | $i$ | $k$ |
| $\beta$ | $k$ | $k$ |

where the top line designates the operand, the extreme left columns – the operators, and the arrow – the direction of the transitions. Prove that these two systems are homomorphous.

*Hint.* Try to find a transformation that would change the elements of the system A into elements of the system B.

*Solution.* For the proof, it is sufficient to construct the following operator, which changes the components of the system A into components of the system B.

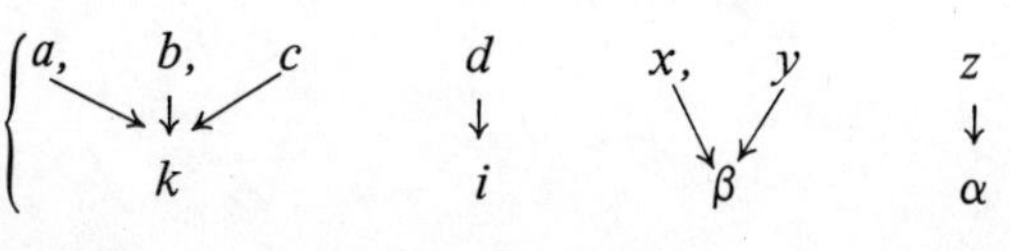

$$\left\{ \begin{array}{l} a,\ b,\ c \rightarrow k \\ d \rightarrow i \\ x,\ y \rightarrow \beta \\ z \rightarrow \alpha \end{array} \right.$$

Then the table of transitions of the system A can be written as follows:

| A′ | | | | |
|---|---|---|---|---|
| $\downarrow$ | $k$ | $k$ | $k$ | $i$ |
| $\beta$ | $k$ | $k$ | $k$ | $k$ |
| $\beta$ | $k$ | $k$ | $k$ | $k$ |
| $\alpha$ | $k$ | $k$ | $k$ | $i$ |

It can easily be seen that after crossing out the repeating rows and columns it repeats exactly the table of transitions of the system B.

# 4 Dynamic Systems

When studying the behaviour of control systems one must consider their motion or variability – the changes in their state. However, there can be no change in the state of any system without conversion and transfer of energy and matter between its component elements. Thus, a change in the temperature of a body entails changes in its internal energy; to change the level in a reservoir it is necessary to change the quantity of liquid contained therein. If an animal wishes to change its position in space during a finite period of time it must move, which in turn requires the accumulation of a reserve of kinetic energy.

If the change in the state of a system could occur instantaneously, this would mean that the reserve of energy or matter in the system would have a finite increment during an infintely short time. If this is to happen, the intensity of the flow of energy or matter through some elements of the system would have to be infinitely large, which is not possible. Consequently the state of a 'real' system cannot change instantaneously and must occur during a finite period of time. This occurs as a result of a certain process which we call a *transient process.*

*Systems whose transition from one state into another cannot be achieved instantaneously and which occur as a result of a transient process are called dynamic systems.* It is clear from what has been said that, strictly speaking, all real systems are dynamic systems. However, when the duration of the transient process is negligibly small compared to the duration of the investigated phenomenon, and where the nature of the transient process does not have an important influence on the behaviour of the system, it is not necessary to take into consideration the dynamic properties of the system under consideration; it can be assumed that the changes in state follow instantaneously the causes which produce them.

## 4.1. Regimes of a dynamic system

Three characteristic types of behaviour of a system must be distinguished – three regimes which are possible for a dynamic system: *equilibrium, transient* and *periodic.*

We will say that a system is in equilibrium if its state does not change with the progress of time. An *equilibrium state* is one in which none of the coordinates of the system change. In the state spaces the equilibrium states of a system are represented by static points, but it is obvious that not all these points in the state space will be points of equilibrium. As an example we will consider the movement of a ship relative to its longitudinal horizontal axis, Fig. 4.1.

The state of the system under consideration is determined by two of its coordinates: the angle $\phi$ between the vertical axis $0' - 0'$ of the ship relative to the line in the direction to the centre of the earth $0 - 0$ (Fig. 4.2), and the speed of rotation $\omega$ of the vertical axis of the ship.

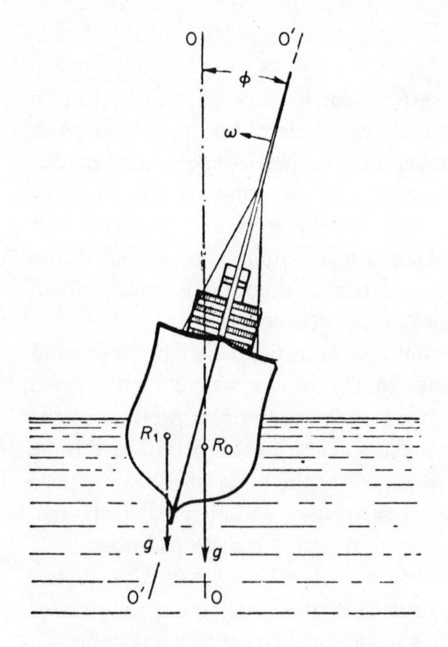


Fig. 4.1. *Movement of a ship relative to its longitudinal axis.*

The equilibrium state in the system under consideration can occur only for such a position of the ship for which its centre of gravity R is on the $0 - 0$ axis (position $R_0$), and the speed of turning $\omega = 0$. The point $a_0$ depicts the equilibrium state of the system and in this case has the coordinates $\omega = 0$; $\phi = \phi_0$. If a part of the cargo is shifted from the left side (port side) of the ship to the right (starboard side), then the equilibrium position of the ship will change and will be depicted for instance by the point $a_1$ with the coordinates

$\omega = 0$, $\phi = \phi_1$. However, for any disposition of the cargo the system will only have one state of equilibrium.

By a 'transient regime' we mean a regime of motion of a dynamic system changing from some initial state to any *steady state regime* – equilibrium or periodic. The transient regime will occur in the system under the effect of changes of external forces or changes in the internal properties of the system. For instance in the system under consideration the transient state may occur as a result of wind, which will change the inclination (list) of the ship, or as a result of changes in the position of its centre of gravity R, due, say, to shifts in the cargo.

If cargo shifts, which may cause changes in the equilibrium state of the ship, are so rapid that the ship does not have time to significantly change its list, then it can be assumed that the movement from the initial state $a_0$ occurs without any change in the position of the centre of gravity of the ship, thus $R = R_1$. The force of gravity, which is in the direction parallel to the 0 – 0 axis and is applied to the centre of gravity $R_1$, will generate a torque that will try to turn the ship with increasing velocity in the direction towards the new equilibrium position. As a result, the point will move along the trajectory 1 – 2, until the centre of gravity $R_1$ is on the 0 – 0 axis. Although the state of the system, depicted by the point 2, corresponds to the equilibrium inclination ($\phi = \phi_1$) and the torque generated by the force $G$ will be equal in this point, the speed of turning will not be zero, the ship will therefore not remain in this position but will continue to turn beyond the equilibrium position. This generates a torque in the direction opposite to that of rotation of the ship and therefore the speed of turning will decrease whilst the representative point moves along the trajectory 2 – 3. The ship will therefore begin to oscillate about its equilibrium position $\phi_1$. Due to the damping effect of the viscous friction between the ship and the water, these oscillations will be attenuated, and the trajectory of the representative point will be a convoluting spiral, as shown in Fig. 4.2.

A *periodic regime* is a regime for which the system will revert to an identical state after equal intervals of time. A periodic regime may occur in the system under consideration in the following two cases: under the effect of waves

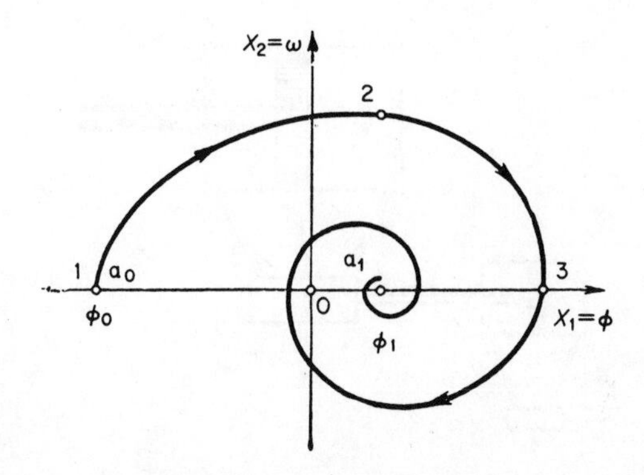


Fig. 4.2. *Trajectory of damped oscillations of a ship relative to its longitudinal axis.*

(forced periodic regime) and in the absence of friction between the water and the ship (regime of free, undamped oscillations.). Undamped, free oscillations in such a system can only be considered possible in theory, when making an abstract, and therefore approximate, study of the process. The trajectories of possible movements of the system being considered under periodic conditions are shown in Fig. 4.3.

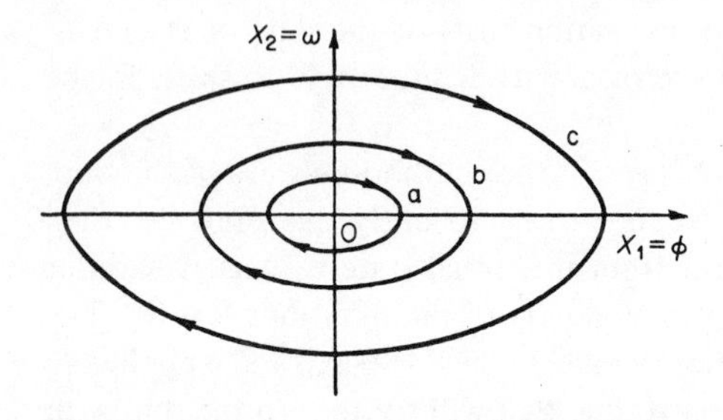


Fig. 4.3. *Trajectories of undamped oscillations of a ship relative to its longitudinal axis.*

## 4.2. Phase space

The behaviour of a dynamic system cannot be studied effectively in any arbitrary state space. If the coordinates in the state space are arbitrarily chosen, the movement of the system may prove to be unpredictable.

Let us observe, for example, the behaviour of the system of a hydraulic drive with an amplifier, as shown in Fig. 4.4. This system consists of the valves $S_1$ and $S_2$ and the cylinders $C_1$ and $C_2$, which are inter-connected in such a way that the position Y of the valve $S_1$ determines the speed of displacement of the piston in the cylinder $C_1$, and the position of the valve $S_2$ displaced by the cylinder $C_1$ determines the speed of movement of the piston in the cylinder $C_2$. The position $X_2$ of the piston of cylinder $C_2$ will be considered as the output value, and the position $Y$ of the valve $S_1$ as the input value.

Let us assume that the observer tries to analyse the relations governing the movements of this system by studying the trajectory of the movement of the

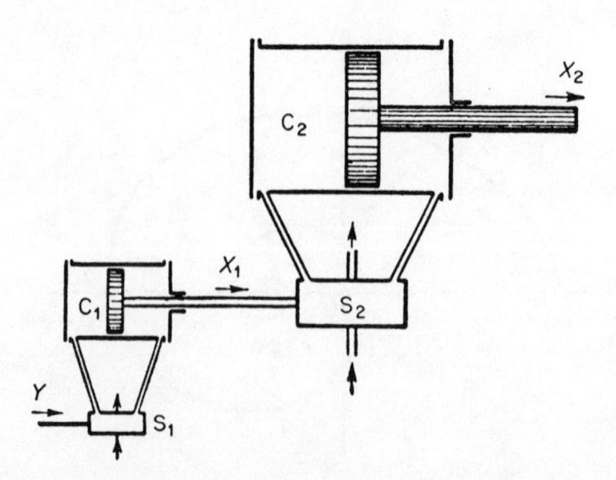


Fig. 4.4. *Hydraulic drive with amplifier.*

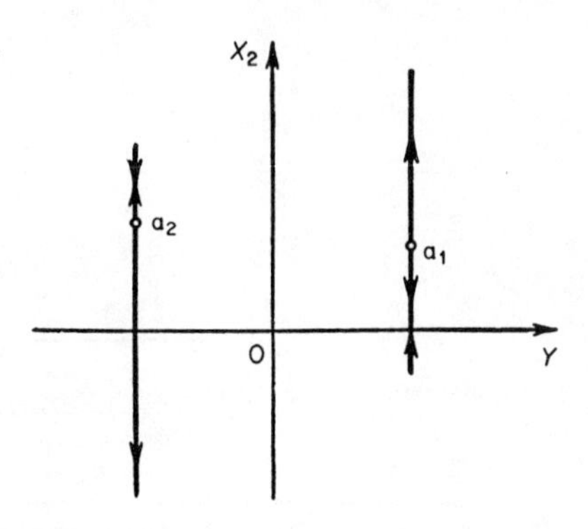


Fig. 4.5. *Space of the input and output variables of a hydraulic drive.*

representative point on the plane $Y$, $X_2$, Fig. 4.5. It will be seen that in many cases the same position at the input $Y$ and at the output $X_2$ of the system, for instance the one represented by the point $a_1$, corresponds to different trajectories of the representative point.

In some cases the piston will move from this position to the right, in other cases to the left, sometimes the piston begins to move to one side and then stops and begins to move in the reverse direction. Such unpredictability in the behaviour of the system deprives the observer of the chance of investigating its properties, and thus provides a real barrier to the study of the phenomenon. In the example given, this is explained by the fact that the observer did not introduce into the investigation the coordinate $X_1$, which characterises the position of the valve $S_2$. In the space $Y$, $X_1$, $X_2$ each fixed input value $Y = Y_j$ corresponds to a family of non-intersecting trajectories that determine unequivocally the movement of the system from any initial state, as is shown in Fig. 4.6.

The space in which the motion of the system is represented by non-intersecting trajectories so that under constant external actions each initial state of the system uniquely determines the progress of the system is called a *phase space*, and the coordinates of this space are called *phase coordinates.* The family of phase trajectories which represent the movement of the system is called its *phase portrait.* The phase space of any dynamic system is completely filled with phase trajectories, i.e. a trajectory passes through each point of this space, although for simplicity and clarity only some of these are shown, for example, in the family of phase portraits in Fig. 4.6. This figure also shows that by changing the external forces on the system it is possible to alter its phase portrait considerably. It can easily be seen that the side-roll of the ship, described in Section 4.1, was effected in the phase space and the trajectory shown in Fig. 4.2 is the phase trajectory, whilst the family of trajectories in Fig. 4.3 is the phase portrait of the system.

The number of dimensions of the phase space of a system is called the *order of the system.* The motion of the system shown in Figs. 4.1 and 4.4 can be uniquely represented in a two-dimensional phase space, therefore these may be referred to as second order systems.

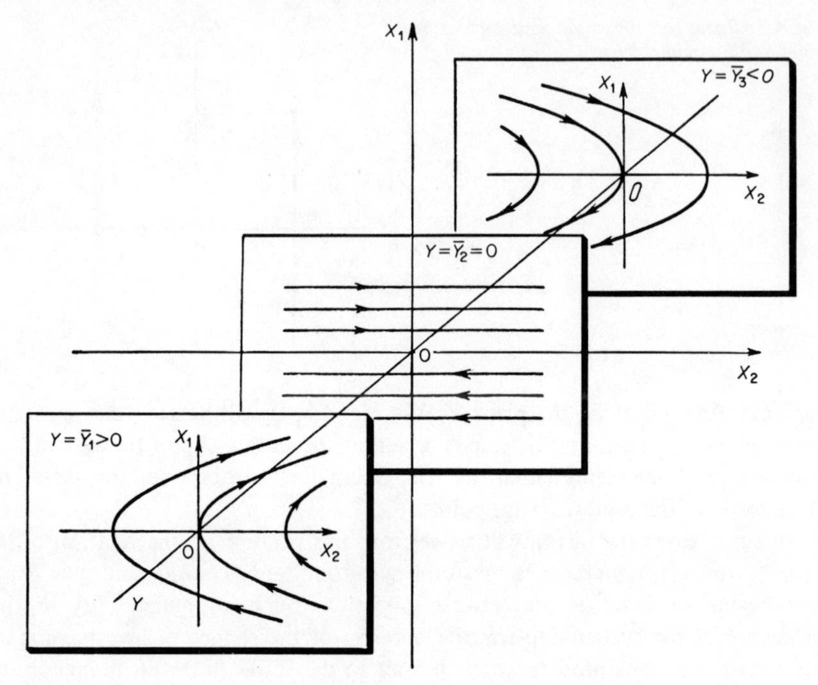


Fig. 4.6. *Space of a hydraulic drive with an amplifier.*

More complicated systems will be of higher order and their phase spaces will necessarily have a higher dimension. Thus, if the hydraulic drive shown in Fig. 4.4 is supplemented by additional amplifier stages, then the order of the system will be increased to a value $n$, which equals the number of cylinders in the drive.

## 4.3. Construction of phase portraits

Phase trajectories of dynamic systems can be plotted on the basis of experimental data. For this purpose we will measure the phase coordinates during movement of the system with fixed input values. The position of the representative point will be determined by the values of the coordinate at a given instant of time. By investigating these processes for various initial states of the system, it is possible to find the relevant set of phase trajectories and from these to construct the phase portrait of the system.

Thus, the three phase trajectories in Fig. 4.3 can be plotted from the results of three experiments to study the roll of the ship by means of gyroscopic instruments which record the inclination angle $\phi$ of the ship's axis and the rate of change $\omega$ of the inclination. Such instrument recordings are reproduced in Fig. 4.7: $a$ – for a small, $b$ – for a medium, and $c$ – for a large oscillation amplitude. For each pair of $\phi(t)$ and $\omega(t)$ curves, phase trajectories $a$, $b$ and $c$ can be plotted (Fig. 4.3), by eliminating time from the functions $\phi(t)$ and $\omega(t)$.

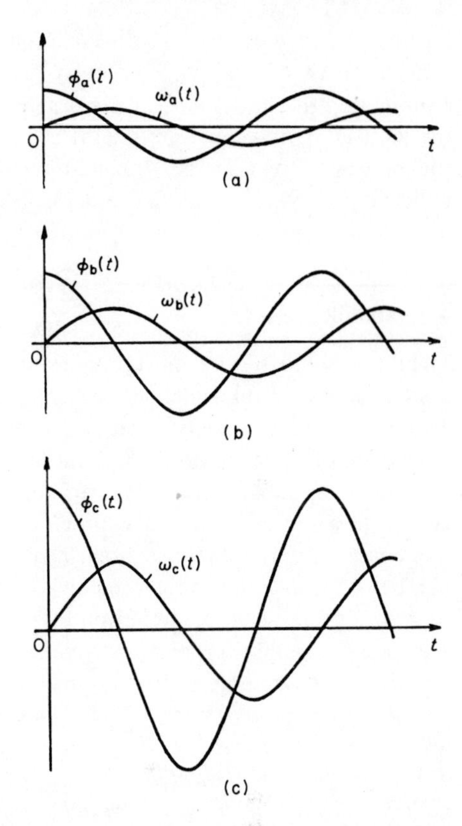


Fig. 4.7. *Record of three possible movements during the roll of a ship.*

For dynamic systems, whose behaviour can be described by appropriate equations, the phase portraits can be obtained analytically. For example, for the hydraulic drives described in Section 4.2 it is possible to construct the family of phase portraits, shown in Fig. 4.6, as follows.

Let $k$ denote the coefficient of proportionality between the speed of movement of the piston and the position of the control valve. Then for a fixed position $\overline{Y}$ of the valve $S_1$, the position of the piston rod (coordinate $X_1$) will change according to the law

$$X_1(t) = X_{\text{lin}} + k\overline{Y}t. \tag{4.1}$$

where $X_{\text{lin}} = X(0)$ is the initial value of the coordinate $X_1$. Similarly, for the coordinate $X_2$ we have

$$X_2(t) = X_{\text{2in}} + kX_1 t. \tag{4.2}$$

Eliminating the time $t$ from equations (4.1) and (4.2) we obtain

$$X_2 = X_{\text{2in}} - \frac{X_{\text{1i}}}{Y} X_1 + \frac{1}{Y} X_1^2. \tag{4.3}$$

The expression (4.3) gives the functional relationship between the coordinates

$X_1$ and $X_2$. In the given case this relation is parabolic. It can be seen from (4.3) that the shape of the phase trajectory depends on the coordinates of the point, representing the initial state of the system ($X_{1\text{in}}$ and $X_{2\text{in}}$), and on the values of the input $\bar{Y}$. Using equation (4.3) it is possible to construct the phase portraits for the cases shown in Fig. 4.6 and to verify the correspondence of these curves with equation (4.3), which the reader is recommended to do.

## 4.4. Stability

Stability is one of the most important characteristics of the behaviour of a system and is a fundamental concept used in physics, biology, engineering, economics and cybernetics. The concept of stability is used to describe the *constancy* of any behavioural feature of a system; using the term 'constancy' in the widest possible sense. This can apply to the constancy of the state of a system (its invariance with the progress of time) or to the constancy of some sequence of states in the system during its movement, or even the constancy of a number of people of a certain type living in the world.

A strict and accurate definition of the concept of stability, as applicable to the state of equilibrium of a dynamic system, was given by the eminent Soviet scientist A. Lyapunov. Let us assume that an immobile point a represents the equilibrium state of a system in the phase space, Fig. 4.8. *This equilibrium state will be stable according to Lyapunov if for any given domain of permissible deviations from the equilibrium state (domain $\epsilon$) it is possible to indicate a domain $\delta$ (which includes the equilibrium state), such that the trajectory of any movement beginning in the region $\delta$ will never reach the limits of the domain $\epsilon$.*

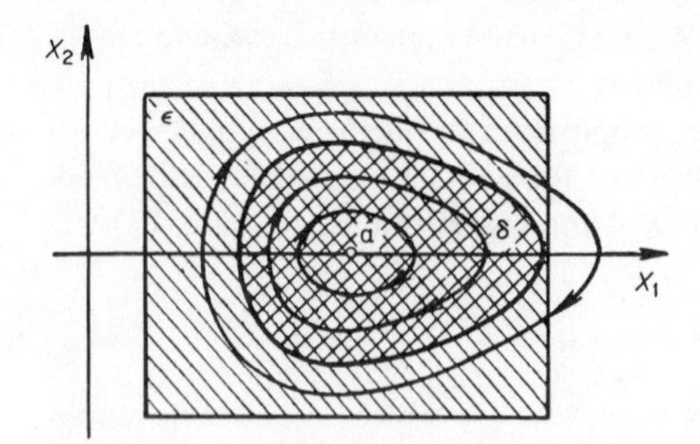


Fig. 4.8. *Graph relating to the definition of stability according to Lyapunov.*

The form of phase trajectories shown in Fig. 4.8 ensures fulfilment of this condition, demonstrating the stability of the equilibrium represented by the point *a*.

It can easily be seen that in this sense the systems represented by the origin of the coordinates, Fig. 4.3 will also be stable, even if the phase trajectories do not contract to this point.

The equilibrium position of a system will not always be stable. Consider, for

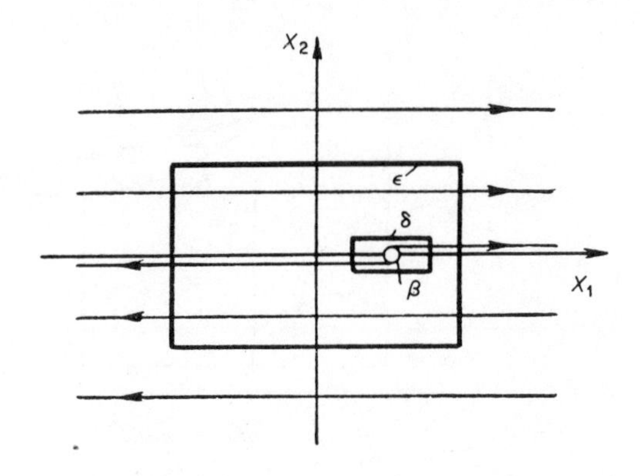


Fig. 4.9. *Unstable equilibrium of a dynamic system.*

instance, the operating regimes of a hydraulic drive with a zero displacement of the input valve ($Y = \overline{Y} = 0$), as shown in Fig. 4.9. Any point on the $X_1$-axis will represent the equilibrium state of the system. For the valves $S_1$ and $S_2$ in the central position, the velocities of the pistons equal zero and the state of the system should not change. In reality the state will not change if it is assumed that the condition $Y = 0$ and $X_2 = 0$ is complied with ideally.

In reality, however, any system is subjected to internal and external disturbing forces. However small such forces may be, they will always cause fluctuations in the state of the system, as a result of which the representative point will wander about its average position in some domain $\beta$.

The representative point will inevitably be above or below the $X_1$ – axis and therefore, as shown in Fig. 4.9, the trajectory of the representative point will sooner or later intersect the boundary of the domain $\epsilon$, whatever the choice of the boundary of the domain $\delta$. Consequently not one of the equilibrium states of this system will be stable.

We will now consider the example of a system of establishing market prices, where there are stable and unstable equilibrium states. Let us assume that the dependence of the demand $D$ and the supply $S$ of some merchandise on the market price $P$ is as shown in Fig. 4.10, and the rate $d$ of price changes is directly proportional to the difference between supply and demand:

$$d = k(D - S). \tag{4.4}$$

Here $k$ is a coefficient indicating the increase in price of the merchandise per unit of time if the difference between demand and supply equals unity. According to the sense of the model the coefficient $k > 0$.

The causes of a drop in demand and increase in supply when the price increases are clearly understood. An increase in supply with a decrease in price below $P_k$ may occur in some particular cases, for instance on changing over to mass production when the prices are reduced and the demand is increased. It can be seen from Fig. 4.10 that the system under consideration has two equilibrium states, $a_1$ and $a_2$, because at these points the demand is equal to the supply and the price, according to equation (4.4), remains unchanged ($d = 0$).

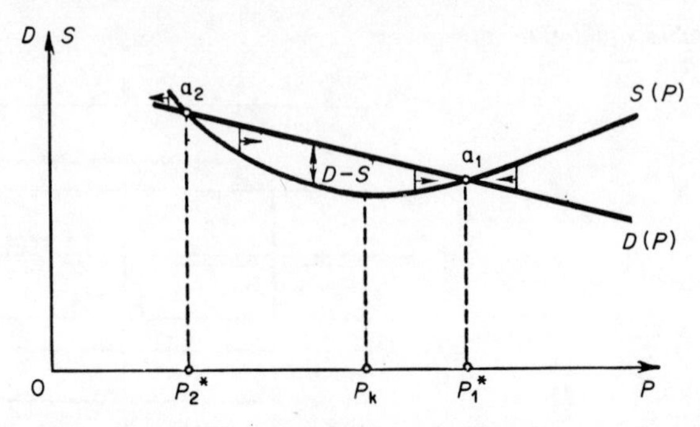


Fig. 4.10. *Dependence of demand* D *and supply* S *on the cost of a product* P.

To elucidate the stability of the equilibrium states, we must first clarify what changes will occur in the price after a small random deviation from the equilibrium values $P_1^*$ and $P_2^*$. From Fig. 4.10 it can be seen that for the point $a_1$ the difference between the price $P$ and the value $P_1^*$ corresponds to the difference $D - S$, producing a change in price which restores the disturbed equilibrium thus the point $a_1$ represents the state of a stable equilibrium of the system. For the point $a_2$, however, any deviation in price from the value $P_2^*$ will cause a further change in the same direction, thus the state of the system at this point is unstable.

## 4.5. Cycles

The 'stability' concept applies not only to estimating the type of *equilibrium state* of a system, but is also of great importance in estimating the type of motion of the system. Thus, it is very important to clarify the problem of stability of cyclic movements (that is, movements along any closed trajectory in the state space) this applying to periodic movements in particular.

Let us consider as an example a pendulum with friction, shown in Fig. 4.11, which is distinguished from the ordinary pendulum by the fact that the shaft on which it is suspended revolves with a constant angular velocity $\omega_0$. During the oscillations of the pendulum the angular velocity $\omega$ of the bushing will change periodically and therefore at some instants of time the direction will coincide with the direction of angular velocity $\omega_0$ of the shaft rotation. During the intervals $\tau_1$ of the lag of the bushing against the movement of the shaft $(\omega < \omega_0)$ the friction forces between them will accelerate the pendulum, whilst during the remaining part of the period $\tau_2$ the friction will decelerate its movement.

The friction work during the intervals $\tau_1$ will increase the reserve of energy in the pendulum by the magnitude $\Delta E_1$ over the period of one oscillation, and during the interval $\tau_2$ it will lower the reserve of energy by the magnitude $\Delta E_2$. Calculations show that the dependence of the energy increment $\Delta E_1$ and the energy decrement $\Delta E_2$ on the amplitude $A$ of the oscillations, is as shown in Fig. 4.12. Accumulation of energy in the pendulum is approximately proportional to the amplitude, whilst the dissipation of energy is proportional to the square of the amplitude.

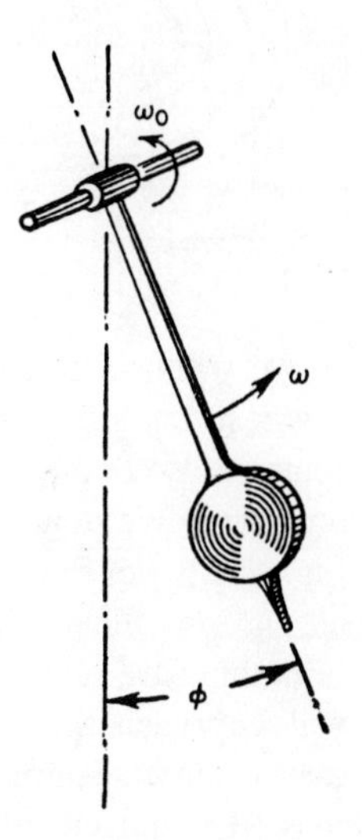


Fig.4.11. *Pendulum with friction.*

It is obvious that the energy stored in the pendulum and consequently the oscillation amplitude, will remain unchanged only if a balance is retained between the increment and the dissipation of energy, i.e. $\Delta E_1' = \Delta E_2'$. As can be seen from Fig. 4.12, such a balanced regime may occur for the amplitude $A^*$. The oscillations of a pendulum with the amplitude $A^*$ are represented in Fig. 4.13 by the closed curve 1 on the plane $\phi$, $\omega$, where $\omega$ is the deflection of the pendulum from the vertical axis.

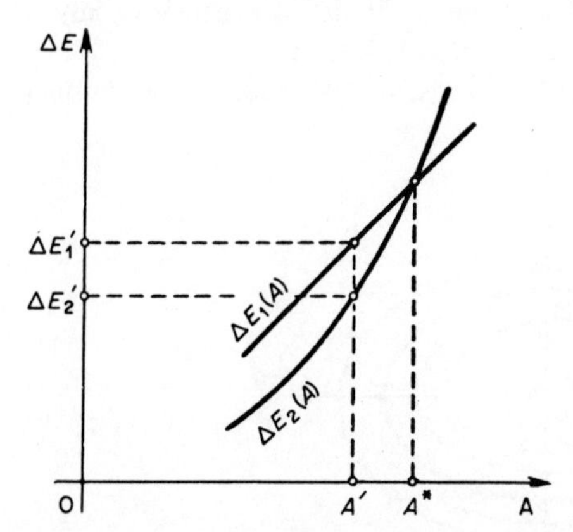


Fig. 4.12. *Dependence of the changes in the energy of the pendulum on the amplitude* A *of the oscillations.*

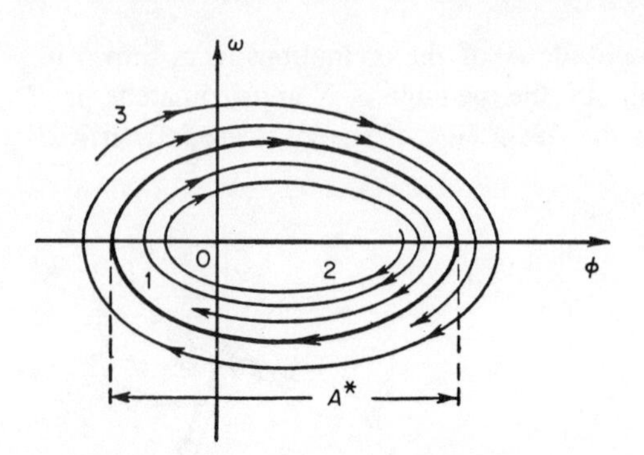


Fig. 4.13. *Phase portrait of a pendulum with friction.*

We will now elucidate the problem of whether the movement along trajectory 1 is restored after random deviations from this trajectory. Let us assume that for some reason the reserve of energy in the pendulum decreases and the representative point is inside the area delineated by the trajectory 1. Then, from Fig. 4.12, we find that $\Delta E_1' > \Delta E_2'$ and the reserve of energy during each oscillation will increase, resulting in an increase of amplitude. In this case the representative point will move along trajectory 2 and with the progress of time will approach the initial trajectory 1. Applying similar considerations for the case when the initial energy reserve exceeds its balanced value, it can be seen that the movement will occur along trajectory 3, which also approaches trajectory 1.

This consideration permits us to conclude that trajectory 1 represents *stable oscillations of the system*, which become established regardless of the initial state of the system. The frequency and amplitude of the oscillations do not depend on the initial conditions. Such oscillations are referred to as free oscillations. The difference between free oscillations and forced oscillations is that the external forces which excite them are non-periodic for the former and periodic for the latter case. For instance, in the given example the shaft revolves at a constant speed and the oscillation frequency is only indirectly associated with this speed.

The trajectory which represents stable oscillations in the phase space is called a *stable limit cycle.*

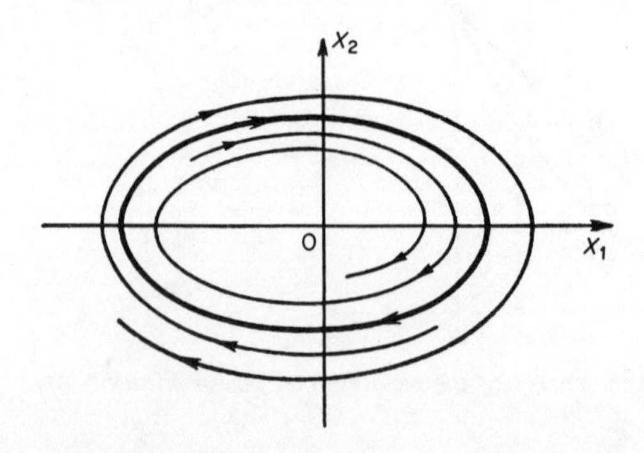


Fig. 4.14. *Unstable limit cycle.*

A closed trajectory will not always represent a stable limit cycle. For instance, the limit cycle shown in Fig. 4.14 is unstable.

Let us consider another example of cyclic movement – the simple system of controlling the traffic at a crossroad. We denote the presence of vehicles waiting for permission to cross by $X_1$, by $X_1 = 1$ if there are vehicles waiting for permission to cross, and by $X_1 = 0$ in the opposite case; we designate by $X_2$ the permission to cross (green traffic light). Here $X_2 = 1$ if there is permission to cross and $X_2 = 0$ if the opposite is the case.

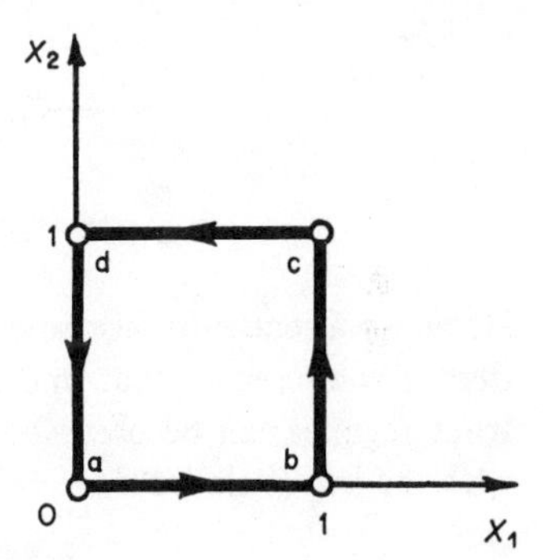


Fig. 4.15. *Diagram of transitions of a system of road traffic control.*

The state of such a system will be determined by the position of the representative point on the plane $X_1$, $X_2$, which can only occupy one of four positions shown in Fig. 4.15. The system can be so constructed that transition from any $i^{th}$ state (characterised by the values of the coordinates of $X_{1i}$, $X_{2i}$) in the $(i+1)^{th}$ state ($X_{1(i+1)}$, $X_{2(i+1)}$ is realised by acting on the operand $X_{1i}$, $X_{2i}$ with the operator $P$

$$\{X_{1(i+1)}, X_{2(i+1)}\} = \{P\}\{X_{1i}, X_{2i}\},$$

where the operator $P$ realises the table of transitions (Table 4.1).

Table 4.1

| | | | | |
|---|---|---|---|---|
| $X_{1i}$ | 0 | 1 | 1 | 0 |
| $X_{2i}$ | 0 | 0 | 1 | 1 |
| $X_{1(i+1)}$ | 1 | 1 | 0 | 0 |
| $X_{2(i+1)}$ | 0 | 1 | 1 | 0 |

It can easily be seen that the system will successively change into each of the four possible states, performing the cycle *a-b-c-d-a*, and the trajectory of its representative point will be a square, as shown in Fig. 4.15.

## Exercises

1. Fig. 4.16 shows a simplified diagram of a tube oscillator with an inductive feedback and an oscillation circuit in the grid loop. The relative disposition of the turns of the coils L and $L_{cb}$ determines the coefficient of mutual inductance

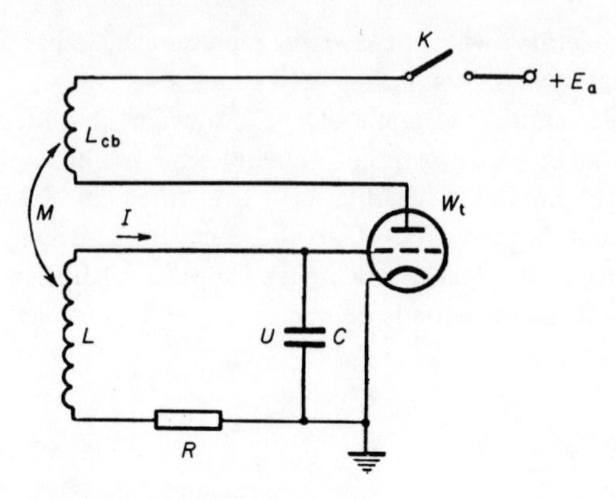

Fig. 4.16. *To Example No. 1.*

*M*, and consequently also the operating regime of the oscillator. Let us assume that *M* is chosen so that undamped oscillations are generated in the circuit LC. What regimes can be observed in this dynamic system with the key K open and with the key K closed?

*Solution.*
(a) Equilibrium – with the key K open.
(b) Transient – on closing the key.
(c) Periodic – from a certain time after termination of the transient process.

2. In Fig. 4.16 the dependence of the amplitude of the voltage oscillations *U* of the condenser C on the coefficient of mutual induction *M* can be characterised by a curve as shown in Fig. 4.17. Draw the phase portraits in the plane *U, I*

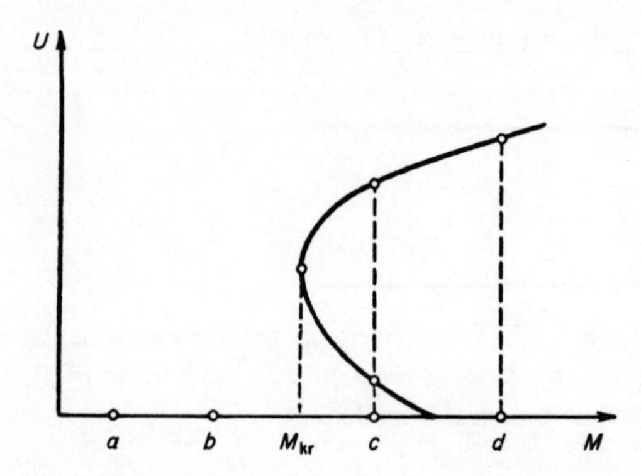

Fig. 4.17. *To Example No. 2.*

(*I* – current in the circuit LC) for values of *M* designated in Fig. 4.17 by the letters *a, b, c, d* under the condition that at the instant of switching on the oscillator a current jump occurs in the circuit.

*Solution.* The phase portraits are shown in Fig. 4.18.

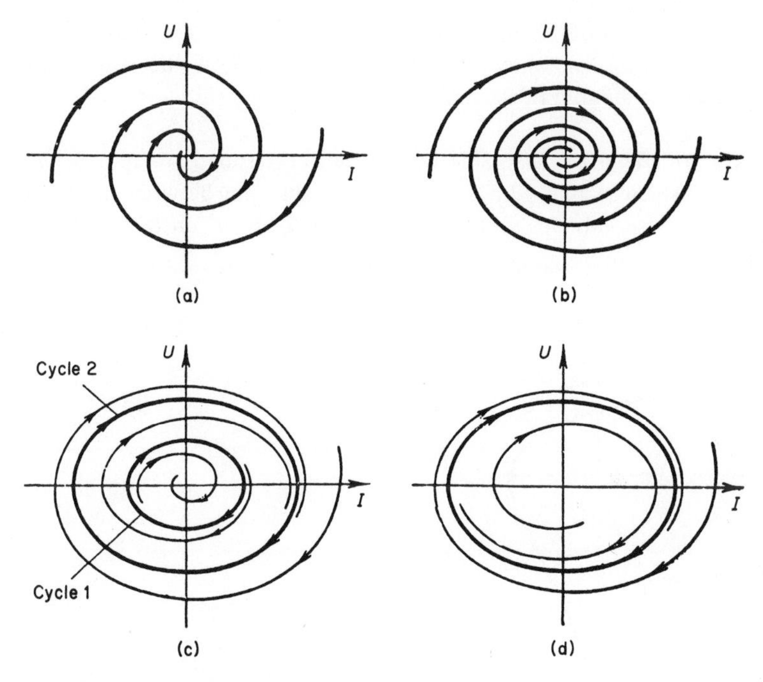


Fig. 4.18. *To Example No. 2.*

**3**. From the phase portrait plotted in Example 2 for the point b, determine which of the cycles is a stable limit and which is unstable.

*Solution.* Cycle 1 – unstable; cycle 2 – stable.

**4**. If the system has a point of stable equilibrium and a stable limit cycle, will an unstable cycle always exist and if so, how will it be positioned on the phase portrait?

*Solution.* Yes, always. On the phase portrait it will be positioned between the stable point and the limit stable cycle.

**5**. Draw the phase portrait of the operation of the heart, selecting as coordinates the rate of filling of the left and right ventricles.

*Solution.* The phase portrait is analogous to the phase portrait shown in Fig. 4.18(d) and represents a stable cycle.

**6**. Can an equilibrium regime occur in the cycle?

*Solution.* No.

7. Find the points of a stable and unstable equilibrium of an induction motor, the characteristics of which are shown in Fig. 3.7, if the load moment is independent of the rpm of the motor.

*Solution.* In Fig. 4.19 the stable equilibrium corresponds to rpm between $n_0$ and $n_k$; unstable equilibrium corresponds to the rpm between $n_k$ and 0. The point 1 corresponds to a stable, and point 2 to an unstable condition of operation of the motor.

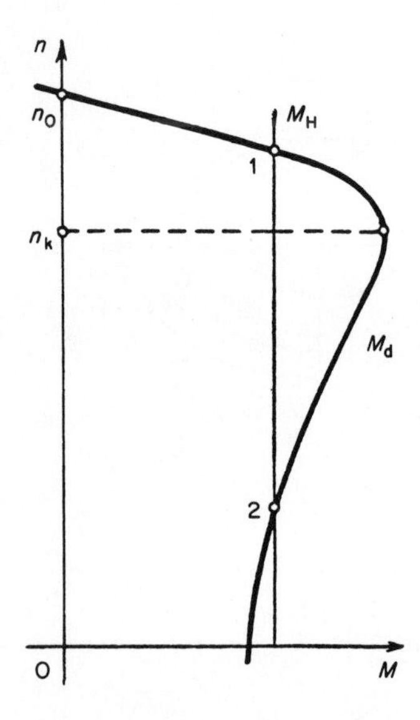


Fig. 4.19. *To Example No. 7.*

8. On a cathode ray oscillograph some periodic process is studied. The voltage $V_x = \sin \omega t$, proportional to the coordinate $X$ of the process, is fed to the horizontal deflecting plate, and the voltage $V_y = \sin 2\omega t$, proportional to the coordinate $Y$, is fed to the vertical plate. What will be the trajectory of the beam on the oscillograph?

*Solution.* The trajectory is shown in Fig. 4.20.

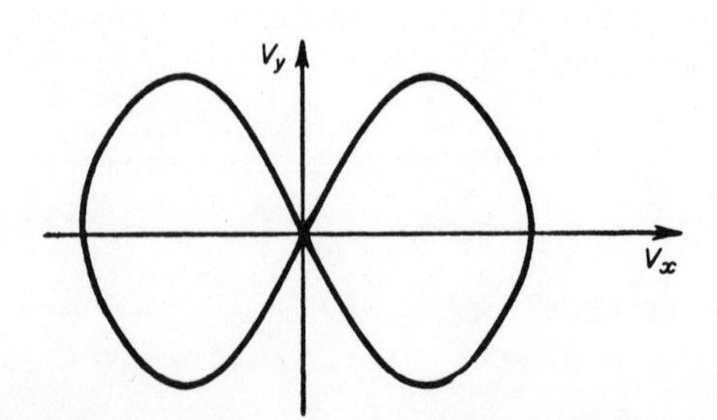


Fig. 4.20. *To Example No. 8.*

**9.** Usually the time-tables of buses in towns are drawn up in such a way that the buses run at regular intervals. However, due to bunching of passengers at any bus stop, a bus may be delayed beyond the time-table time, and the next bus will follow behind at too short a time interval and will take fewer passengers than usual. Is the resulting non-uniform movement stable or unstable?

*Solution.* Unfortunately, such a non-uniformity is stable and represents a continuous source of complaints by passengers and a headache for the transport organisers.

# 5 Signals

Between the elements of any system and between the systems as a whole there are connections through which they interact. These inter-connections may consist of exchange of energy or matter between the interacting objects. However, there may also be connections in which the predominant factor is the *information* content of the inter-connections, i.e. information sent to one object on the state of the other object. The material form of this information is of secondary importance. Such information links are realised by means of *signals* which circulate in cybernetic systems.

A signal can be transmitted over distances and as a result connections can be established between objects which are isolated in space. Storage of signals allows their reproduction and transmission with time delay, and thereby the connection of objects which are separated in time.

## 5.1. Coding

We have so far considered systems for which the state space can be discrete or continuous. In the discrete state space we assumed the possibility of existence of states represented only by separate points, whilst in the continuous space we assumed the possibility of existence of states represented by *any* point in some permitted range of the state space.

If the state of a system is to be observed, there must be some possibility of estimating the values of its coordinates. It appears that not one of the methods of observation can provide the observer with absolutely accurate data on the values of the system coordinates, although the visual observation and any measurement are characterised by a certain finite resolution.

If for instance we use a ruler to measure the length of two rods which differ from each other by less than the width of a scale division on the ruler, then we will obtain equal lengths for both rods. Errors, noise and fluctuations of the measured values limit the accuracy of any measurement. This means that two observations will usually be approximately the same if the measurements are distinguishable but will not necessarily be precisely the same.

Therefore, even systems which are described mathematically as being continuous, will under certain concrete procedures appear to the observer as discrete systems with a finite number of possible states.

We will consider the *set of states* of a system which includes all the distinguishable states. Let us assume, for instance, that we are interested in the state of a man as characterised by his body temperature. If we exclude the case of artificial hypothermy, then it can be assumed that the temperature of a living man will remain within the range 34 to 42°C. It is obvious that the temperature may assume any value within this interval. In practice, however, we measure the temperature with an ordinary medical thermometer whose resolution equals 0.1°C. Therefore the set of states in the given case will consist of a total of 81 elements (the element is a certain value of the temperature), namely: (1) 34.0°C; (2) 34.1°C; (3) 34.2°C; . . . (81) 42.0°C. If the state of the system is represented by a vector, whose components can assume independently of each other: $X_1 = r_1$ values, $X_2 = r_2$ values, . . . $X_n = r_n$ values, then the number of elements (i.e. the number of all possible sets of values $X_1, \ldots, X_n$) belonging to the sets of states of the system will equal

$$N = r_1, r_2, \ldots r_n.$$

We will call an *event* the state of a system at a given instant of time. If in any instant of time $t$ the system may be in any of its states from the set $X$, then $(X, t)$ will represent a set of possible events for each instant of time.

It is possible to specify for each state of the system a corresponding definite value or sequence of values of any physical quantity. By means of this quantity it is possible to realise the transmission of messages (information about events) from one object to another.

*A physical process which is the material embodiment of a message is called a signal. The system or the medium in which the transmission of the signal takes place is called a communication channel.*

Since each state of the system $X$ corresponds to a certain message $X_c$, then the set of possible events corresponds to the *set of messages* transmitted by means of signals. Thus, for instance, information on the state of the atmosphere at a given point is transmitted to the meteorological office by radio or telegraph, using a certain sequence of electric signals.

Formation of a message can be considered as a transformation of the state of the system $X = (X_1, X_2, \ldots, X_n)$ into $X_c$ – one of the set of possible messages $X_c = (X_c^1, X_c^2, \ldots, X_c^n)$ produced by a certain operator $P$:

$$X_c^i = (P)(X_i).$$

The operator $P$ which transforms any operand into its transform (message) is called the *code*, and the operation of transformation is called *coding*. Not only the state $X$ of the system or the events $(X, t)$ but also the message $X_c^i$ can be considered as the operand of such a transformation. The transcoding of a message proves necessary when for convenience or for secrecy of the transmission it is necessary to transform the message $X_{c_1}^i$ coded by one method into a message $X_{c_2}^i$ coded by a different method. Such a transformation of messages can be represented as a sequence of applications to $X$ operators of $P_1$, $P_2, \ldots, P_l$ in accordance with the scheme

$$X_{c_1}^i = P_1 X_i, X_{c_2}^i = P_2 X_{c_1}^i, \ldots X_{c_l}^i = P_l X_{c_{l-1}}^i .$$

To re-establish the transmitted message, even when it has been transcoded several times, it is not necessary to re-establish all the intermediate codes. It is sufficient to carry out on the message $X_{c_1}^l$ a single transformation

$$X_i = P^{-1} X_{c_l}^i ,$$

where

$$P^{-1} = P_l^{-1} P_2^{-1} \ldots P_l^{-1} .$$

The sign '$-1$' designates operators which realise that transformation, which is the inverse of the transformation realised by an operator without such a sign. Thus, the operators $P_k$ and $P_k^{-1}$ are interrelated as follows:

$$P_k^{-1} P_k X_i = X_i .$$

One example of a complex system of transmission, its coding and decoding, is given by the televised transmission of images. There, the following chain of transformation occurs: the state of the system (i.e. the brightness distribution of the transmitted picture) – optical image on the iconoscope screen – electrical signal in the form of varying current in the circuit formed by the electron beam, scanning the screen – radio signal in the form of electromagnetic waves of varying frequency radiated by the transmitter – image on the tube of the receiver produced as a result of the inverse transformation (decoding) of the message received by the television receiver. These are all stages of the total process.

## 5.2. Information

From what has been said above, it can be seen that a signal may contain some information on the state of a system, on events or on a process, but it is still not clear whether it is possible to measure quantitatively the data contained in the signal or to calculate the amount of information carried by the signal. Such a measure is necessary, however, for calculating the capacity of channels over which signals are transmitted. This may be necessary in order to determine the characteristics of the equipment which transforms for example the signals and suitable methods of coding.

The problem of quantitative measurement of information processing is not yet fully resolved, but for a rather broad class of problems, associated with the transmission of information from one object to another, with coding, recording and storage of information, has been achieved in formulating quantitative theory. This was primarily achieved by the American scientist Claude Elwood Shannon. If we dissociate ourselves from the semantic aspects of the information, from is value to the receiver and the form in which it has been preserved, then any message can be considered as information on a certain event $X_i$, $t_i$, containing data on which of the set of possible states the system S is at the instant $t_i$.

Let us consider in greater detail so-called discrete messages. *Discrete messages* are sequences of symbols taken from some set of symbols – the *alphabet*. Each individual symbol is called a *letter* of the alphabet. An example of discrete messages is the ordinary text in any language (for instance, in English). Note that the alphabet of such a message is larger than the English alphabet proper: in addition to letters it contains spaces between words (a very important symbol!) and punctuation marks. The finite sequence of symbols taken from some alphabet is called a *word in the given alphabet*.

Use of discrete messages permits the transmission of data about some state, chosen from any large number of possible states, by means of a few different symbols taken from the alphabet (the number of these symbols is called the *basis of the code*). For instance, any message, however complicated, can be transmitted by a sequence of only two different symbols, say the symbols 0 and 1, which can respectively correspond to: 0 – absence of signal, 1 – presence of signal. If the system can be in one of $N$ different states $X_i$, the set of which $X_1$, $X_2, \ldots, X_N$ is known to the receiver (recipient) of the message, then in order to transmit the information on the state of the system it is sufficient to indicate the number $i$ ($i = 1, 2, \ldots, N$) of the state in which the system is. This number is a word in the alphabet, and the letters are figures.

The quantity of the various figures forming the number depends on the code used. In the binary system each number is expressed by a combination of zeros and ones, which form the digits of the number. Any number $i$ can be written as follows:

$$a_m \, a_{m-1} \ldots a_1 \, , \tag{5.1}$$

where each $a$ can assume only two values: 0 and 1, and the expression (5.1) denotes

$$i = a_m 2^{m-1} + a_{m-1} 2^{m-2} + \ldots + a_1$$

If, for instance, the number $i$ in decimal form of writing is

$$i = 27$$

then in the binary form of writing it will be

$$i = 11011$$

$$(1.2^4 + 1.2^3 + 0.2^2 + 1.2 + 1). \tag{5.2}$$

The word (5.2) can be transmitted by a sequence of signals consisting of current pulses and pauses, in a sequence as shown in Fig. 5.1.

A message about any event can therefore be written as a word in the two-letter alphabet. It can easily be seen that the number of various binary sequences of the length $m$ is written as $2^m$, because each symbol can assume two values independent of any others. Therefore, using the binary sequence of the length $m$ it is possible to transmit a message on an event selected from $N$ possible events, where $N = 2^m$, which may also be written as $m = \log_2 N$. If this message were not transmitted by a binary, but by a decimal, code (i.e. using ten different symbols instead of two), we would require a sequence of the length $m' = \log_{10} N$. In this case $m' = m \cdot \log_{10} 2$, i.e. $m'$ would differ from $m$ by a constant factor which does not depend on $N$.

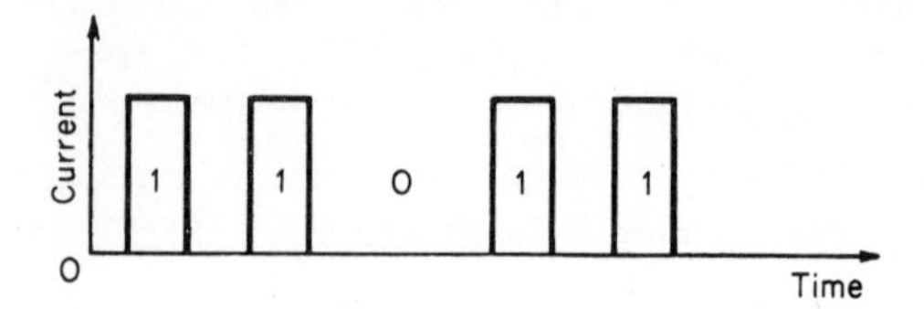


Fig. 5.1. *Signal (sequence of current pulses) corresponding to the binary message 11011.*

Generally speaking, whatever the basis of the code we use, the length of the sequence which is necessary for transmitting some message is proportional to the logarithm of the number of possible events. It is reasonable to assume that the maximum amount of information $H_{max}$ contained in a message is proportional to its length, i.e.

$$H_{max} \sim m \sim \log N.$$

In selecting the coefficient of proportionality we choose the basis of the logarithm, and this means choosing a unit for the amount of information. Most frequently the logarithm to the base 2 is used for this purpose. In this case, as a unit of information we take the information contained in one binary digit, i.e. in the choosing of one of two possible messages. Such a unit of information is referred to as a *bit* (from the words binary digit).

We obtain

$$H_{max} = \log_2 N. \tag{5.3}$$

This measure of the maximum information which can be obtained in a message was proposed well before the theory of information was elaborated in 1928 by the American scientist L. Hartley. The measure proposed has two important properties: it increases monotonically with increasing $N$ and is additive. The additive property means the following. Let us assume that a message a is selected from $N_1$ possible messages, and independently of this the message b is selected from $N_2$ possible messages. What information will be contained in the combined message consisting of the messages a and b? It is obvious that the number of all possible messages will equal $N_1 \cdot N_2$. Consequently

$$H_{max}(N_1 N_2) = \ldots$$

i.e. the amount of information in the two independent messages equals the sum of the amounts of information in each of these messages, which is in agreement with our intuitive notion of information.

The magnitude of $H_{max}$ indicates the upper boundary of the amount of information which can be contained in a message. Indeed, the amount of information depends not only on the number of possible messages but also on their probabilities. For instance, the amount of information contained in the message which says whether a newborn baby is a boy or a girl is considerably greater than in the message which says whether or not your friend gave birth to triplets. Indeed the cases of triplets being born are rare and you can almost forecast with certainty that this will not happen.

It is essential to emphasise that we do not discuss the actual amount of information in a particular statement (for instance in the message 'triplets were born'), but the *average* amount of information contained in a message, selected with a given probability, from a set of possible messages. Since the unexpected message 'Triplets were born' is very rarely encountered and in most cases of childbirth we get the expected message 'Triplets were not born', the average amount of information will be small. In the limit case, when the probability of all messages, except one, becomes zero, the amount of information equals zero – in this case it is known in advance what message will be received and, hence, the message does not include anything new; it contains no information. On the other hand, when *a priori* (before obtaining the message) all possible messages are equally probable, the amount of information must be a maximum. This intuitive concept helps us in understanding the sense of the quantitative measure of information, introduced by Shannon in 1947.

Let us now assume that an experiment was carried out, the result of which was not known in advance. It was only known that the set of possible results was $x_1, x_2, \ldots, x_n$ and the probabilities of the results were $p(x_1)$, $p(x_2)$, ..., $p(x_N)$.* According to Shannon the amount of information in the message (in bits) which depends on the result of such an experiment equals

$$H(X) = -\sum_{i=1}^{i=N} p(x_i)\log_2 p(x_i). \tag{5.4}$$

* Using the terminology from the theory of probability, we can say that $X$ is a *random variable*, which assumes the values $x_1, \ldots, x_N$ with the probabilities $p(x_1), \ldots, p(x_N)$.

If all the results were equally likely, then $p(x_i) = 1/N$ for all $x_i$ and

$$H = \log_2 N = \max.$$

If $p(x_k) = 1$ for any result $X_k$, and for the other results $p(X_i) = 0$ $(i = k)$, then $H = 0$. (the value 0 log 0 is assumed to equal zero). In other cases the following equality is valid:

$$0 < H < \log_2 N.$$

The value $H$ is called the *entropy* of a random experiment (the random variable) $X$. It is a measure of the initial uncertainty of the outcome of the experiment, a measure of the statistical variety of its possible results. The message on the outcome of the experiment fully 'removes' this uncertainty and therefore yields the amount of information $H$.

In theoretical investigations, it is usually convenient to use natural logarithms rather than logarithms to base 2. The appropriate unit of information is called a natural unit (abbreviated to 'nit' or 'nat'). The entropy in natural units equals

$$H(X) = -\sum_{i=1}^{i=N} p(x_i) \ln p(x_i). \tag{5.5}$$

The expression (5.5) is identical with the expression for the entropy in statistical physics (if we interpret the outcome $x_1, \ldots, x_n$ as various states of a physical system). This coincidence is not only formal. As long ago as the beginning of the present century, the German physicist Soltzman wrote: 'Entropy is a measure of lack of information about the state of a physical system'. By now, a deep interrelation has been established between the theory of information and statistical physics, which permits the development of a theory of information as a physical theory.

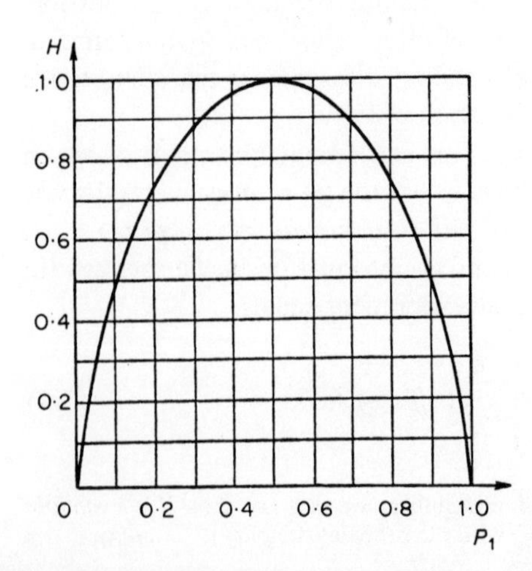


Fig. 5.2. *Entropy of an experiment with two possible results:* $p_1$ *– probability of one of the outcomes.*

Fig. 5.2 shows the change in entropy of an experiment with two possible results $x_1$ and $x_2$, as a function of the probability $p_1$ of the result of $x_1$ (probability of the result $x_2$ will then obviously be $p_2 = 1 - p_1$).

### 5.3. Transmission of signals

The transmission of signals is always realised through some communication channel: a telephone line, a radio channel, the air, a magnetic recording channel, etc. Communication channels usually also include coding, transcoding and decoding equipment. A communication channel can be considered as a system in to which the input signal $Y$ is fed, whilst the output signal $X$ is obtained, as shown in Fig. 5.3. In the same way as any other system, a communication channel will be modified by internal and external perturbations $M$. The sources of these perturbations are random external and internal noises, and in some cases represent faults of components which make up the channel. Under the effect of the perturbations, signals transmitted along the channel may become distorted or may not reach the receiving end at all.

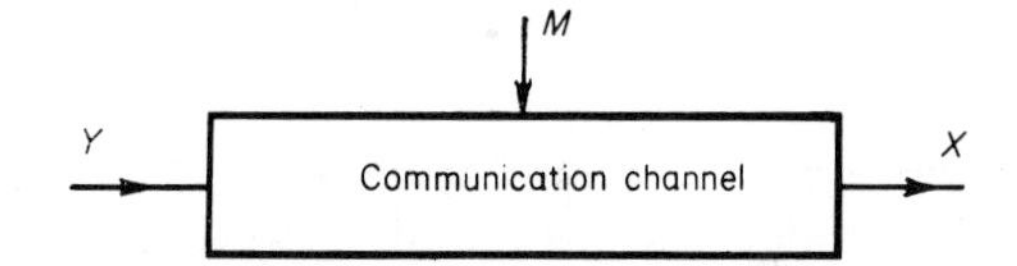


Fig. 5.3. *Communication channel.*

Let us assume, for example, that you receive a message in Morse Code, using radio-telegraphy under conditions of strong atmospheric perturbations. Then some pulses of the signal (dots and dashes) will be 'drowned' by the noise and conversely the noise pulses could be mistaken for the signal. Therefore, if noise is present, the output signal $X$ is not a one-to-one function of the input signal $Y$, but associated with it only by probabilistic or statistical relationships. The question arises of how to measure quantitatively information content derived from the knowledge of one random variable (of the output signal $X$) on the value assumed by another random variable (the input signal $Y$). This most generalised measure of the amount of information was also introduced by Shannon. In order to explain this concept, we must first be acquainted with the notion of *conditional probability*.

Let us consider as an example the production of components from the raw materials of two different qualities I and II. Let us assume that 70% of the components are manufactured from the material I and 30% from the material II. Of each hundred components made from the material I, 83 are satisfactory, and of each hundred components made from the material II, 63 are satisfactory.

The probability $p$ of the event $A$, that a randomly chosen component will be satisfactory (let us denote this component by $p(A)$), can easily be calculated as the mean value of the satisfactory components from the entire production

$$p(A) = \left(\frac{83}{100} \cdot \frac{70}{100}\right) + \left(\frac{63}{100} \cdot \frac{30}{100}\right) = 0.77.$$

Let us assume now that it is known that a component is produced from the raw material I. Then the probability that this component will be satisfactory is 0.83. Obviously, the addition of a condition (in the given case information on the quality of raw material) changes the probability of the event. Let us denote by $B$ the condition on which $A$ depends, and the probability of the event $A$ under the condition $B$ – the *conditional probability* of this event – by $p(A/B)$; we can now write for the given example

$$p(A/B) = 0.83.$$

Generally, if the random variable $Y$ assumes the value $y_1, y_2, \ldots, y_N$ and the random value $X$ assumes the values $x_1, x_2, \ldots, x_M$, then the *conditional probability* $p(y_i/y_j)$ is the probability that $Y$ will assume the value $y_i$, if it is known that $X$ has the value $x_j$. The unconditional probability $p(y_i)$ equals the conditional probability averaged for all possible values of $x$:

$$p(y_i) = \sum_{j=1}^{M} p(x_j)p(y_i/x_j),$$

where $p(x_j)$ is the probability of the $j^{\text{th}}$ value $X$. The value $p(x_j)p(y_i/x_j)$ is the probability that $X$ will assume the value $x_j$, and $Y$ the value $y_i$. This is called the *joint probability* of the event $(x_j, y_i)$ and is denoted $p(x_j, y_i)$.

We will now try to find out what information is received at the receiving end of the communication channel about the signal which has been sent. The initial (*a priori*) uncertainty of the signal $Y$ equals its entropy $H(Y)$

$$H(Y) = \sum_{i=1}^{N} p(y_i)\log_2 p(y_i).$$

If the received signal $X$ were unequivocally associated with the sent signal, the uncertainty would cease after reception of the signal and we would obtain an amount of information equalling $H(Y)$. In reality, however, after receiving for example a signal $x_j$ the uncertainty of the signal $Y$, which was actually sent, becomes:

$$Hx_j(Y) = -\sum_{i=1}^{N} p(y_i/x_j)\log_2 p(y_i/x_j).$$

since at that time only the conditional probabilities $p(y_i/x_j)$ of various values of $Y$ become known.

The received signal $X$ can assume any value $x_1, \ldots, x_M$ with the probabilities $p(x_1), \ldots, p(x_M)$, then the *average uncertainty of the transmitted signal, if the received signal is known*, equals:

$$H(Y/X) = -\sum_{j=1}^{M}\sum_{i=1}^{N} p(x_j)p(y_i/x_j)\log_2 p(y_i/x_j).$$

This is the *conditional entropy* of the random variable $Y$ for a given random variable $X$. The conditional entropy is always smaller, or at least is never larger, than the unconditional one:

$$H(Y/X) \leqslant H(Y).$$

the equality taking place only if knowledge of $X$ does not change the probabilities of the values of $Y$, i.e. if $P(y_i/x_j) = P(y_j)$, whatever the value of $x_j$.*

It is natural to use as a measure of the *amount of information in the random variable X about the random variable Y the value by which the uncertainty of Y* decreases (on average) if the value of $X$ becomes known, i.e. the difference between the unconditional and the conditional entropies $I(X, Y) = H(Y) - H(Y/X) =$

$$\sum_{j=1}^{M}\sum_{i=1}^{N} p(x_j, y_i)\log_2 \frac{p(x_j, y_i)}{p(x_j)p(y_i)} \tag{5.7}$$

It is proved in the theory of information that the measure of the amount of information, introduced here, is very useful, and, in particular, it permits a full solution of the problem of what the communication channel should be like in order that messages generated by some message source can be transmitted over it, regardless of the real nature and meaning of these messages, whether they be human speech, music, visual images, meteorological data, nerve pulses in the living body or anything else.

An important property of the amount of information is its uniqueness and symmetry: $I(X, Y) = 0$ and $I(X, Y) = (Y, X)$. The latter means that the amount of information in the received signal about the transmitted signal equals the amount of information in the transmitted signal about the received one. The amount of information equals zero if the input and output signals are independent, i.e. if they are not in any way associated, even statistically. The amount of information reaches a maximum – the value $H(Y)$, if the received signal uniquely determines the transmitted one, i.e. $H(Y/X) = 0$ which is the uncertainty of the transmitted signal for a known received signal equals zero.†

* In this case the random variables $X$ and $Y$ are called *independent* variables.

† It is important that a uniquely determined value of $Y$ should correspond to each known value of $X$. However, several values of $X$ may correspond to one and the same value of $Y$.

In the general case the amount of information satisfies the inequalities

$$I(X, Y) \leqslant H(Y) \text{ and } I(X, Y) \leqslant H(X).$$

Let the input of the channel be a signal $Y$ which can assume only two values, $y_0 = 0$ and $y_1 = 1$, with the probabilities $p(y_0) = p(y_1) = 0.5$. Then, as can be seen from Fig. 5.2, $H(Y) = 1$ bit for each letter. Let us assume that on applying the signal $y_1 = 1$ to the input the output signal $x_1 = 1$ will appear in 90% of the cases, i.e. the conditional probability $p(y_1/x_1) = 0.9$ and the conditional probability that for the input signal $y_0 - 0$ we will obtain at the output the signal $x_0 = 0$, also equals $p(y_0/x_0) = 0.9$. Then we find from Fig. 5.2 that $H(Y/X) = 0.5$, and the amount of information contained in each letter of the output signal will not be one bit but, according to 5.7, will decrease to the value

$$I = 1 - 0.5 = 0.5 \text{ bits.}$$

The concepts and the relations introduced above for the quantitative measuring of information, contained in messages, is extensively used in cybernetics. They are used not only for calculating communication channels, but also for evaluating the capacity of memories, for finding quantitative characteristics of information processing in cybernetic systems and for choosing the structure of information networks in complex systems.

In later chapters we will also meet problems where the amount of information transmitted is of considerable importance.

## 5.4. Memory

For realising connections between systems which are separated in time, the communication channels must contain 'reservoirs' where information can be stored during the intervals between the instant of generation of a message from the instants when the message may be required.* The existence of such 'reservoirs' for storing information in biological systems is obvious from practical observation. If they were not available it would be impossible to explain the undeniable fact of the existence of memory in man and animals. Organisation of a memory in artificial systems is one of the most important and most difficult problems of communication and control engineering.

For storing information, it is obviously necessary to transform the real physical signal, which in the general case varies with time and space $X_c = X_c(x, y, z, t)$, where $x, y, z$ are spatial coordinates, and $t$ is time, in some sequence of signal $F_c = F_c(x, y, z)$, which is independent (or at least only slightly dependent) on time. For realising such a transformation, it is obviously necessary to have some medium (information carrier) whose state can be changed in the desired manner by means of the signal $X_c$. This medium should

* Due to the properties of the physical world in which we live, transmission of information in time can be realised only in *one direction*: from the past to the future but not from the future to the past.

retain, in the form of sequence $F_c$ without appreciable charges over a sufficiently long time after the actuation has ceased, the original state in which it arrived.

There can be a correspondence between the signal $X_c$ and its sequence $F_c$, if $F_c$ is a homomorphic model of $X_c$, so that to each element of the communication contained in the signal $X_c$ at the time $t$ there will be a one-to-one correspondence to a state of a specific element of the sequence, or trace $F_c$. An accurate reproduction of the initial signal from its sequence, or trace, is not possible, as was explained in Chapter 3 when discussing the properties of homomorphic models, (absence of one-to-one correspondence between the original and its homomorphic model).

The loss of information during the process of memorising and reproduction also occurs as a result of internal and external perturbations M, which are present in any communication channel, as mentioned earlier.

If the stored signal $X_c$ is a one-, two- or three- dimensional function: $X_c = X_c(t)$, $X_c = X_c(x, t)$ or $X_c = (x, y, t)$, it can be mapped respectively by a one-, two- or three-dimensional distribution of the state of the carrier:

$$F_c = F_c(x), F_c = F_c(x, y) \text{ or } F_c = (x, y, z).$$

The time coordinate of the signal will correspond to one of the space coordinates of the carrier. Thus, for instance, recording of sound on a magnetic tape is by transforming $P$ of the signal $X_c(t)$ into a function of magnetic induction $B$ of the tape along its length $x$

$$B(x) = P\,X_c(t).$$

The two-dimensional signal $X_c(x, t)$ or $X_c(x, y)$ can be represented on a two-dimensional carrier, for instance on a photographic film.

It is much more difficult to visualise the artificial memory when it is necessary to memorise a three-dimensional signal $X_c = X(x, y, t)$. Theoretically, such a signal can be memorised by means of a three-dimensional carrier, one of the space coordinates of which $z$ will represent time whilst the other coordinates $x$ and $y$ of the state $F_c$ will represent the distribution of the states in the original at the appropriate instant of time.

The main difficulties occur when we try to visualise the analogue mechanism of memorising a four-dimensional signal $X_c = X(x, y, t)$ by means of a three-dimensional carrier, because in this case the sequence of signals should have four dimensions whilst the carrier will be a material one and consequently will be three-dimensional. It is known that man is able to memorise three-dimensional images which change with the progress of time. This phenomenon can be explained, for instance, by the hypothesis of parallel memorising of two three-dimensional images which, due to their stereoscopic effect, will give as a whole a picture of the changing three-dimensional image. The striking relationship between the capacity of the memory and the

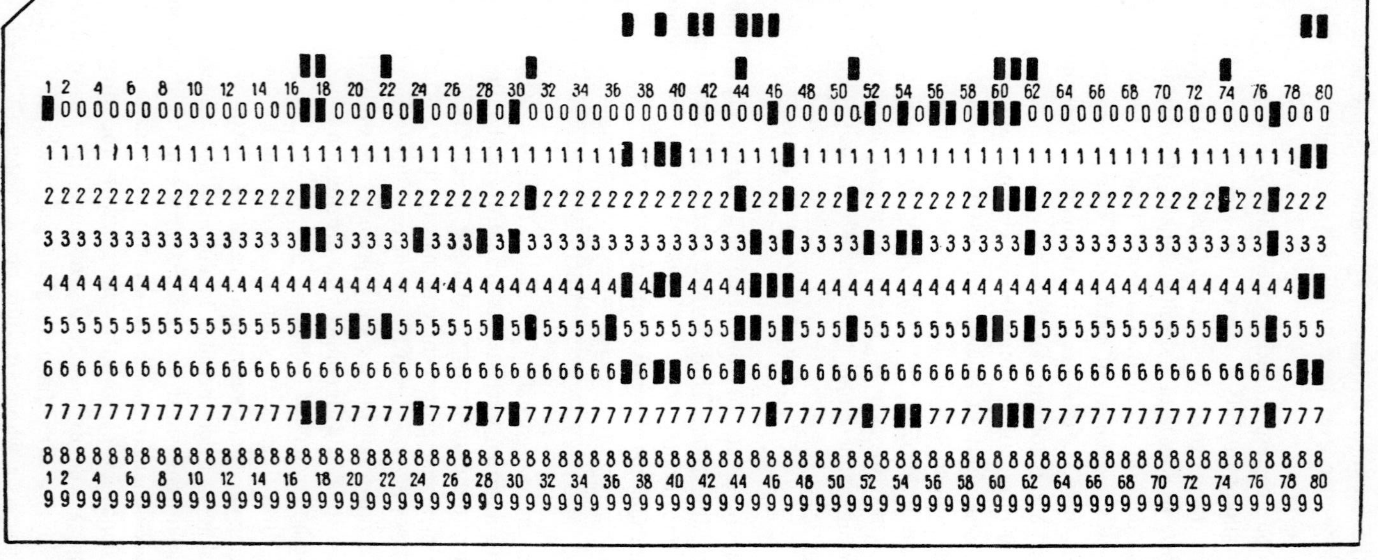

Fig. 5.4. *Standard punched card.*

dimensions of the carrier is also one of the many mysteries of memory in living systems which so far has not been resolved by science.

In the artificial memories used in control engineering, methods are frequently applied which permit reducing the problem of storing complex signals to storage of simpler ones, usually one-dimensional signals, by changing over from the initial continuous signals to discrete ones. For instance, transmission of moving pictures in television by means of a one-dimensional radio signal is based on scanning the images.

The storage, or memorising, of discrete signals (sequence consisting of binary signals) extensively used in automation, computer engineering and communications, is achieved by discrete changes in the state of elements of the memories, such as on punched cards or tape, magnetic drums or discs, sets of ferrite cores, etc.

Fig. 5.4 shows a standard punched card with punched holes representing coded binary numbers.

The principles of design of the most widely used artificial memories are shown in Table 5.1.

**Exercises**

**1.** Rolling stock assembled at a marshalling yard with 7 tracks is channelled into the main track by means of, say, green light signals, each track having its own set of signals. In principle a sequence of 3 signals would suffice, if for each track a code could be made out of a combination of green lights operating over three instants of time. Write this combination for each of the seven tracks. In the given case what is the operand, the image and the operator?

*Solution.* The code combinations are shown in Fig. 5.5. The black dots are the green signals The operand in the given case is the rolling stock standing at the marshalling yard, the operator is the code of green signals, and the image is the rolling stock which has been channelled onto the main track.

**2.** Which of the two transformations $y = \log x$ and $y = \cos x$ is in a one-to-one relationship with regard to coding for x 70?

*Solution.* The first. For the second various values of $x$ may give one and the same value $y$.

**3.** Let us assume that one of your acquaintances gave birth to a child and you ask: 'Did you have a boy or a girl?'. What is the amount of information contained in the answer?

*Solution.* $N = \log_2 2 = 1$ bit. (We consider it is equally likely that the birth is that of a boy or girl).

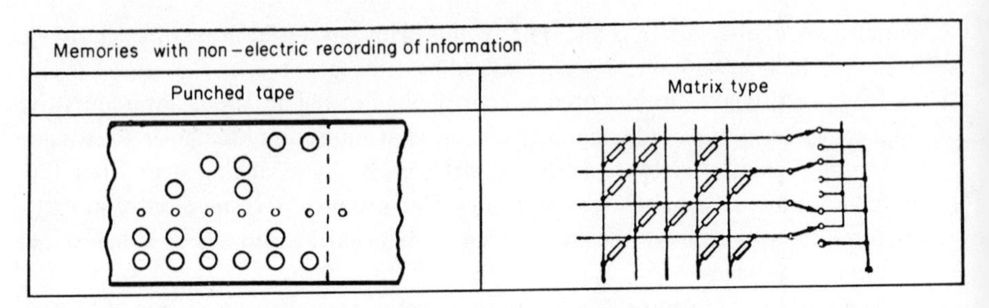


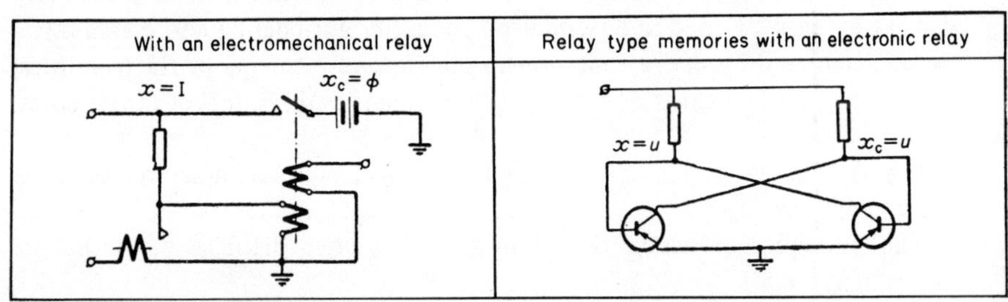


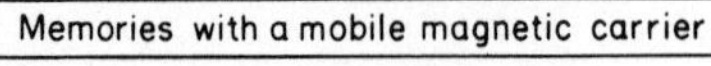


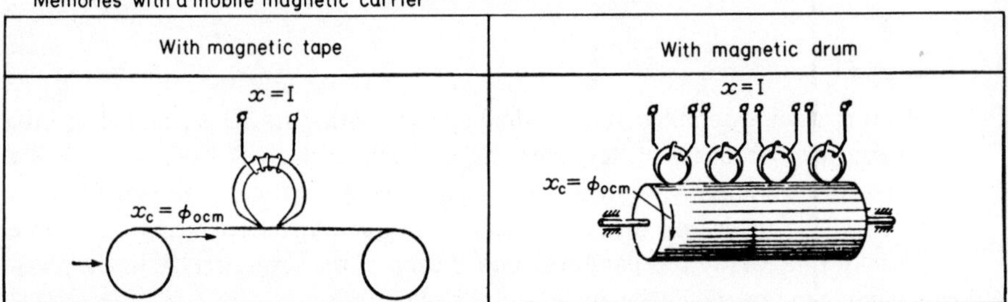


Memories with static magnetic elements

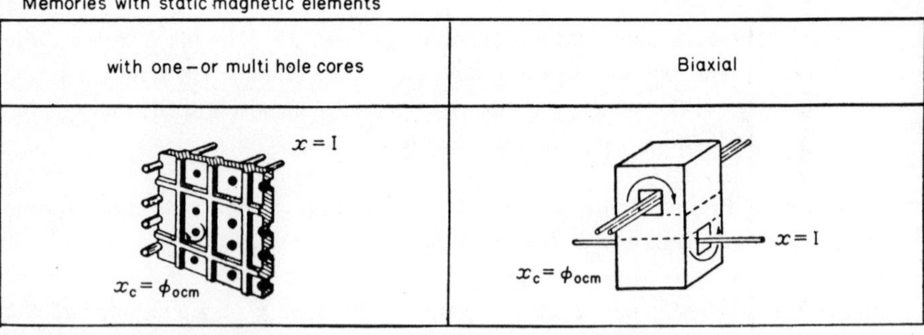


Cryogenic type memories

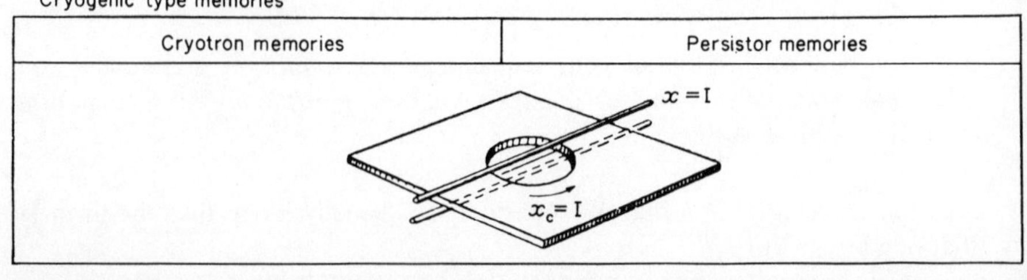

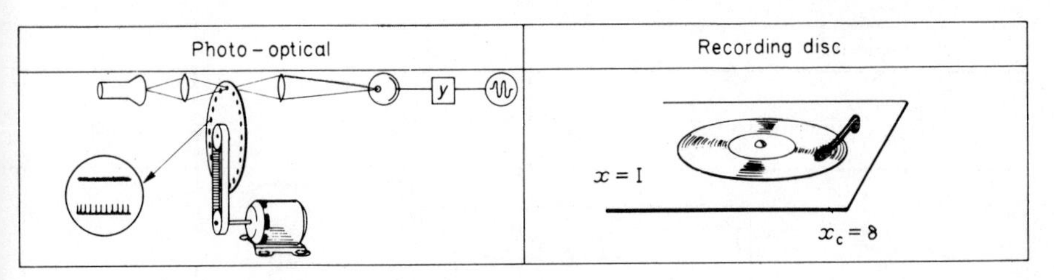
Photo-optical
y
Recording disc
x = I
x_c = 8

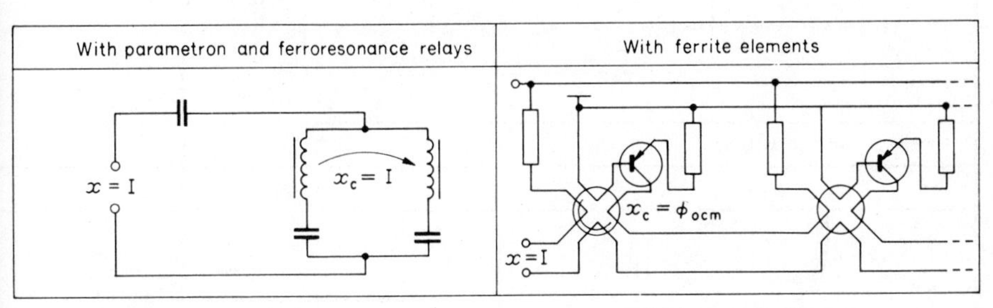
With parametron and ferroresonance relays
x = I
x_c = I
With ferrite elements
x_c = ϕ ocm
x = I

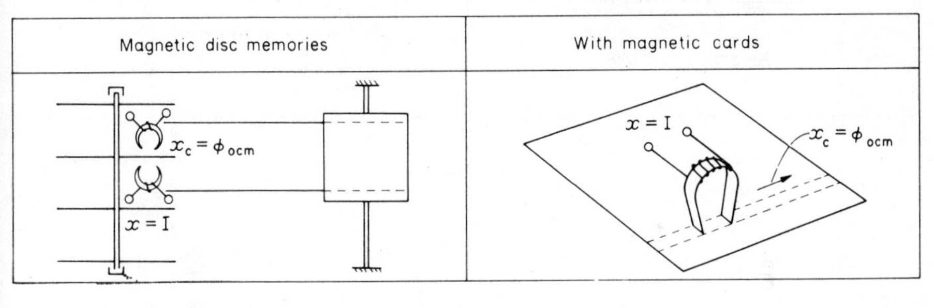
Magnetic disc memories
x_c = ϕ ocm
x = I
With magnetic cards
x = I
x_c = ϕ ocm

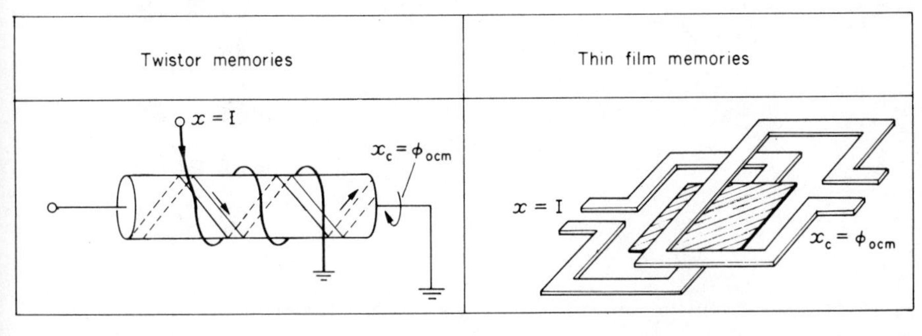
Twistor memories
x = I
x_c = ϕ ocm
Thin film memories
x = I
x_c = ϕ ocm

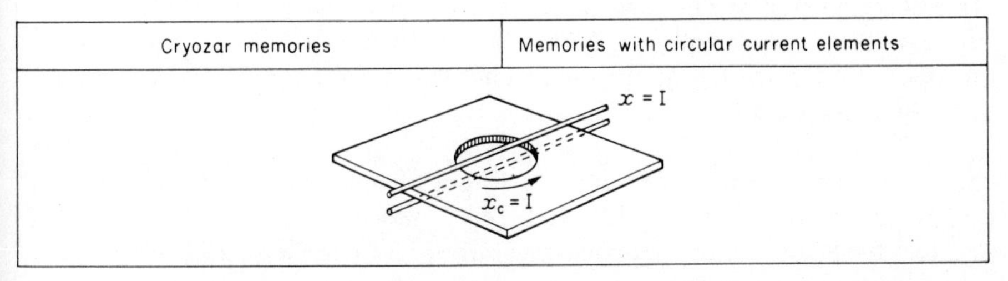
Cryozar memories
Memories with circular current elements
x = I
x_c = I

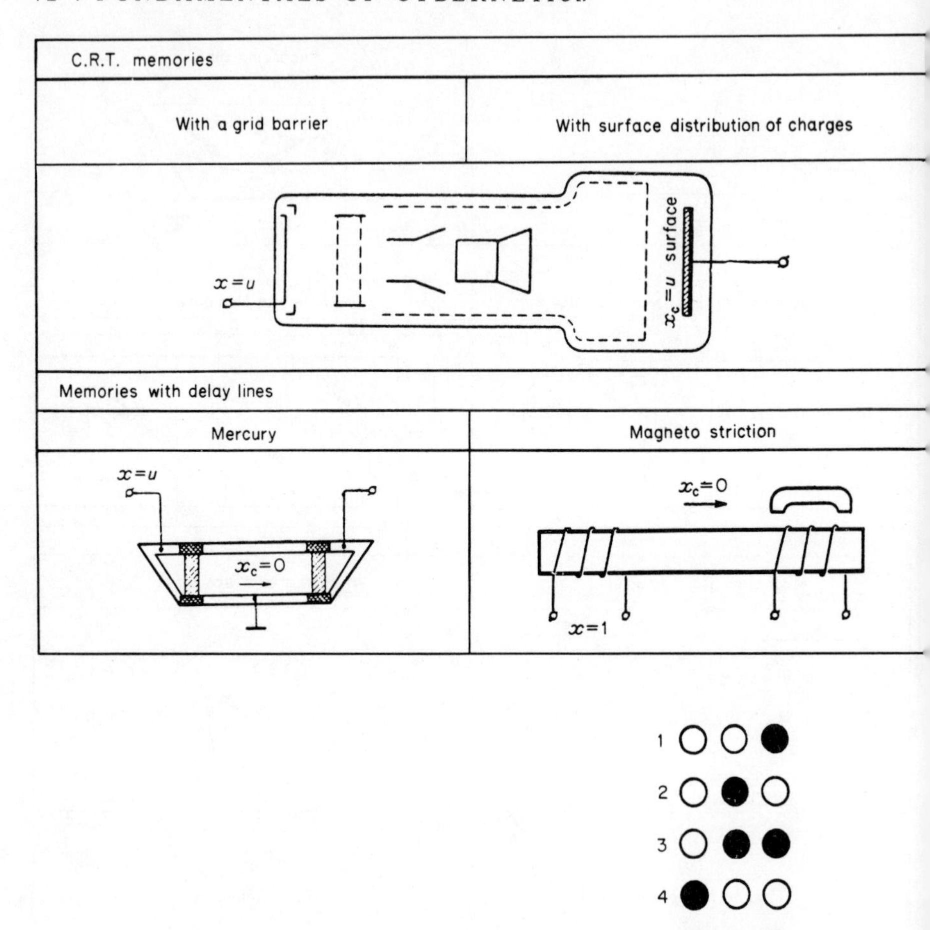


Fig. 5.5. *To Example No. 1.*

**4.** A message is written in the form of a decimal number made up of 5 digits (zeros are allowed in any digits), and it is assumed that all the numerals are equally likely and independent. What is the amount of information in this message? How much less information would the message contain if it consisted only of 5 binary digits?

*Solution.* The quantity of information in the decimal number consisting of 5 digits: $I = 5 \cdot \log_2 10 = 5 \cdot 3.32 = 16.6$ bits, whilst in the 5 digits: $I = 5 \cdot \log_2 2 = 5$ bits, i.e. the latter message carries 3.32 times less information than the 5 digits decimal number message.

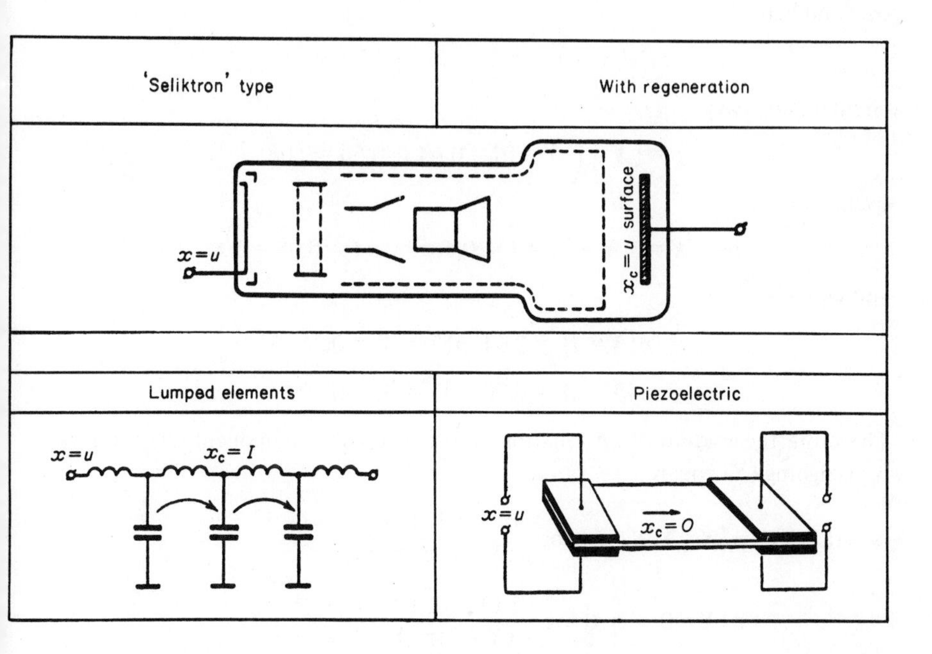


5. A trolleybus motor can operate in one of 5 regimes. The probability that it operates in the first regime $p_1 = 0.08$, in the second regime $p_2 = 0.12$, in the third regime $p_3 = 0.15$, in the fourth regime $p_4 = 0.28$, and in the fifth regime $p_5 = 0.37$. Find the entropy of the set of possible operating regimes of the motor.

*Solution.*

$$H = -(0.8 \times \log 0.08 + 0.12 \log 0.12 + 0.15 \log 0.15 + 0.28 \log 0.28 + 0.37 \log 0.37) = 1.565 \text{ bits.}$$

6. Depending on accuracy of manufacture, a number of components of the same type are divided into circular and oval, and according to weight into light and heavy. 70% of all the components belong to the light ones, and of these 80% are circular. Altogether 64% of the total number of components are circular. What amount of information about the shape of the component can be obtained by weighing?

*Solution.* We compile a table of distribution of a system with two random variables: $X$ – the number of circular components, $Y$ – the number of light components. For one component each of these random variables can assume the values 1 and 0. The probability that a randomly chosen component is circular and light is

$$p = \{X = 1, Y = 1\} = 0.7 \times 0.8 = 0.56;$$

oval and light

$$p = \{X = 0, Y = 1\} = 0.7 \times 0.2 = 0.14;$$

circular and heavy

$$p = \{X = 1, Y = 0\} = 0.64 - 0.56 = 0.08;$$

oval and heavy

$$p = \{X = 0, Y = 0\} = 1 - 0.56 - 0.14 - 0.08 = 0.22$$

and where

$$p\{X = 1\} = 0.64, p\{X = 0\} = 0.36,$$

$$p\{Y = 1\} = 0.70, p\{Y = 0\} = 0.30.$$

Therefore the amount of information on the shape of a component ($X$) obtained by weighing ($Y$) equals

$$1(X/Y) = p\{X = 1, Y = 1\} \log \frac{p\{X = 1, Y = 1\}}{p\{X = 1\}p\{Y = 1\}}$$

$$+ p\{X = 0, Y = 1\} \log \frac{p\{X = 0, Y = 1\}}{p\{X = 0\}p\{Y = 1\}}$$

$$+ p\{X = 0, Y = 0\} \log \frac{p\{X = 0, Y = 0\}}{p\{X = 0\}p\{Y = 0\}}$$

$$= 0.56 \log \frac{0.56}{0.64 \times 0.7} + 0.14 \log \frac{0.14}{0.36 \times 0.7} + 0.08 \log \frac{0.08}{0.64 \times 0.3}$$

$$+ 0.22 \log \frac{0.22}{0.36 \times 0.3} = 0.0561 \text{ dec. units} = 0.186252 \text{ bits.}$$

# 6 Control

As was mentioned in Chapter 2, the required behaviour of a control system is achieved by control actions. The effect of these actions will be to make a system assume a 'better' state than it would in the absence of such control actuations.

We will explain in what sense the word 'better' is used.

If we speak of a man-made control system built to achieve a certain goal, the behaviour of such a system is evaluated by the designer, and the word 'better' means better with respect to the objectives of the subject – the system designer. Biological control systems were formed during the process of evolution of the animal world, and for these it is not possible to determine the specific aims, which the control must realise. In spite of this, the concept of 'better behaviour' has a meaning for biological systems. It is that the behaviour of the organism in its natural environment influences its survival and reproduction. Therefore the evaluation of the behaviour of the organism as a controlled system is determined by its interaction with the medium, so that *better* is the behaviour which improves the chances of the given organism's survival and production of offspring.

It was mentioned in Chapter 2 that some external forces, particularly those available for controlling the system, are control forces. The effect on the behaviour of the system can be achieved by forces acting on its coordinate and by changing the parameters of the controlled system, i.e. the *controlled object*. Thus, for instance, control of the rpm of a turbine can be effected by changing the water head h which is a force acting on the coordinate representing the turbine torque. However, the same effect can be achieved if without changing the water head, we change the angle $\phi$ of the guide vanes (GV) and influence the direction of flow relative to the turbine rotor (TR), as shown in Fig. 6.1. A change in the angle $\phi$ produces a change in the parameters which determine the characteristics of the turbine, its internal properties.

The possibilities of control are greater and the control is the more effective

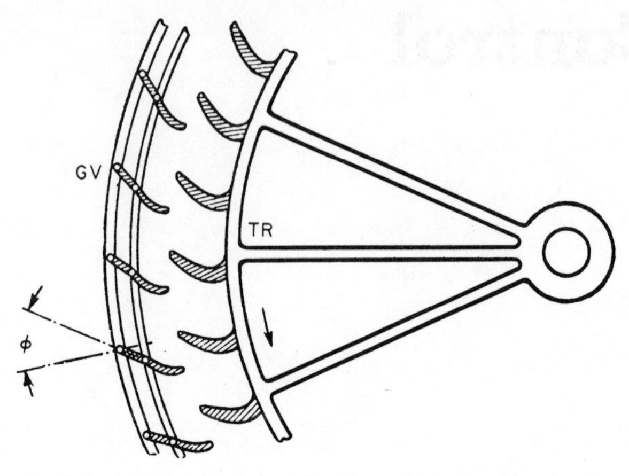


Fig. 6.1. *Control of the speed of movement of a turbine by changing the angle $\phi$ of the guide vanes.*

according to the range of values available for controlling forces, during the control process. However, we must take into account the fact that in real systems the range of variation in each control force is limited. In the given example of controlling a turbine, both the head h and the angle $\phi$ can change, but only within certain limits.

$$h' \leqslant h \leqslant h'', \quad \phi' \leqslant \phi \leqslant \phi''.$$

Since the control of any object can be realised by several control forces, each of which is limited by certain limiting values, it is possible to separate in the *space of the control forces* $Y_1, Y_2, \ldots, Y_m$, an area $\Omega$ which satisfies the conditions

$$Y_i' \leqslant Y_i \leqslant Y_i'' \; (i = 1, 2, \ldots, m),$$

within which there are points which represent all the sets of control forces, Fig. 6.2. This area is called the area of *possible (control) forces.*

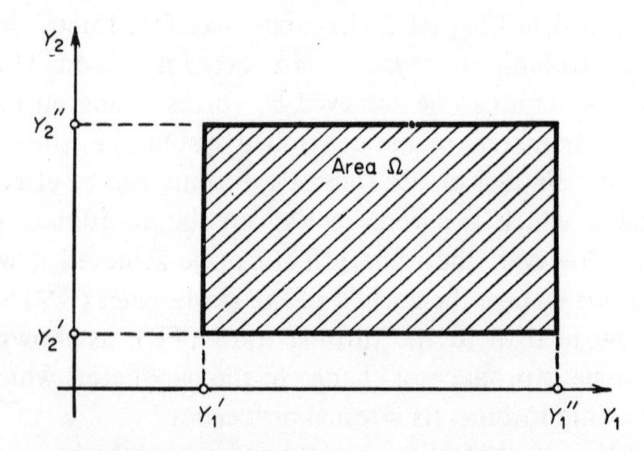


Fig. 6.2. *Area of possible control forces $\Omega$.*

Control forces can often only assume finite fixed values or can be considered as such values. In this case the area of possible control forces will contain a finite number of possible sets of control forces, which we will call a *set of possible (control) forces*.

The temperature in a refrigerator, for instance, can be maintained near to a given value by switching on or off the refrigeration unit. The set of possible forces in such a system consists of two control forces: 'on' and 'off'.

We will now explain how the set of controlling variables or actions required under the given conditions for the given controlled plant can be selected.

In order to control any object, it is necessary to change the controlling variables in a certain manner. Such a change can be brought about by means of *control signals* which carry the message on the required values of the control variables. The set of elements of a system which generates control signals is called a *control device*. If the behaviour required and the relevant properties of the plant are known in advance, then it is possible to introduce into the control device information about the sequence of control forces in the form of a control programme.

In other cases, when all the necessary data for a control programme are not known in advance, we can acquire the necessary information during the operation of the system. Such information may be data on the state of the controlled system, on its desired state, on disturbances or on characteristics of the controlled system. Processing this information in the control system, in accordance with certain rules, allows it to be used for generating the necessary control forces. The set of rules according to which information passing into the control system is changed into control signals is called the *control algorithm*.

Control is necessary not only in the normal functioning of a system, but also to ensure its development in the required direction of change; for example, for the development of an organism from a nucleus or for the development of a transport system.

Development control depends on a plan of development and in the realisation of that plan. The plan of development of living organisms is based on hereditary information expressed in the form of macromolecules entering into the composition of the cell nucleus. The plan of development of any economic system is a document containing information on the activities (capital investments, reconstruction work, etc.) which bring about the required changes in time of its functions and structure.

Taking the above described concepts into consideration, we can define 'control' as follows.

*Control represents selected actions on some object or objects, on the basis of information obtained and used in order to 'improve' the functioning or development of the given object.*

If the task of control is to stabilise the state of the object, then the control can be interpreted as an active protection against disturbances.

Passive protection, as opposed to active, consists in giving the object properties such that its output values (i.e. the functions of its state which are of interest to us) depend very little on disturbances.

Here are a few examples of passive protection: the position of a ship at anchor does not depend on the wind and on currents; a Dewar vessel, which limits the heat transfer between its charge and the ambient medium; high duty barriers which ensure that the industry of a given country is competitive in its internal market; the lipoid shell of a tuberculosis Koch rod, which protects it from phagocytes and medicaments; the creeping shapes of plants in the polar regions and the small evaporation surfaces of plants in zones; the anti-corrosion coatings of machine parts.

In contrast to passive protection, control systems can organise control forces which act against such disturbances. Thus, in the above examples, the stabilisation of the position of the ship can be achieved by suitable manoeuvring; the temperature of any arbitrary body can be maintained by controlling the flow of heat from the external source; the competitiveness of an industry can be ensured by increasing the productivity and reducing the production bottlenecks; bacteria can decompose medicaments or change their metabolism in such a way that the medicaments will be harmless to them; plants can control evaporation by stoma opening and closure, folding and shedding leaves, etc.

## 6.1 Control systems.

A controlled plant and the control devices connected to it form a *control system*. In order that a control signal U generated by the control equipment on the basis of the processing of the information Z can change the control action Y, organs are necessary which transform the control actions in accordance with the control signals – *actuating units*. For a simple case with a single actuating unit a block diagram of the interaction between the controls (YY) and the plant O through the actuating unit (AO) is given in Fig. 6.3.

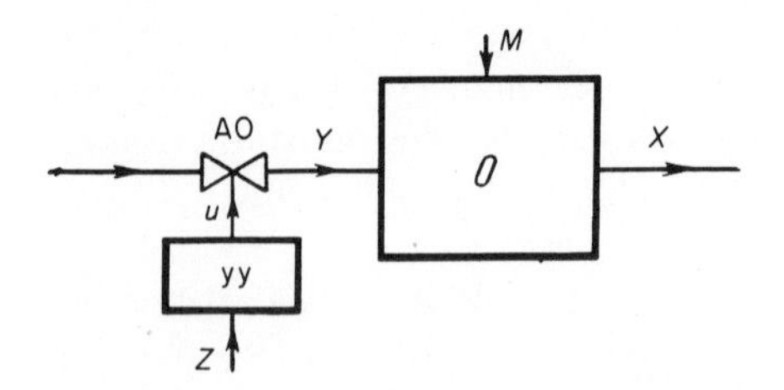


Fig. 6.3. *Diagram showing the interaction between the control equipment and the controlled plant.*

As actuating units we use, for instance, a solenoid operated valve (Fig. 6.4) which opens or closes off the flow of liquid in the valve K under the effect of an electric signal I, the current through the solenoid C. The role of the actuator can also be played by man, for instance the helmsman steering his boat in accordance with commands received.

In control systems four basic types of control are solved: stabilisation, fulfilment of a programme, following (tracking) and optimisation.

The task of stabilising a system is the task of maintaining some of its given values $X_y$, regardless of disturbances M which act on the values $X_y$. Thus, to ensure normal living of a warm-blooded animal body, it is necessary to stabilise the body temperature, the composition and pressure of the blood, regardless of changes taking place in the ambient medium. In power supply systems it is necessary to stabilise the voltage and frequency in the system so that it remains independent of changes in the power consumption.

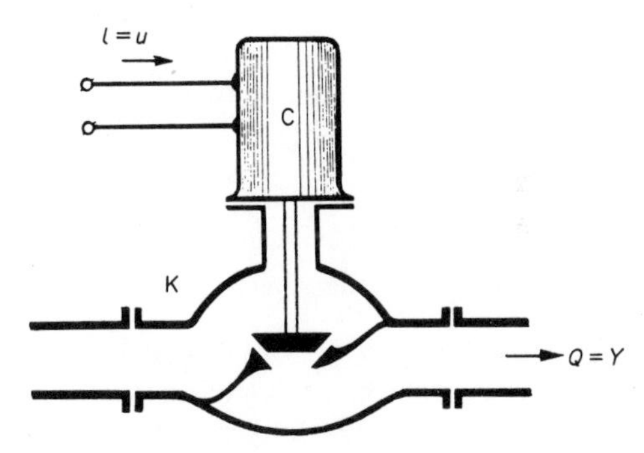


Fig. 6.4. *Solenoid control valve.*

The task of fulfilling a programme arises when given values of the controlled quantities $X_{y0}$ change with time in a manner which is known in advance. For instance, when controlling a ballistic rocket, guidance of the rocket into a predetermined trajectory must be in accordance with a programme $X_{y0}(t)$ of changes in its position in space and its velocity, known in advance. When controlling the position of a telescope in order to compensate for the rotation of the Earth, it is necessary to move it in accordance with a predetermined programme. A similar problem arises in production, when working in accordance with a predetermined production schedule.

Clear examples of action of the type of 'execution of a programme' in biology are the development of an organism from the egg cell, seasonal migration of birds, metamorphosis of insects, etc.

In cases when changes in the given values of controlled quantities are not known in advance, the task of *tracking* arises, i.e. to observe as closely as possible the correspondence between changes in the state of the system $X(t)$ and the values $X_0(t)$. The necessity of tracking arises for instance when controlling the production of an article under conditions of unpredictable changes in demand; the rhythm and depth of breathing must follow the changes of the physical exertion of the body; a radar antenna must follow the unpredictable movements of a maneouvring aircraft.*

* To prevent misunderstanding, it is emphasised that, for instance in the latter case, the controlled quantity is not the position of the aircraft but the position of the radar antenna. $X_0(t)$ is such a sequence of antenna positions which will always be aimed at the aircraft. In contrast to the problem of 'fulfilling a programme', this sequence cannot be given in advance and is determined by the external factor – the movement of the aircraft.

In a number of cases, control cannot be formulated as a problem of ensuring correspondence between the state of a system and a specified state since information on the specified state cannot be introduced into the control system in advance or obtained during operation. Such a system arises for instance when controlling a power generating set which operates under complicated variable conditions, when the aim of the control is to assure maximum efficiency of the generating set under any regime of operation.

The problem of optimisation – of establishing conditions which are optimal in a certain sense – is very often encountered. Examples are, for instance, control of an economic system in order to maximise profit, control of technological processes in order to minimise losses of raw materials and semi-finished goods, and many others. These problems will be considered again in Chapter 11.

## 6.2 Direct link and feedback

The properties of a control system depend largely on the sources of information used in the control equipment generating control signals. Let us first consider systems in which the information Z, obtained by the control equipment, does not include information on the state X of the controlled plant. In this case Z may contain a programme of sequences of changes of control actions $Y_0(t)$ or information on disturbances $M(t)$. In the latter case, to obtain a control signal $u$ the control equipment should contain data on what the value of $Y$ should be for each value of $M$, in order to achieve the desired goal. The control algorithm in such systems will consist of the transformation

$$u = PM \tag{6.1}$$

In this tranformation the operator $P$ is fed (inserted) into the control equipment in advance, on the basis of data on the aims of control and the properties of the controlled object.

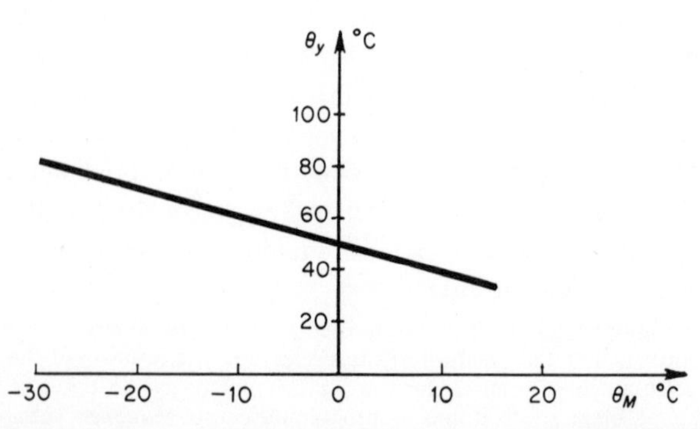


Fig. 6.5. *Graph for determining the required water temperature in a heating system.*

Assume, for example, in a heating system the main disturbance is the outside temperature $\theta_M$, and the aim of the control is to maintain the temperature $\theta_x$ in the heated rooms near to a pre-given temperature $\theta_0$. This can be achieved with the required accuracy by controlling the temperature of the water $\theta_y$ of the heating system in accordance with the graph Fig. 6.5. The persons operating the heating system will observe the thermometer readings $\theta_M$ and $\theta_y$, and operate the control units accordingly, so as to ensure the required functional dependence $\theta_y$ ($\theta_M$). A characteristic feature of the described control system is that in order to generate the control actions we do not use data on the controlled quantity itself – the temperature $\theta_x$ in the rooms being heated.

*Control systems in which information on the values of the controlled quantities is not used to form control actuations during the control process are called open-loop systems.* The structure of a control system of this type is shown in Fig. 6.6.

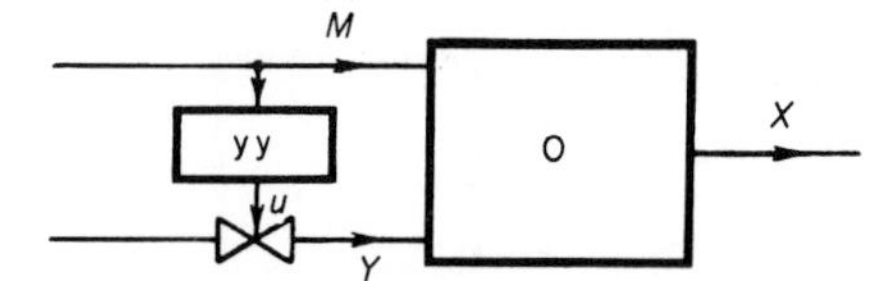


Fig. 6.6. *Diagram of an open-loop system.*

The control algorithm (6.1) realised by the control equipment YY is based on the idea of compensating the disturbances: for each disturbance $M$, using the transformation (6.1), a value $Y$ is chosen which compensates the effect of $M$ on the controlled quantity $X_y$. Here, the control action $Y$ should be so chosen that the sum of the deflections: $\Delta X_y(M)$ which occur under the effect of the disturbance, and $\Delta X_y(Y)$, arising due to a control action, would equal zero, so that

$$\Delta X(Y) = - \Delta X(M). \tag{6.2}$$

It can be seen from (6.2) that for selecting a control force, it is important to have available information on the influence of disturbances on the controlled quantity rather than information on the disturbances themselves. Consequently, a control can be organised without directly measuring the disturbances, monitoring only deviations of the controlled system caused by these disturbances. Information on the deviations of the controlled quantity $X_y$ from the pre-determined value $X_{y0}$ can be considered as an indirect method of obtaining information on the disturbances.

The above considerations show that control signals can also be generated from information about the deviations of the controlled quantity from its predetermined values.

For instance, it is possible to organise the control of a heating system by requiring the personnel to increase the temperature $\theta_y$ of the water in the heating system, when the temperature $\theta_x$ in the heated rooms drops below a given value $\theta_0$, or to reduce the water temperature in the opposite case. In this case the control algorithm will contain information on the value of the controlled quantity, because the control equipment realises the transformation

$$u = P(X, X_0) \tag{6.3}$$

where $P$ is the operator, realising the correspondence between each combination of values $X$ and $X_0$, and the value $u$.

*Systems in which information on the value of the controlled quantity is used to generate control forces are called closed-loop systems.* The structure of a closed-loop system is shown in Fig. 6.7. A system with such a structure is called 'closed' due to the presence of a closed loop in the circuit for transmitting the control forces of the system. Starting from any point of the circuit and moving in the direction of transmission of the control forces, we come back to the starting point, for instance along the loop $u \to Y \to X \to u$.

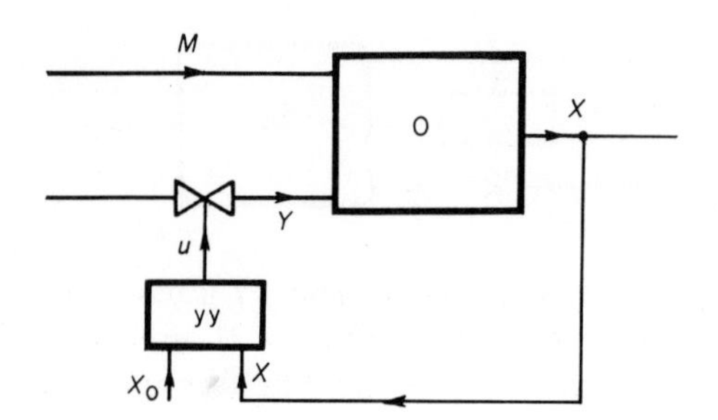


Fig. 6.7. *Diagram of a closed-loop system feedback.*

*The relation between the output forces $X_i$ of the $i^{th}$ element of the system and the input $Y_j$ of any other $j^{th}$ element is called a direct link. The link between the output $X_i$ and the input $Y_i$ of the same element is called feedback.*

Feedback can either be direct from the output of the system to the input, or it can occur via other elements of the system. It can be seen from Figs. 6.6 and 6.7 that in open control systems only direct links are used, whilst in closed systems feedback is also used. Thus in Fig. 6.7 the input – control forces $Y$ on the object O, depends on its output value $X$, due to the feedback through the control equipment YY. The link between the output and input elements of the system is referred to as feedback because in this case the transmission of forces is in the direction opposite to that of transmission of the forces in this element.

Feedback is one of the most important concepts of cybernetics, it helps us to understand many phenomena which occur in control systems of various types. Feedback can be detected when studying processes in living organisms, in economic structures and in automatic control systems.

*Feedback which increases the influence of the input on the output of the elements of a system is called positive, and a feedback which reduces this influence is called negative.*

Negative feedback ensures, for example, a narrowing of the range of changes in the brightness of an image on the retina, as compared to the range of brightness in the field of vision. This is done by changing the diameter of the pupil in accordance with the brightness on the retina. Due to negative feedback it is possible virtually to eliminate the effect of the parameters of electronic amplifiers on their performance. Positive feedback is used in many types of engineering equipment to increase the transmission coefficient.

Generally speaking, negative feedback brings about re-establishment of stability in a system when it has been disturbed by external forces, whilst positive feedback will produce a still greater deviation than would be caused by the external forces alone.

It should be noticed than any system with feedback is a system with a closed loop of transmission of forces.

The advantage of closed-loop control systems is that they can achieve the necessary control under conditions when disturbances are considerable, and not all of the disturbances measurable. This also applies under conditions where the effects of disturbances on the controlled variables are not known in advance.

The advantage of open-loop control systems is that the control forces change in accordance with changes in the disturbances immediately before the disturbances have time to influence, to any extent, the values of the controlled system.

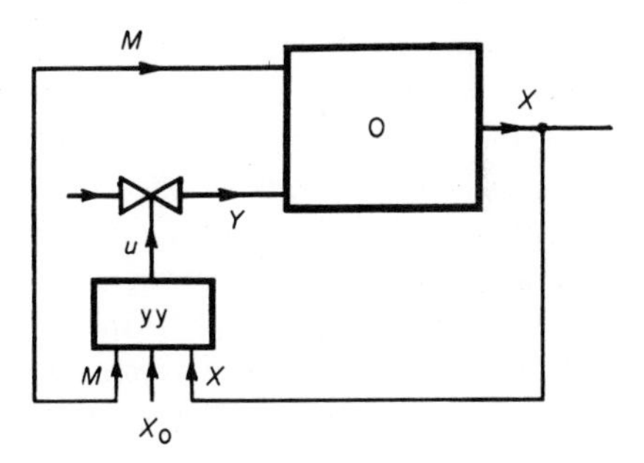


Fig. 6.8. *Diagram of a combined control system.*

A mixture of open and closed control systems can be achieved in what we shall call *combined control systems*; the structure of such a system is shown in Fig. 6.8. Here, information on the basic disturbances as well as information on the values of the controlled quantities are used to generate control signals, and the algorithm of operation of the control equipment will consist in realising the transformation.

$$u = P(M, X, X_0). \tag{6.4}$$

Such a method of control is realised in the example under consideration by means of the following instruction to the people concerned: establish the temperature $\theta_y$ in accordance with the graph 6.5, then observe the deviations in temperature $\theta_x$ from a predetermined value $\theta_0$, change $\theta_y$ so that this deviation is reduced to a minimum. A fast but approximate compensation of the perturbations is achieved owing to the first component of the control action, which depends on the deviation of the controlled quantity and which reduces this deviation to a permissible value, whatever its cause.

### 6.3. Limitations on control.

In order to evaluate the limitations on possible control, and to reveal the quality of that control which in principle can be achieved, let us consider a control system as a singular system of transmitting information, Fig. 6.9. For simplicity and clarity, we shall limit ourselves to considering a controlled object such as a factory with one output value $X$ and one control actuation $Y$, which is under the effect of a single perturbation $M$.

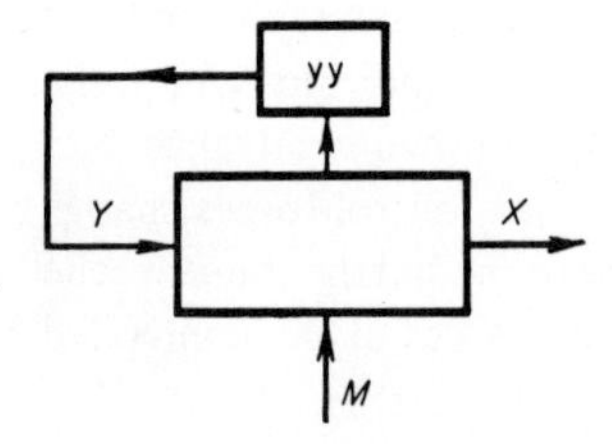


Fig. 6.9. *Control system.*

Let the aim of the control be to maintain $X$ at the constant level $X_0$. The quality of control in the given case can be evaluated from the degree of uncertainty $H(X)$ of the controlled quantity $X$. If the control were ideal: $X = X_0$, then the degree of uncertainty – the entropy $H(X)$ – would equal zero.

Let us assume that under the effect of a random disturbance $M$, the value $X$ fluctuates, exhibiting a random deviation from its specified value, and its entropy $H(X) = H(X(M)) = 0$. In this case the control system should play the role of a correcting device which lowers the uncertainty of the value of $X$. In the presence of control signals the uncertainty of the controlled quantity is $H(X/Y)$. Thus, the degree of decrease of the uncertainty in the state of the system is expressed by

$$H(X) - H(X/Y) = I(Y, X) \tag{6.5}$$

i.e. by the amount of information in $Y$ (considered as a random value, i.e. as a set of all the possible control forces used with definite probabilities) with repect to the quantity $X$. In order to ensure such a decrease in uncertainty, the control system should possess a sufficient variety of control forces in accordance with relation (see Chapter 5)

$$H(Y) \geqslant I(Y, X) \tag{6.6}$$

From (6.5) and (6.6) we find that the uncertainty in $X$ in the presence of control satisfies the inequality

$$H(X/Y) \geqslant H(X) - H(Y). \tag{6.7}$$

The inequality (6.7) expresses the limitations of the control. It is emphasised that in (6.7), equality can be achieved only if there is one-to-one correspondence between the control signal $Y$ and the controlled quantity $X$, i.e. if the control system determines precisely what the deviation of $X$ should be under the effect of random disturbances $M$ and generates very accurately the required correction signal $Y$. In practice, however, neither the random disturbances nor their relations with the changes in the quantity $X$ can be taken into consideration sufficiently. In the same way, it is not possible to produce absolutely accurate changes in $X$, whose values determine the control forces in closed control systems. Furthermore, the control system itself is subject to random disturbances and therefore the signal $Y$ may not accurately equal the required correction signal. As a result of all these factors there will be no unequivocal relationship between the values of $X$ and the control signal $Y$. In other words, the condition entropy $H\ (Y/X)$ does not equal zero. The amount of information $I(Y, X) = H(Y) - H(Y/X)$ (see Chapter 5). Therefore the uncertainty of $X$ is determined by

$$H(X/Y) = H(X) - H(Y) + H(Y/X). \tag{6.8}$$

The expression (6.8) shows that to improve the quality of control (reducing $H\ (X/Y)$) it is necessary to increase the number of control forces $H\ (Y)$, which are trying to achieve $H(X)$. More simply, for each possible deviation of $X$ it is necessary to have 'in reserve' an appropriate correction signal $Y$, also to be able to use it each time the given value of $X$ is encountered. However, this alone is not enough. It is necessary to ensure maximum adequacy of the control action to deviations in the controlled quantity. In other words, it is necessary to achieve a control action $Y$ which is required to correct the actually occurring deviations in the value $X$. This means that it is necessary to aim at reducing the nonuniqueness of the control signal $H\ (Y/X)$. For this purpose, very accurate and detailed information is required on the controlled system and the perturbations acting on it, and on the controlling system itself.

In a number of cases the possibility of control is also limited by some other factors, for instance by a limited speed of transmission of information along the direct link and feedback channels. Furthermore, one must bear in mind that usually the conditional entropy $H(X/Y)$ cannot serve as an exhaustive characteristic of the control quality, because the entropy depends only on the probability distribution but not on the magnitudes of the random values.

In the given case, however, the *magnitude* of the random deviations in the control system are more important than their probabilities. Let us assume, for instance, that it is necessary to maintain $X - X_0$ and the control system ensures in one case the values $X$ and their probabilities as given in the left-hand side of Table 6.1. whilst in the other case the values and probabilities as given in the righ-hand side of Table 6.1. Obviously in a real situation we would prefer the first alternative, although in the second the entropy is considerably smaller. So far no general complete solution is available to the problem of limitations of control in one or the other set of conditions.

Table 6.1

| $X_i$ | $0.99X_0$ | $X_0$ | $1.01\,X_0$ |
|---|---|---|---|
| $p_i$ | $1/3$ | $1/3$ | $1/3$ |

| $X_i$ | $0.5X_0$ | $X_0$ | $1.5X_0$ |
|---|---|---|---|
| $p_i$ | $1/6$ | $2/3$ | $1/6$ |

**Exercises**

1. A controller must maintain a certain temperature and level of the solution in a bath made for chromating components. The temperature in the bath may increase, with the level of the solution remaining unchanged, or decrease if the level rises, etc., and in such cases the control device must respond in a suitable manner. How is the set of input forces expressed? Write the control algorithm for the system, assuming for simplicity that the controlled quantities are independent.

*Solution.* The set of input actuations: (a) a fall in the level, (b) a rise in the level, (c) a drop in the temperature, (d) an increase in the temperature.

The control algorithm of the system:

(a) when the level of the solution falls, open the valve (gate) for feeding in the solution and close the drainage gate; (b) when the level rises, close the valve (gate) for feeding in the solution and open the drainage gate (valve); (c) when the temperature falls, increase the quantity of heat fed to the heater; (d) when the temperature rises, lower the quantity of heat fed to the heater.

2. Which is the controlling (A) and which the controlled (B) system in the following processes? What control tasks are solved in the process?

(1) The vegetative cycle in annual plants.
(2) Migration of fish to spawning grounds.
(3) Feeding of their nestlings by birds.
(4) Pairing of males and females.
(5) Sexual selection.
(6) Control of the animal population of a given species in a given location.
(7) Production control under conditions of private enterprise manufacture.
(8) Legal regulation of the life of a community.
(9) Moral control of the behaviour of Man.
(10) Democratic elections.

*Solution.* (1) A – genotype*, B – phenotype†; fulfilment of a programme.
(2) A – genotype, B – phenotype; fulfilment of a programme.
(3) A – genotype and nervous system, B – the body. The task of the genotype is fulfilment of a programme, the task of the nervous system – following.

* *Genotype* – set of heredity factors (genes) of one or another organism.
† *Phenotype* – set of all the individual characteristics of an organism.

(4) A – genotype and nervous system, B – the body. The task of the genotype is fulfilment of a programme, the task of the nervous system – following.
(5) A – genotype, B – population*. Task – optimisation.
(6) A – biocoenosis†, B – population of a given species. Task – stabilisation and follow-up (during changes of external conditions).
(7) A – market, B – production. Task – following and optimisation (in the sense of approximating in the best way to objective economic laws and tendencies of evolution).
(8) A – the State, B – the individual. Task – stabilisation (of a given social-political system).
(9) A – ruling class, B – the individual. Task – stabilisation (of given social relationships and forms).
(10) A – society (a system of social classes and groups), B – members of the executive. Task – optimisation (in the sense of reflecting in the best way the political wishes of most voters).

3. What control tasks are fulfilled by the following systems: (a) a windscreen wiper; (b) a system for controlling anti-rocket missiles; (c) aqualung; (d) a system for switching on street lighting; (e) a thermostat; (f) a system for controlling the movement of the slide of a copying milling machine; (g) the respiratory system of an animal.

*Solution.* (a) Stabilisation of the coefficient of transmission of the windscreen; (b) tracking and execution of a programme; (c) stabilisation of air pressure entering into the mask; (d) fulfilment of a programme; (e) temperature stabilisation; (f) following the template; (g) optimisation under varying external and internal conditions.

4. (Continuation). Which of the above systems are closed and which are open? Draw block diagrams.

*Solution.* (a) open; (b) closed; (c) closed; (d) open; (e) closed; (f) closed; (g) closed.

5. The entropy $H(X/Y)$ of a controlled quantity $X$ of a system as represented in Fig. 6.9 must not exceed 0.35 bits. A perturbation M acts on the system which is expressed by the following table (Table 6.2).)

Table 6.2

| $M$ | 1 | 2 | 3 | 4 |
|---|---|---|---|---|
| $p_i$ | 0.2 | 0.3 | 0.1 | 0.4 |

* *Population* – set of specimens of a given species who at a given instant of time inhabit a certain area in space.
† *Biocoenosis* – The complex of organisms related to each other and to the surrounding medium.

where the top row defines the type of perturbation and the bottom row the probability of its occurrence. Based on the assumption of an ideal control device, calculate the amount of information which it must introduce to achieve the required quality of control, and say what the multitude (variety) of control actions should be.

*Solution.* The entropy of the quantity $X$ in the absence of control action is

$$H(X) = H(M) = -\sum_{i=1}^{i=4} p_i \log_2 p_i = 0.2 \log_2 0.2 - 0.3 \log_2 0.3 - 0.1 \log_2 0.1 - 0.4 \log_2 0.4 = 1.85 \text{ bits.}$$

Since according to the condition $H(X/Y) \leqslant 0.35$ bits, $I(Y, X) = H(X) - H(X/Y) \leqslant 1.85 - 0.35 = 1.5$ bits. The necessary variety of the input actions is determined by the expression

$$H(Y) = I(Y, X) + H(Y/X).$$

For the ideal system $H(Y/X) = 0$ and consequently

$$H(Y) = I(Y, X) \geqslant 1.5 \text{ bits.}$$

# 7 Automatic Control

By automatic control we usually understand a control without the direct participation of man. This definition, however, does not give an accurate idea of the full sense of this term. In the first place, it is necessary to specify more accurately what we understand by 'direct participation'. In order that a control should be implemented in the interest of man, he must participate in the control process. If this is not done, sooner or later the controlled process will deviate from that which the human being desires it to be. The causes of these deviations may be: the inevitable ageing and wear of the elements of the system, changes in the environment or changes in man's purposes. Therefore the *participation of human beings in controlling systems which function in his interest is inevitable.*

As control techniques improve, the participation of man in the control process becomes increasingly indirect. First he became relieved of direct forces on the control elements, transferring this to a technical actuator on which man operated by generating control signals. The next step was for him to relieve himself of the function of generating control signals by transferring this function to a *technical control* device into which he has fed setting points to determine the desired values, which was realised by *technical* devices on the basis of *tasks of control* imposed by man. The process of removing humans from the direct forces means extending the functions of technical equipment, and this will continue with the development of control engineering.

In this way the degree of participation of humans in controlling engineering systems changes continuously in nature, and there will be no sharp boundary between systems that include man as opposed to those that exclude him.

If we speak of biological control systems, then it is quite senseless to subdivide them into automatic and nonautomatic, depending on whether man does or does not participate in the control. For instance, he does not participate in control of the flight of a bird, but birds can hardly be classified as automatic systems. This criterion is even less applicable to systems controlling individual organs in the human body.

Therefore in this chapter we will consider only control of *engineering, artificial systems designed by man* and we will call automatic systems those which essentially are only *relatively* automatic.

## 7.1. Elements of automatic control systems

At least three basic elements are required to realise an automatic control system: a measuring device or element (MET), a control (CONT) and an actuating (ACT) element, Fig. 7.1. These elements are connected to the object (plant) O and together form an automatic control system.

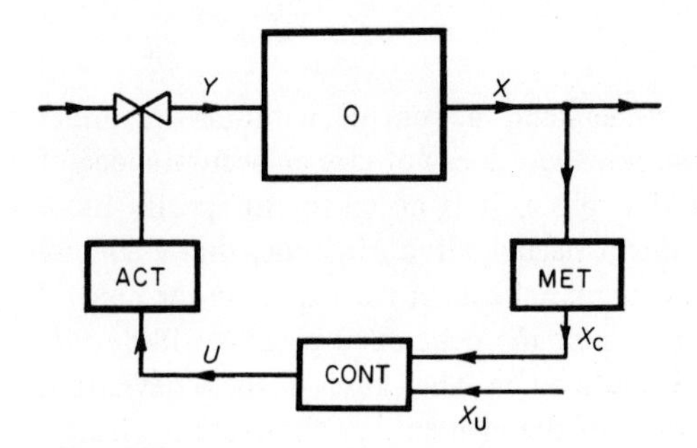


Fig. 7.1. *Automatic control system.*

The purpose of the measuring element is to generate the signal $X_c$, which characterises the magnitude of the controlled quantity $X$. As measuring elements we usually apply transducers which convert the controlled quantity $X$ into a unique corresponding quantity – the signal $X_c$, which is convenient for transmission to the control element and for further utilisation. The measuring element realises the functional transformation

$$X_c = f_c(X), \tag{7.1}$$

in particular the linear transformation

$$X_c = k_c X, \tag{7.2}$$

where $k_c$ is *transmission coefficient* of the metering element. Diagrams of a number of widely used measuring elements are given in Table 7.1.

The simplest control element can be the device for comparing two signals: the signal $X_c$ which characterises the value of the controlled quantity and the signal $X_0$ which characterises its desired value. Such a control element generates the control signal u, which is the result of comparing the signals $X_0$ and $X_c$ and is determined by the difference between them

$$u = f_u(X_0 - X_c). \tag{7.3}$$

Table 7.1 *Measuring elements*

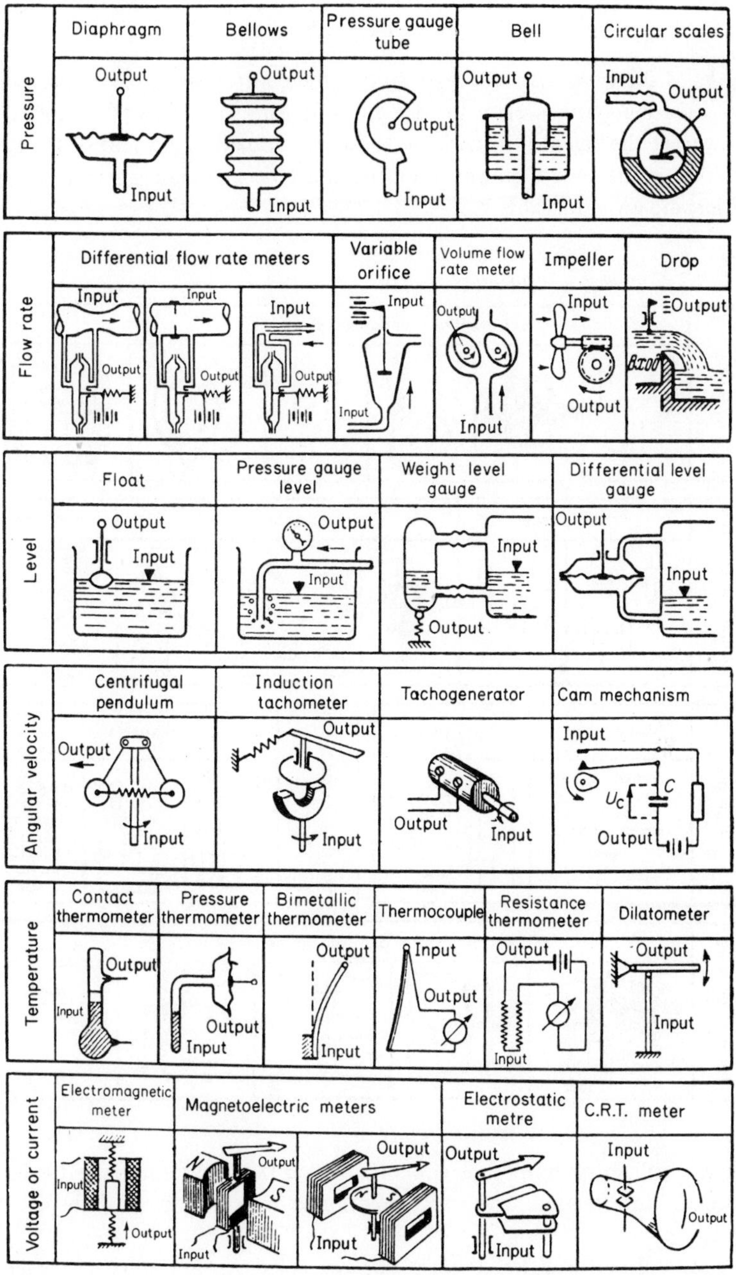

| Quantity | Measuring elements | | | | | |
|---|---|---|---|---|---|---|
| Pressure | Diaphragm | Bellows | Pressure gauge tube | Bell | Circular scales | |
| Flow rate | Differential flow rate meters | Variable orifice | Volume flow rate meter | Impeller | Drop | |
| Level | Float | Pressure gauge level | Weight level gauge | Differential level gauge | | |
| Angular velocity | Centrifugal pendulum | Induction tachometer | Tachogenerator | Cam mechanism | | |
| Temperature | Contact thermometer | Pressure thermometer | Bimetallic thermometer | Thermocouple | Resistance thermometer | Dilatometer |
| Voltage or current | Electromagnetic meter | Magnetoelectric meters | Electrostatic metre | C.R.T. meter | | |

In the particular case the control signal is a linear function of the deviation of the controlled quantity from the pre-assigned value

$$u = k_u (X_o - X_c), \tag{7.4}$$

where $k_u$ is the *transmission coefficient* of the controlling element. Diagrams of widely used comparison elements are given in Table 7.2. It is often found that the power of the signal u is inadequate for the actuation element of the system and if this is the case a power amplifier must be added. Diagrams of typical amplifiers are given in Table 7.3.

Table 7.2 *Comparison elements*

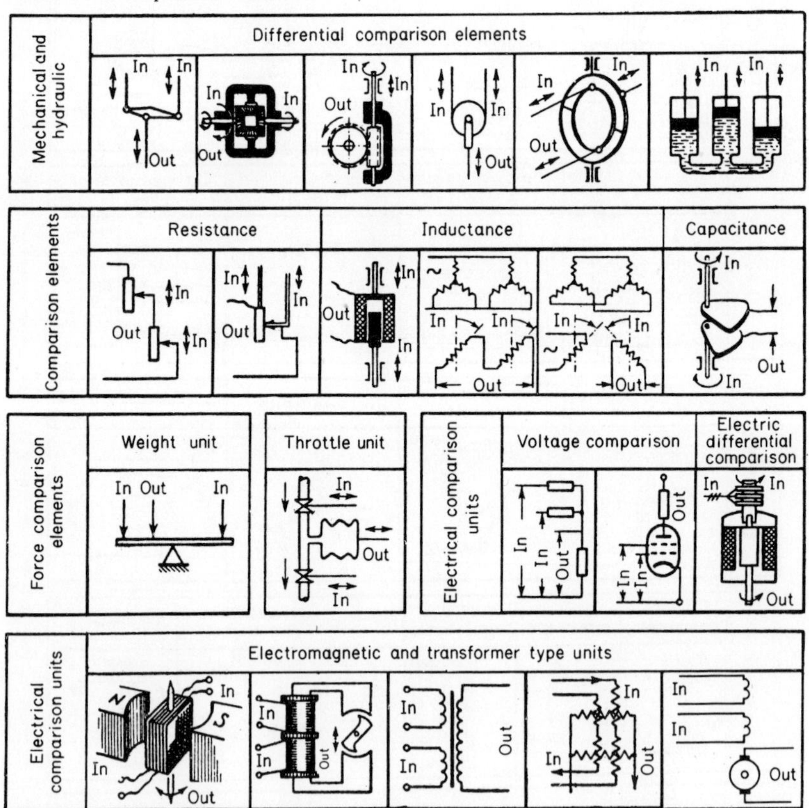


We will call *static* units such units for which the input and output values are functionally interrelated. The above mentioned units can obviously be classed as static.

Usually units of a different class – *astatic* (or dynamic) units, i.e. where the input value is functionally interrelated with the *rate of change of the output value*, are used as correcting servo elements.

Table 7.3 *Amplifiers*

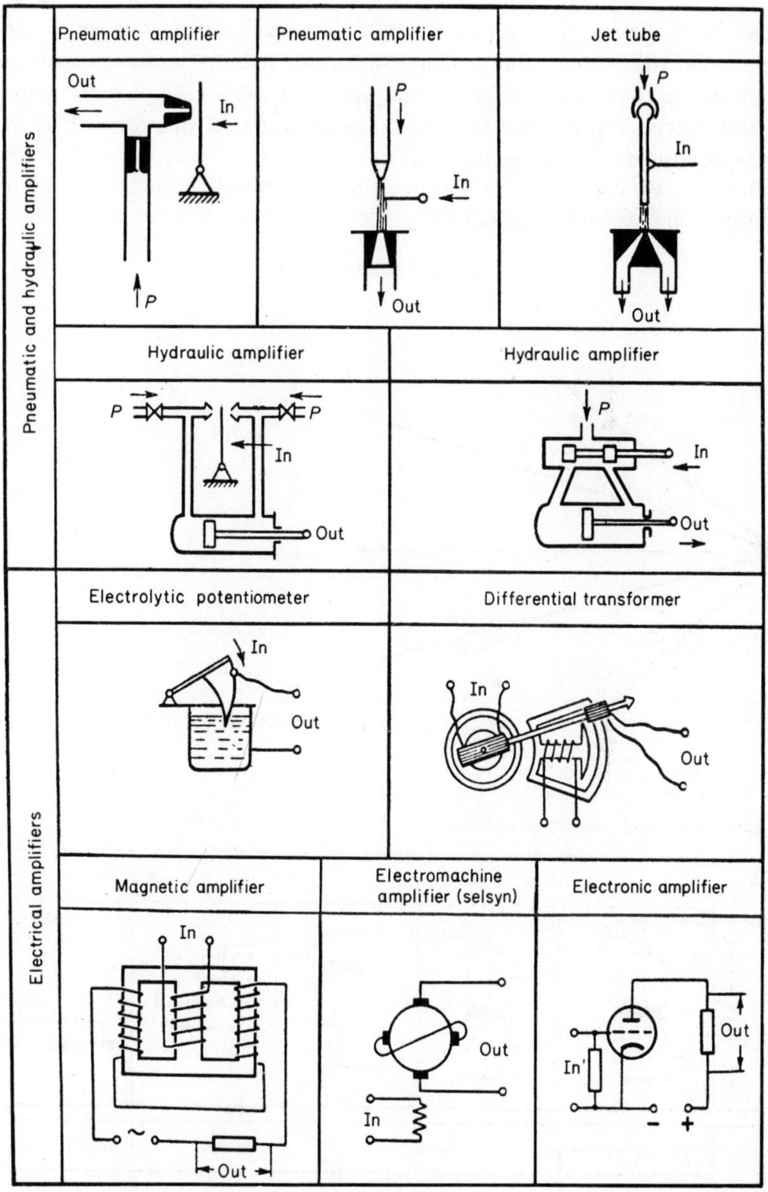

| | | | |
|---|---|---|---|
| Pneumatic and hydraulic amplifiers | Pneumatic amplifier | Pneumatic amplifier | Jet tube |
| | Hydraulic amplifier | Hydraulic amplifier | |
| Electrical amplifiers | Electrolytic potentiometer | Differential transformer | |
| | Magnetic amplifier | Electromachine amplifier (selsyn) | Electronic amplifier |

Thus the characteristics of transformation for the most widely used types of correcting servo devices can be written as follows:

$$v_Y = f_Y^*(u), \tag{7.5}$$

where $v_Y$ is the rate of change of the control actuation for the linear case

$$v_Y = k^*_Y u, \tag{7.6}$$

where $k_Y$ is the transmission coefficient. The characteristic feature of astatic units is that the output value remains unchanged only for certain (usually zero) input values. The changes in the output value of an astatic unit with the characteristic as expressed in Eq. (7.6), as a function of time for various values of the input $u$, are plotted in Fig. 7.2.

In some automatic control systems static correcting elements may be used, with the transformation characteristics:

$$Y = f_Y(u), \tag{7.7}$$

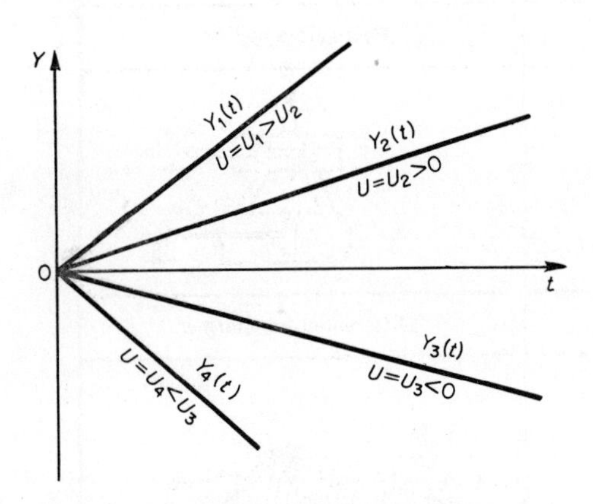


Fig. 7.2. *Changes with time of the output of an astatic unit.*

Table 7.4 *Final control elements*

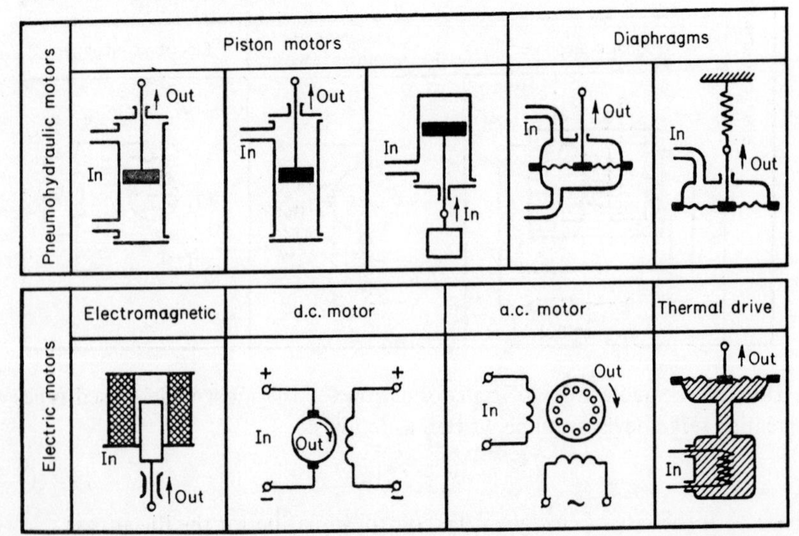

and for linear elements with the transmission coefficient $k_Y$:

$$Y = k_Y \, . \, u. \tag{7.8}$$

The most widely used final control units are given in Table 7.4.

## 7.2. Automatic control systems

As an example we will consider an automatic system for controlling the pressure $P$ of the gas in a gas reservoir GR, as shown in Fig. 7.3.

The measuring element here is the bellows B (elastic metallic chamber) which generates a force $F_c$ that is proportional to the pressure $P$. In the control element $C_E$ the force $F_c$ is compared with the force $F_o$ which characterises a pre-assigned gas pressure in the reservoir and this is set by means of the tension of the spring S. Deflection of the weight lever WL changes the size of the gap through which compressed air flows from the nozzle N, and accordingly changes the air pressure P. The final diaphragm element D puts the butterfly valve V$-_u$ into a position which depends on the pressure $P_u$ and consequently changes the throughput $R_1$ of the flap V and thus also the flow of gas into the reservoir. The correction is chosen in such a way that a drop in the pressure in the reservoir will open the throttle flap, and vice versa. The control $R_1$ will always change in such a way as to prevent deviations in the gas pressure caused by changes in the flow rate from the reservoir due to load changes ( the total throughput capacity $R_2$ of the gas users).

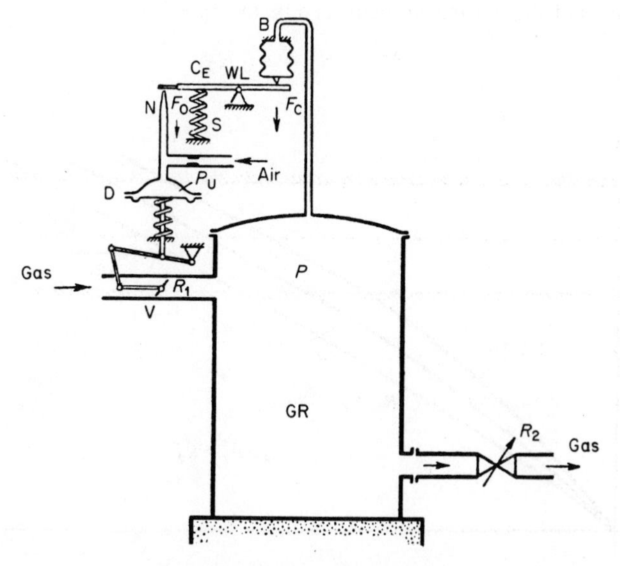


Fig. 7.3. *System for controlling gas pressure.*

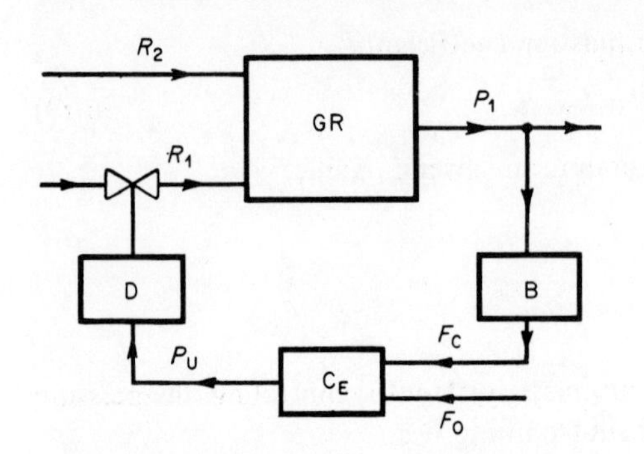


Fig. 7.4. *Block diagram of the system for controlling gas pressure.*

A simplified model of the control system under consideration is the block diagram 7.4, and it can be seen that its structure does not differ from the general diagram Fig. 7.1.

The dependence of the controlled value $P$ on the control $R_1$ and the perturbations $R_2$ is expressed by a family of characteristics of the plant, which in the given case are the curves 1, 2 and 3 in Fig. 7.5. For pressures which are near to the pre-assigned pressure $P_0$ it can be assumed approximately that these characteristics are linear and have the same coefficient of slope

$$k_0 = \frac{\Delta P}{\Delta R_1}$$

A shift in the characteristics caused by a change in load $R_2$ can be reflected by means of the coefficient $k_M$, which characterises the influence of the load $R_2$ on the pressure $P$. Taking into consideration what has been said above, the static characteristics of the controlled plant can be expressed by

$$P = k_0 R_1 - k_M R_2. \tag{7.9}$$

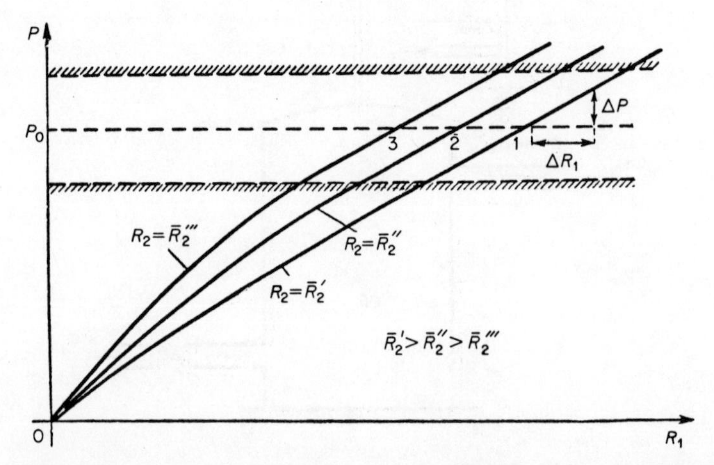


Fig. 7.5. *Dependence of the controlled value* P *on the control* $R_1$ *and the perturbations* $R_2$.

Taking into consideration that all the elements used in the system under consideration are static, and assuming that they are linear, we use the expressions given in Section 7.1. In this case we obtain the following set of equations which describe the static properties of the system:

$$\left.\begin{aligned} F_c &= k_c P, && \text{(a)} \\ F_0 &= k_c P_0, && \text{(b)} \\ P_u &= k_u(F_0 - F_c), && \text{(c)} \\ R_1 &= k_y P_u, && \text{(d)} \\ P &= k_0 R_1 - k_M R_2 && \text{(e)} \end{aligned}\right\} \tag{7.10}$$

Eliminating from the system (7.10) the intermediate coordinates by successive substitutions of all the other equations into equation (e), we obtain

$$P = \frac{KP_0 - k_M R_2}{K+1}, \tag{7.11}$$

where $K = k_0,\ k_y,\ k_n,\ k_c$ are transmission coefficients of the system. The expression (7.11) is called the static characteristic of the considered closed control system.

It can be seen from (7.11) that the pressure $P$ in the reservoir can be considered as consisting of the two components $P'$ and $P''$.

$$P = P' + P'' = \frac{K}{K+1} P_0 - \frac{k_M}{K+1} R_2 .$$

The component $P'$ depends on the pre-given pressure $P_0$ and will be the nearer to that pressure the higher the transmission coefficient $K$. The second component $P''$ characterises the influence of perturbations $R_2$ on the controlled value $P$; it decreases with increasing $K$. The system possesses such properties because the manipulated variable is proportional to the deviation of the controlled quantity from the pre-assigned value. Systems of this type are called linear *static systems.*

What has been said can be generalised to all linear static systems. The static characteristics of such systems can be written as follows

$$X = \frac{KX_0 - k_M M}{K+1} . \tag{7.12}$$

The graphs of these characteristics are given in Fig. 7.6.

The task of the control in the class of systems under consideration is to maintain the controlled variable as near as possible to the pre-assigned value (regardless of the changes in the perturbations), and therefore the greater the magnitude of the transmission coefficient $K$, the better will be the solution of the problem.

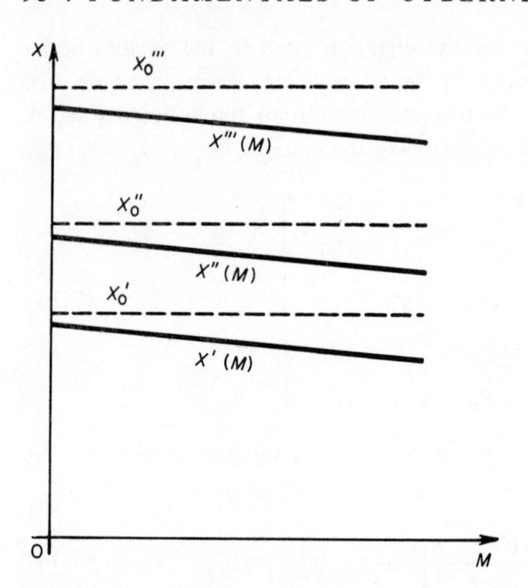


Fig. 7.6. *Characteristics of a linear static system.*

This conclusion was arrived at without taking transient phenomena in the system into consideration. Further consideration (Section 7.3) shows that when considering the dynamics of control processes, considerable corrections must be introduced in results which are based on analysis of all the static properties of the system.

We will now describe the changes in the properties of a system when the static correcting element is replaced by an astatic (dynamic) one (for instance, a piston). In this case the magnitude of the manipulated variable $Y = R_1$ will no longer be determined by the control signal $u = P_u$, but by its rate of change $v_Y = v_R$, and according to Eq. (7.6) the characteristic of the correcting element can be written as

$$v_R = k_y^* P_u. \tag{d'}$$

All the other elements of the system remain unchanged, and therefore the series of equations of the new system will be the same as in (7.10) except that equation (d) will be replaced by equation (d′).

By definition, not one of the coordinates of the system should change in the steady state, which means that the rates of change of all the coordinates should equal zero. It can be seen from (d′) that this will occur only if $P_u = 0$. However, it follows from equation (7.10) that the manipulated variable will equal zero only if the value of the controlled quantity $P$ equals its pre-assigned value $P_0$. In this case the equilibrium value of the controlled quantity will no longer depend on the perturbations and it will exactly equal the pre-assigned value. The system has this property because the *rate of change* of the correcting action is proportional to the deviation of the controlled quantity from the pre-assigned value. A control system of this type is called an *astatic system.*

It would appear that astatic systems ideally solve the problem of control, but their application is complicated because there are many drawbacks associated with the dynamic properties of such systems (see Section 7.3).

## 7.3. The dynamic of control

Some properties of automatic control systems which relate to the steady-state of operation have now been explained.

However, it is risky to judge the properties of such systems without taking into account transient regimes which occur during the control process. The dynamic properties of the system may be such that the transient operation is not completed during the time between subsequent changes in the perturbation. Furthermore, the steady-state may not occur at all (if, for instance, the equilibrium is unstable). Therefore, to estimate the working ability of the control system it is very important to consider not only the static but also the dynamic properties.

Some important dynamic properties of automatic control systems can be explained on the example of the control of the course of an aircraft by means of a relay type automatic pilot. Such a system is schematically drawn in Fig. 7.7.

Here the metering device is the gyroscopic compass GC whose position relative to the longitudinal axis of the aircraft characterises its course $\phi$. The control mechanism is represented by the setting ring SR, fitted in a position

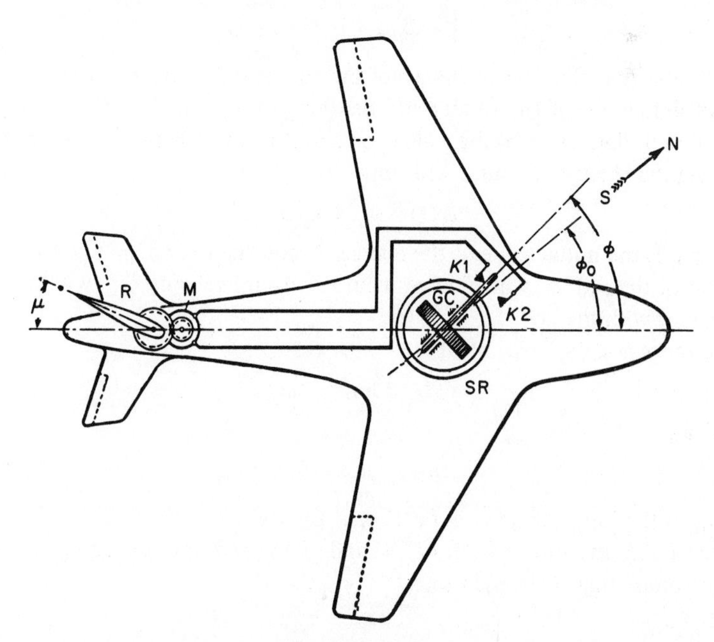


Fig. 7.7. *Relay type automatic pilot.*

corresponding to the pre-assigned course $\phi_0$ and equipped with a set of contacts. The link between the compass and the control device is so designed that when the course deviates to one side the contact K1 closes, and if it deviates to the other side, the contact K2 closes. These contacts switch on the servo motor M of the rudder R,, so that the rudder turns to port or starboard, depending on which of the contacts is closed. The connections are so designed that a deflection of the aircraft's course to the left produces a turn of the rudder to the right, and vice versa, producing a control action aimed at a re-establishment of the pre-assigned course.

Since the only possible state of equilibrium of such a system is movement of the aircraft exactly along *its course* (since only in that case will the rudder not turn), static considerations lead to the conclusion that the tasks of the system are accurately fulfilled in the steady-state.

Let us now, however, also consider the transient process in the system, and verify whether the equilibrium conditions will set in. To do this we construct its phase portrait.

We will denote by the angle $\mu$ the turning angle of the rudder with respect to its mean (central) position and by $\omega_\mu$ we denote its rate of turning. In a relay type of automatic pilot the rate of turning of the rudder may assume the following values:

$$\omega_\mu = \begin{cases} +\bar{\omega}_\mu, & \text{if} \quad \phi < \phi_0, \\ 0, & \text{if} \quad \phi = \phi_0, \\ -\bar{\omega}_\mu, & \text{if} \quad \phi > \phi_0 \end{cases}$$

Assuming that the rate of change of the course is directly proportional to the angle of deflection of the rudder and denoting by $k_\mu$ the rate of change of the course due to deflection of the rudder on a unit measurement, we obtain the law governing the change of course with time

$$\phi\,(t) = \phi_{in} + k_\mu\,\mu t, \tag{7.14}$$

where $\phi_{in}$ is the initial value of the course. According to (7.13) the change in the position of the rudder, which moves with a constant velocity $\bar{\omega}_\mu$, will proceed in accordance with the law:

for the range $\phi < \phi_0$

$$\mu(t) = \mu_{in} + \bar{\omega}_\mu t, \tag{7.15}$$

for $\phi < \phi_0$

$$\mu(t) = \mu_{in} - \bar{\omega}_\mu t. \tag{7.15'}$$

here $\mu_{in}$ is the initial angle of deflection of the rudder.

Eliminating the time $t$ from (7.14) and (7.15) and introducing $k_\mu/\bar{\omega}_\mu = r$, we find the phase trajectory equations:

for $\phi < \phi_0$

$$\phi\,(\mu) = \phi_{in} - r_\mu\,\mu_{in}\,\mu + r_\mu\,\mu^2. \tag{7.16}$$

for $\phi > \phi_0$

$$\phi(\mu) = \phi_{in} + r_\mu \mu_H \mu + r_\mu \mu^2. \tag{7.16'}$$

The phase trajectory equations (7.16) represent parabolas, similar to the trajectories of the hydraulic drive considered in Chapter 4. Fig. 7.8 shows the phase portrait of the system considered, from which it can be seen that once oscillations start in the system they will not die out, which is in itself sufficient to recognise that the system is not viable. If, in addition, we take into consideration that the rate of change of course does not accurately follow the changes in the position of the rudder, but due to the inertia of the aircraft will lag behind, it will be seen that the fluctuations in the course will not only not be attenuated, but will even increase with each oscillation.

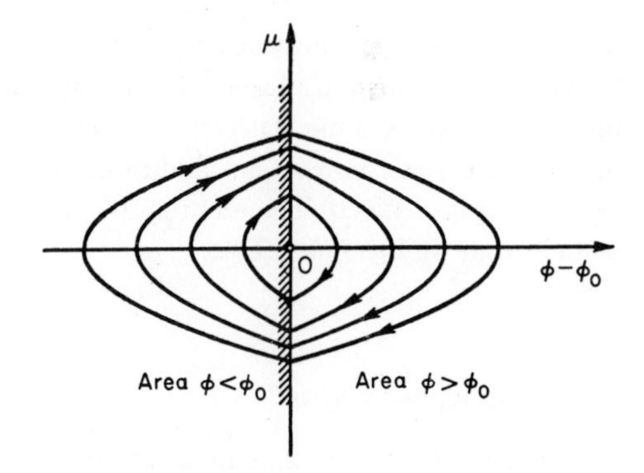


Fig. 7.8. *Phase portrait of 'hunting' with respect to the course of an aircraft in the absence of a link to the position of the rudder.*

The tendency of the system under consideration to oscillate about its steady-state ($\mu = 0$, $\phi - \phi_0 = 0$) is due to the successively late switching of the correcting element (for $\phi = \phi_0$). Furthermore, during the time of return of the rudder to the mean (central) position, an excessively large deviation of the controlled value from the pre-assigned value ($\mu = 0$) will build up in the direction opposite to the initial deviation. To eliminate this phenomenon, various types of *stabilising devices* are used which produce a lead in the switching of the correcting element that compensates at least partly for the above mentioned time lag. One such means of stabilisation is the introduction into the control element of an additional link (po) to the site of the final positioning element, in the given case to the position of the rudder R. This can be achieved, for instance, by turning the setting ring SR by an angle proportional to the rudder turning angle $\mu$.

In this case the dependence of the rate of turning the rudder $\omega_\mu$ will no longer be determined according to (7.13), but will assume the value

$$\omega_\mu = \begin{cases} +\overline{\omega}_\mu & \text{if} \quad (\phi + k_c\mu) < \phi_0 \\ 0 & \text{if} \quad \phi + k_c\mu = \phi_0 \\ -\overline{\omega}_\mu & \text{if} \quad (\phi + k_c\mu) > \phi_0, \end{cases}$$

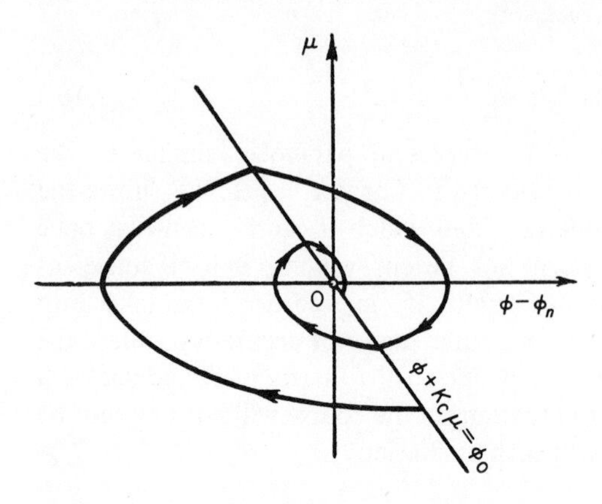


Fig. 7.9. *Phase portrait of the same system on introducing an additional link connection to the position of the rudder.*

where $k_c$ is the proportionality coefficient between the turning angle of the rudder and the turning angle of the setting ring. In this case the phase portrait of the system will assume the form as shown in Fig. 7.9, from which it can be seen that the introduction of an additional link to the position of the correcting element will lead to attenuation of the oscillations in the system and the equilibrium will become stable.

## 7.4. Programmed control

The automatic control systems considered in the previous paragraphs are, as the reader has probably noticed, intended for solving one of the four basic types of control problem enumerated in Chapter 6, namely the problem of stability or stabilisation.

The second frequently encountered control problem is to maintain the state of a controlled system near the state $X_0(t)$, which changes in time secondary to a pre-determined law, and not near an invariant (stationary) state $X_0$ = const., as was the case in the systems considered above. This is a problem of following a programmed control.

It is pointed out that the programme for changing a given state of the controlled system can be given not only as a function of time but also as a function of any other quantity which changes with the progress of time.

The law governing the change of the given state is known in advance, therefore information on it can be fixed in some type of storage unit SU which is connected to the control system, forming a system of programmed control as shown diagrammatically in Fig. 7.10.

Realisation of the programme can be achieved by an open (a) or closed (b) control system.

Let us consider as an example a closed programme-controlled system of a mine-shaft winding machine. The simplified sketch of such a system is shown in Fig. 7.11. The task of the control in the given case is to comply with a given

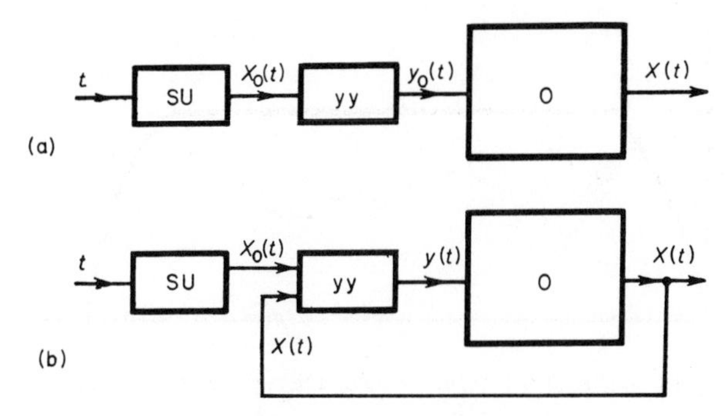


Fig. 7.10. *Block diagram of a system of programmed control.*

diagram of the velocity $\omega_0$ of the lifting cage as a function of its position $S$, Fig. 7.12. The form of the diagram is chosen so that the speed of the cage reaches the limit permissible value $\omega_{0\ \lim}$ as quickly as possible, and then reduce to zero when the cage reaches the reception platform.

In the lifting installation shown in Fig. 7.11, the winch W on which the cage K is suspended is moved by a motor M which is fed from the controlled rectifier CR. A control signal in the form of the differences between the signals $X_0$ and $X$, characterising respectively the pre-assigned and the real values of the speed of

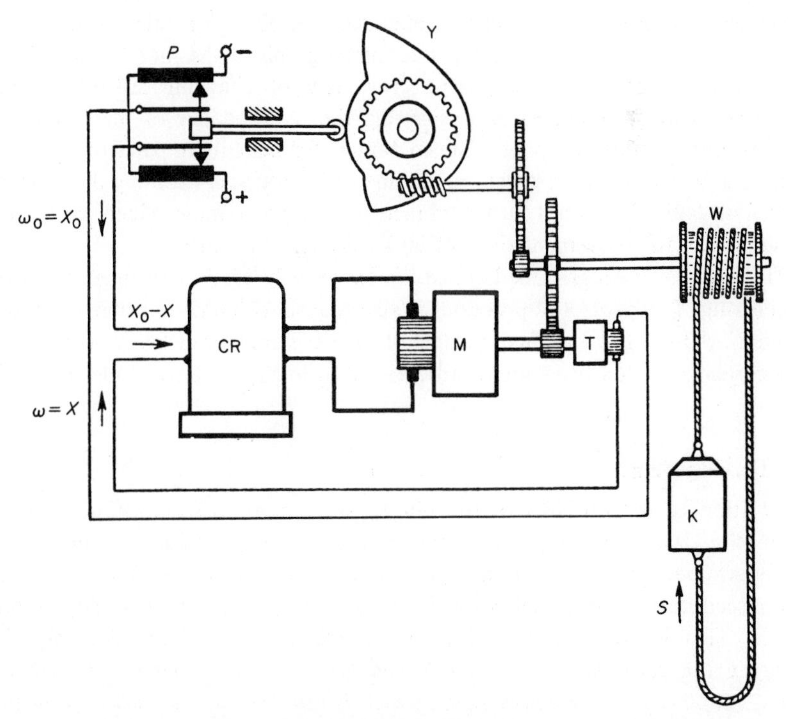


Fig. 7.11. *System of programmed control of a mine shaft winding machine.*

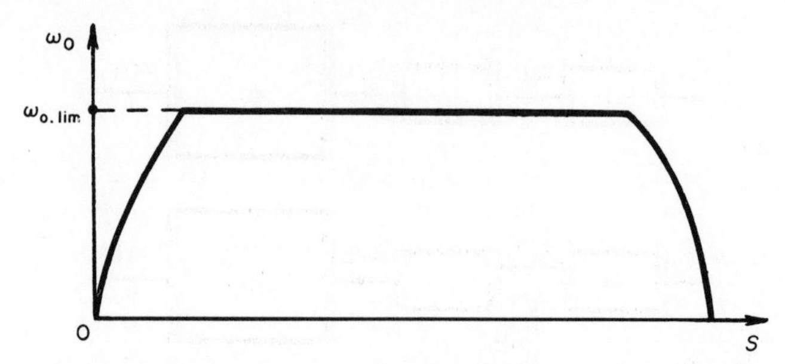


Fig. 7.12. *Dependence of the speed of the cage on its position.*

the motor (cage), is applied to the input of the controlled rectifier. The signal representing the real value of the speed is obtained by means of a tachogenerator T, which is coupled to the drive shaft. The signal corresponding to the pre-assigned speed is formed by means of the potentiometer P, whose contacts are moved by a profiled cam of the setting device Y. The cam profile reflects the required cage speed diagram. When the cage moves along the shaft the cam turns and changes the pre-arranged speed value in accordance with the programme. Deviations of the speed from the pre-determined speed produce a change in the signal on the input of the controlled rectifier, in a direction so as to reduce the deviation.

Systems of programmed control operating on this principle are used, for example, for complying with a pre-determined graph of the metal temperature during heat treatment; satisfying a given law of changing altitude during automatic landing of an aircraft; ensuring the required law of increasing the pressure during starting up of a steam boiler; for machining parts on machine tools in accordance with a given programme of contours; controlling the drive of optical or radio telescopes in accordance with a programme which takes into consideration the movement of celestial bodies and the earth.

This system of programmed control is extensively used in automatic control engineering. The control of processes in accordance with a given programme also occurs in nature. For instance, control of the development of living organisms is in accordance with a programme which is given in the nucleus of the germ cell.

## 7.5. Servo systems

In addition to the problem of fulfilling a programme, a situation often arises when the law governing the changes with the progress of time of a pre-determined state of a system is not known in advance, and has to be determined during the actual control process in accordance with some external signals. This is the problem of tracking. *The control system, intended for changing the state $X(t)$ of the controlled system in accordance with the law $X_0(t)$, supplied by an external signal which is not known in advance, is called a tracking servo system.*

Servo systems are being extensively used in engineering: for controlling the drives of radar antennae which track the movement of an aircraft, for controlling the fuel-to-air ratio in combustion systems, for driving mechanisms of lifting cranes, etc.

As in control systems considered in Section 7.2, similarly in control devices with servo systems the correcting signals are based on information about the state of the controlled plant, transmitted along the feedback channel.

In a servo system the task of the control is that the controlled quantity $X(t)$ should reproduce a law of change of the state $X_0(t)$ generated by an external signal, and therefore investigation of the statics of the system leads to a conclusion as to the expediency of increasing as much as possible the transmission coefficient of the system K in order to lower the steady state value of the tracking error, as was shown in Section 7.2.

Due to inevitable time lags in changes of the state of the system after changes of the correcting action, transients in the system have a considerable influence on its properties and this influence increases with increasing transmission coefficient K of the system. Fig. 7.13 shows typical graphs of the transition of a system from the initial state to a given state, for various $K$ values. It can be seen that with increasing $K$, damping of the oscillations is slower, and for some critical value $K = K_{cr}$ the oscillations in the system will not die out, and for $K > K_{cr}$ the system becomes unstable.

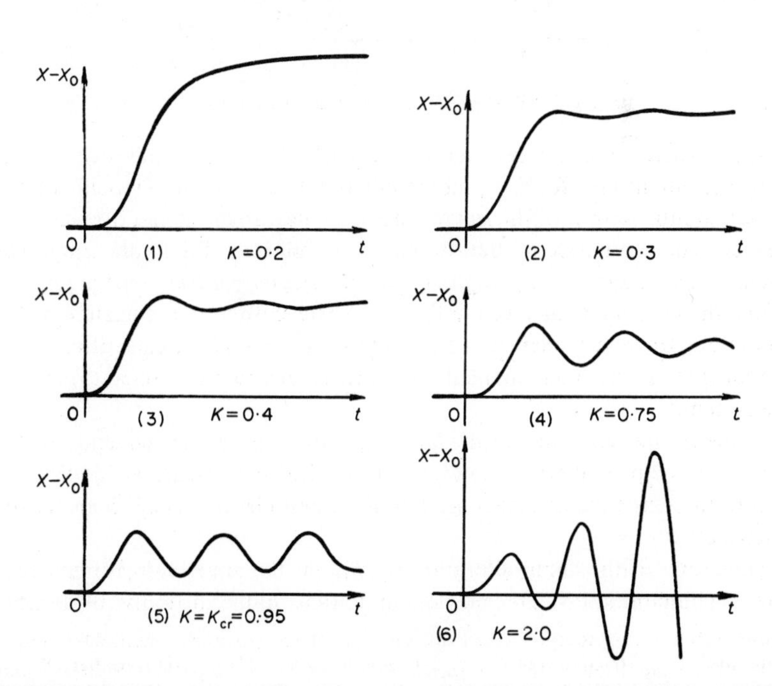


Fig. 7.13. *Transient characteristics of a system for various values of the transmission coefficient* K.

The dynamic properties of a linear servo system* can be judged from its *frequency characteristics.*

Let a setting signal which changes in accordance with the sinusoidal law $X_0(t) = X_{0a} \sin(\omega t)$, where $\omega$ is the oscillation frequency measured in numbers of radians per second, be applied to the input of the system. Then after a sufficiently long time from the beginning of the process a periodic regime will establish itself in the system and the output values will also change according to a sinusoidal law with the same frequency $\omega$, but with an amplitude $X_a$ differing from the amplitude of the given input $X_{0a}$ and with some shift in the phase $\phi$ relative to the phase of the input.

An example of the periodic regime caused in a servo system by the effect of a sinusoidal change in the input given is given in Fig. 7.14.

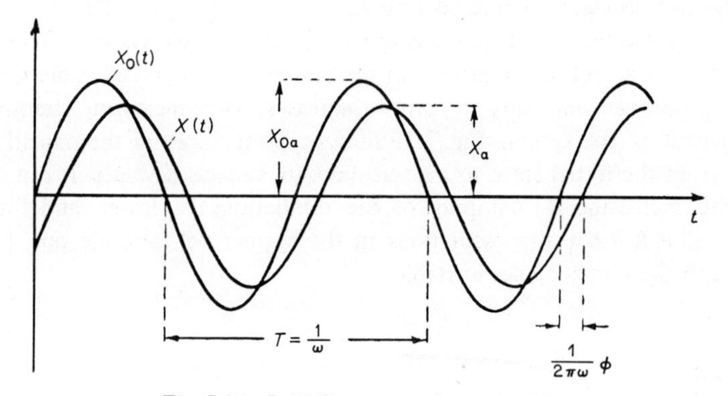


Fig. 7.14. *Periodic regime of a servo system.*

If such experiments are repeated for various frequencies $\omega$ and each time the ratio of the amplitude $X_x/X_{0a}$ and the phase shift $\phi$ between the oscillations at the input and the output of the servo system are measured, the results enable the frequency characteristics of the system to be plotted: this is the amplitude-frequency characteristic. $X_a/X_{0a}(\omega)$ and the *phase-frequency characteristic* $\phi(\omega)$ for which examples are given in Fig. 7.15. The existence of a maximum M in the amplitude-frequency characteristic demonstrates the occurrence of resonance at a frequency of the input oscillations which is near to the resonant frequency of the system.

The higher the resonance peak M, the greater will be the tendency of the system to develop oscillations, and the slower will the oscillations be damped. Therefore the existence of a high peak in the amplitude-frequency characteristic is undesirable.

A reduction of the amplitude ratio $X_a/X_{0a}$ in the range of high frequencies (in the neighbourhood of $\omega$) evidences limitations in the capability of the servo

*A *linear* system is a system for which the principle of superposition is applicable, namely: if to an input $X_{1inp}(t)$ an output $X_{1outp}(t)$ and to an input $X_{2inp}(t)$ an output $X_{2outp}(t)$ correspond, then the sum of inputs $X_{inp}(t) = X_{1inp}(t) + X_{2inp}(t)$ will produce the output $X_{outp}(t) = X_{1outp}(t) + X_{2outp}(t)$.

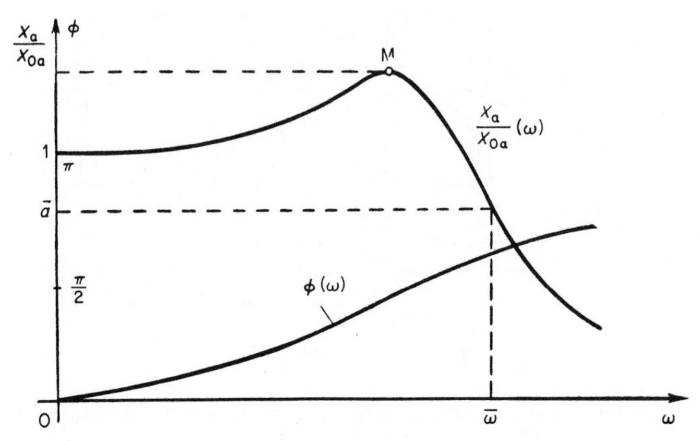


Fig. 7.15. *Amplitude-frequency and phase-frequency characteristics of a servo system.*

system to reproduce rapidly changing signals. If it is conditionally assumed that the ratio $X_a/X_{0a} < a$ is inadmissible, then a frequency range $0 - \omega$ can be distinguished on the amplitude-frequency characteristic, which is called the *pass band of the system.* The wider the pass band, the more accurately can a system reproduce rapidly changing signals.

The phase-frequency characteristic permits judging the phase lag in reproducing the input signal. Obviously, the accuracy of reproduction will be the higher, the smaller the phase lag $\phi$.

If analysis indicates that the dynamic properties of a servo system do not correspond to the required properties, then an attempt can be made to improve them. For this purpose it is possible to use so-called *correction elements.* These generate signals of the rate of change or the speed and acceleration of changes in the input or output values, and as a result it is possible to improve the operation algorithm of the control devices. By this means it is frequently possible to improve appreciably the accuracy of reproduction of the input signals of a servo system over a wide range of frequencies.

## Exercises

1. Draw the functional diagram of an automatic system for controlling the voltage of a d.c. generator (Fig. 7.16) and name its main elements.

The circuit operates as follows. The voltage $V_u$, which is proportional to the controlled quantity $V_g$, is compared with a voltage standard $V_{av}$, which the operator can change as desired. The difference voltage is fed to the input of an electronic amplifier which feeds the control winding CW of amplidyne A. The amplidyne generates a direct current for feeding the excitation winding of the generator EWG which is proportional to the voltage in the control winding CW.

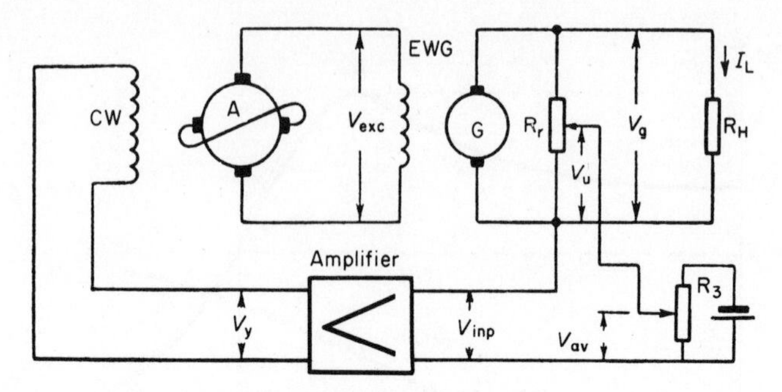


Fig. 7.16. *To Example No. 1.*

*Solution.* Subdividing the automatic control system into its functional elements, we first distinguish the generator G which will be considered as the controlled plant. The excitation voltage $V_{exc}$ which acts on it is the control variable and the disturbance the load current $I_L$. The quantity to be controlled, $V_g$, is transformed into the voltage $V_u$ of the metering element and compared with the voltage $V_{av}$. The amplifier in the control equipment is electronic, and the selsyn will be considered as the final control element.

The enumerated subdivision is relative, because we would be equally justified in considering the electronic amplifier together with the selsyn as the amplifier, and the excitation circuit of the generator as the final control element, taking the armature of the generator as the controlled plant. Of the above enumerated subdivision of the voltage control system, the functional elements of the diagram can be represented as shown in Fig. 7.17.

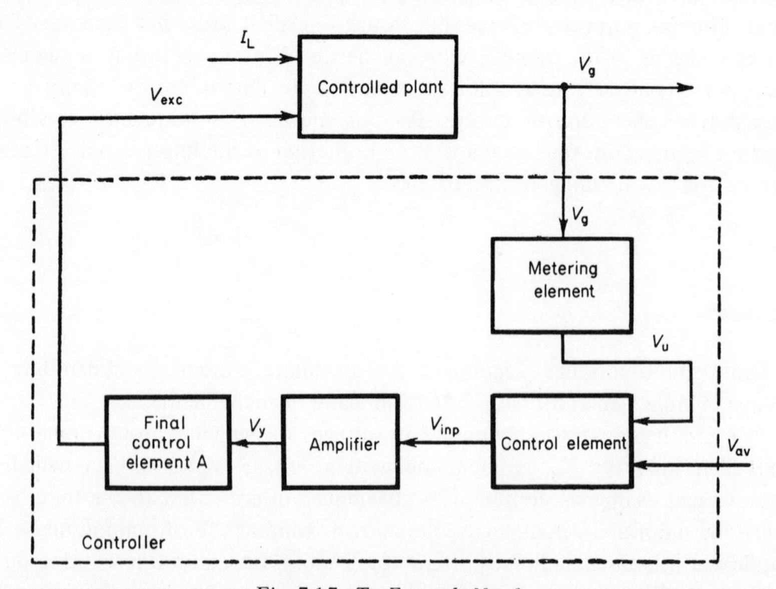


Fig. 7.17. *To Example No. 1.*

**2.** Assuming linearity (within certain limits) of the individual elements, write the equations of the static characteristics of the system.

*Solution.* The static properties of the system can be described by the following system of equations:

$$V = k_u V_g \qquad (1)$$

$$V_{av} = k_u V_0 \qquad (2)$$

$$V_{inp} = k_c(V_{av} - V_u); \qquad (3)$$

$$V_Y = k_Y V_{inp}; \qquad (4)$$

$$V_{exc} = k_A V_Y; \qquad (5)$$

$$V_g = k_0 V - k_M I_L \qquad (6)$$

The equations of the static characteristic of the system are derived from equation (6) by substituting into it the values from equations (1 to 5):

$$U_g = \frac{K V_0 - k_M I_L}{1 + K}$$

where $K = k_0\ k_A\ k_Y\ k_c\ k_u$, and $V_0$ is the given voltage which the controller is required to maintain.

**3.** Fig. 7.18 shows a system for controlling the level of a solution in a tank.

When the level changes, the float F switches the contacts in such a way that the reversible motor M either opens or closes the gate valve B in the solution feeding line, compensating the change in level. Draw the phase trajectories on the plane $h$, $l$, where $h$ is the level in the tank and $l$ is the magnitude of opening the feed flap of the solution (it is assumed that the flap is so designed that the quantity of solution flowing into the tank will be directly proportional to the value of $l$). Will the given system be stable, and if so, why?

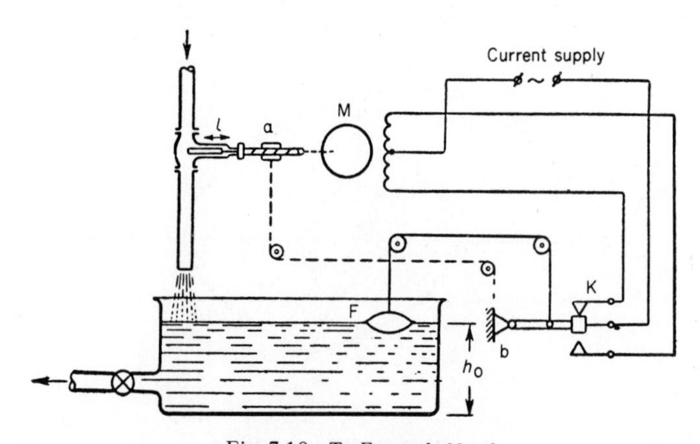


Fig. 7.18. *To Example No. 3.*

*Solution.* The phase trajectories of the system are plotted in Fig. 7.19. Each curve of the family of phase trajectories represents a second degree parabola. Due to the symmetry of parabolic phase trajectories, the representative point will always return to the initial position after passing all the four quadrants of the phase plane. Therefore, oscillations in the regulated quantity which occur in the system will not attenuate, as is evidenced by the closed phase trajectories surrounding the equilibrium point.

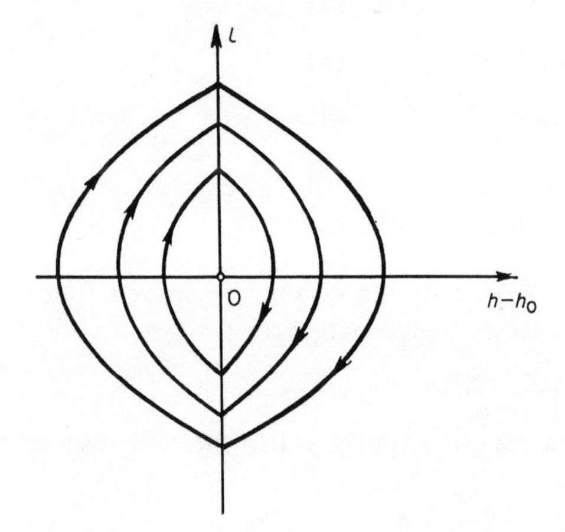


Fig. 7.19. *To Example No. 3.*

**4.** How can the system be stabilised and what will the phase portrait of the system be like?

*Solution.* The system can be stabilised, for instance, by introducing an internal negative feedback consisting of a mechanical connection between the drive of the controlled organ and the contact system K of the float relay. In Fig. 7.18 this feedback (ab) is indicated by a dotted line. If there is feedback the phase trajectories will change as shown in Fig. 7.20.

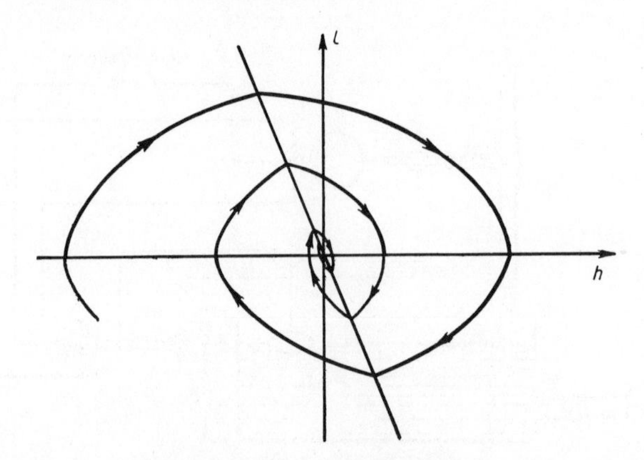


Fig. 7. 20. *To Example No. 4.*

**5.** Is the programmed control of switching street lighting a closed system?

*Solution.* No.

**6.** In Fig. 7.21 (a) and (b) amplitude-frequency and phase characteristics are given for two control systems A and B. Determine from the characteristics which of the systems will terminate the transient process more rapidly.

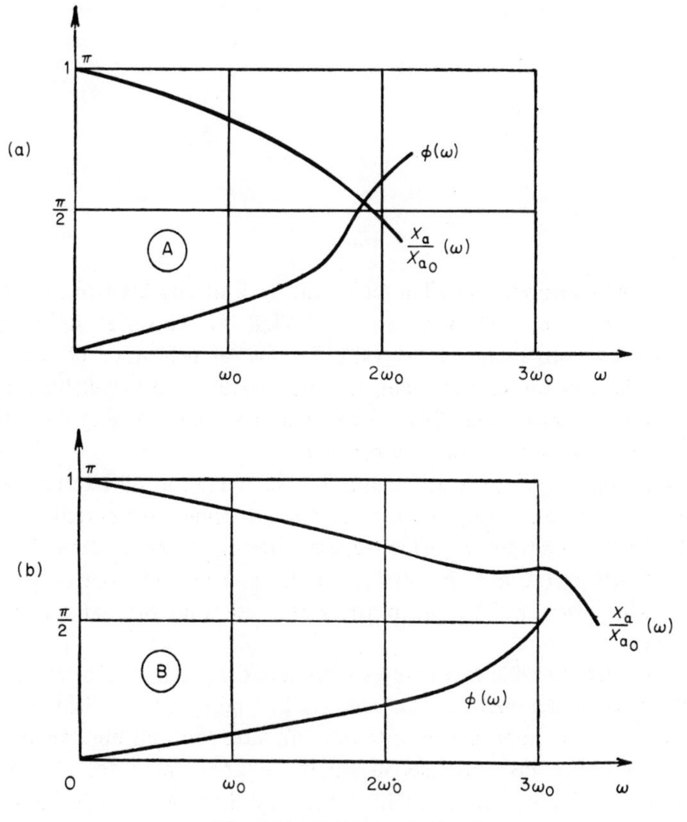


Fig. 7.21. *To Example No. 6.*

*Solution.* In the system B the transient process will terminate more rapidly, because this system has a wider pass band and 'better' (shallower) phase-frequency characteristic.

# 8 Optimal Control

As was already mentioned in Chapter 6, the task of any control process is to actively influence the object to be controlled to 'improve' its behaviour. However, in order to compare the various types of behaviour of the control system and discover the 'better' ones, it is necessary to have available a suitable measure for the purpose and this implies quantity characterising the effectiveness of control – the *effectiveness criterion* $J$.

Depending on the purpose and conditions of operation of a system, various quantities can serve as a criterion of effectiveness. Thus, for a system of control of the movement of a train the effectiveness criterion can be the time $T$ of travel from the departure station to its destination; for an irrigation control system the effectiveness criterion will be the profit $g$ realised from the harvest from the irrigated fields.

To each control variable there corresponds a definite criterion of effectiveness $J$, and the task of *optimal* control consists in finding and realising a control variable at which the appropriate criterion will have the optimum value. In the examples given, the task consists in finding a programme for changing the tractive effort of locomotive for which the travelling time would be minimal, $J = T = \min$; or an irrigation programme from which the maximum profit would be made from the harvest, $J = G = \max$. It is essential to bear in mind that the controlled variable can be changed only within certain limits, i.e. cannot exceed the limits of certain permissible values. Furthermore, additional limitations may be imposed on the system: limitations as regards the phase coordinates, limitations on the complexity of the control algorithm and on the volume of the information used.

*By optimal control we understand a set of control actions which are compatible with the limitations imposed on the system and which will ensure the 'best' values of the effectiveness criterion.*

## 8.1. Optimal process

Let us assume that the task consists in changing the state of the system from the initial state $X_{in}$ to a given state $X_0$ by means of the control action (manipulated variable) $Y$. In the phase space the states $X_{in}$ and $X_0$ correspond to the points $a_{in}$ and $a_0$ and the transition of the system from $a_{in}$ to $a_0$ will correspond to some trajectory which joins the two points, Fig. 8.1. It is possible to select many control variables which will satisfy the requirements for changing the state of the system from $a_{in}$ to $a_0$ by means of the manipulated variable $Y$. Each of these control variables corresponds to a certain trajectory which joins $a_{in}$ with $a_0$. However, as regards the effectiveness criterion $J$ these trajectories are not equivalent because each of them corresponds to a specific value of $J$, equalling $J_1, J_2, \ldots$. The problem of finding an optimal control can in this case be treated as the problem of selecting from a multitude of possible trajectories joining $a_{in}$ with $a_0$, the particular trajectory for which the effectiveness criterion will be the most favourable.

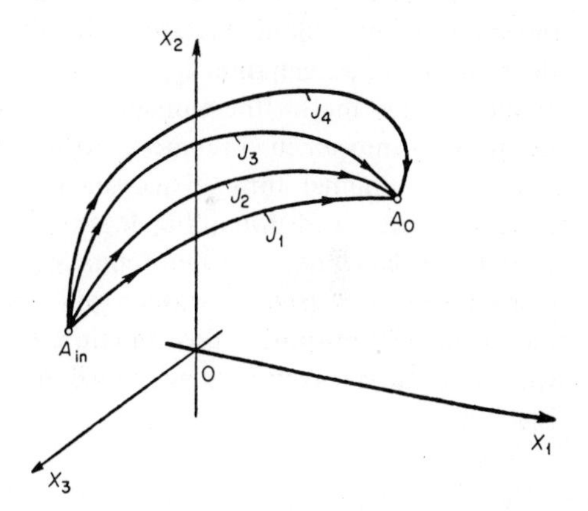


Fig. 8.1. *Family of possible trajectories of transition of the system from the state* $a_{in}$ *to the state* $a_0$.

Let us consider as an example the problem of finding the optimal control of a locomotive, which should move the rolling stock in a minimum period of time from one point $S_{in}$ to another $S_0$. Of the factors which limit the possible control, we must take into consideration the fact that accelerations $u$ during starting and decelerations during braking should not exceed certain limits

$$-\bar{u} < \bar{u} < +\bar{u}.$$

Furthermore the velocity $v$ should not exceed the maximum permissible velocity $\bar{v}$.

In order that the train should travel from $S_{in}$ to $S_0$ in the shortest possible time, it is necessary that its average rate of travel should be as high as possible. For this purpose the speed must increase at the beginning as quickly as possible and that means that the rolling stock must be accelerated with the maximum permissible acceleration $u$. As a result, the velocity $v$ will increase in accordance with the linear law

$$v = \overline{u}t.$$

If this regime were to be maintained until the moment at which the train arrives at the point $S_0$, then its speed at the moment of arrival would differ from zero and in view of the limitations imposed on the intensity of braking, it would not be possible to reduce the speed instantaneously, and as a result the train would overshoot its destination.

In order then that the train should stop at the desired destination it is necessary to start braking well ahead, before the system reaches the position $S_0$. The more intensive the braking, the later it can start, which means that the higher will be the average rate of travel. This means that deceleration should be with the maximum permissible intensity $-\overline{u}$.

During deceleration with a constant value $-\overline{u}$ the velocity will decrease according to the linear law $v = v_M - \overline{u}t$.

Accordingly, the fast movement of the train from $S_{in}$ to $S_0$ will consist of two stages: start-up with the limit acceleration $+\overline{u}$ and braking with the maximum permissible deceleration $-\overline{u}$. By suitable selection of the duration of the intervals of acceleration $t_{acc}$ and deceleration $t_{dec}$ it is possible to meet the condition such that at the moment of termination of the braking process (when the speed of movement becomes zero) the train will be exactly at the point $S_0$.

Such an optimal form of movement – optimal from the point of view of travelling time – is shown in Fig. 8.2.

If the distance between the points $S_{in}$ and $S_0$ is so great that for such a form of movement the speed of travel $v$ will reach its limit value $\overline{v}$, then the optimal movement will obviously also contain a third interval $t_L$ during which the train will travel at a constant speed equal to the limit permissible $\overline{v}$, as shown in Fig. 8.3.

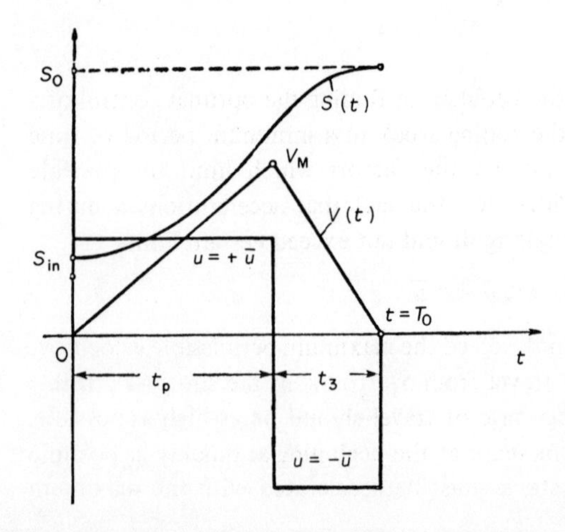


Fig. 8.2. *Optimal graph of the movement of a train with limited acceleration.*

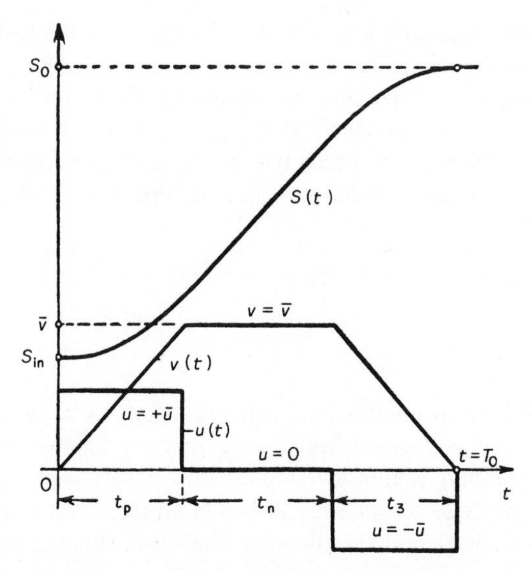


Fig. 8.3. *Same graph with an additional limitation on velocity.*

During the periods of uniform acceleration or deceleration the distance travelled by the train will change in accordance with a parabolic law, and when the train moves at a constant speed it will change in accordance with the linear law.

If we consider now the actions of the train driver on the acceleration or deceleration as control actions $Y$, and the distance $S$ travelled as the output value, then an analogy between the laws of motion of the considered system and the system of hydraulic drives considered in Chapter 4, is easy to see.

Therefore the phase portraits of these systems will be similar and represent a family of parabolas whose parameters change as a function of the value $Y$, as is shown in Fig. 8.4.

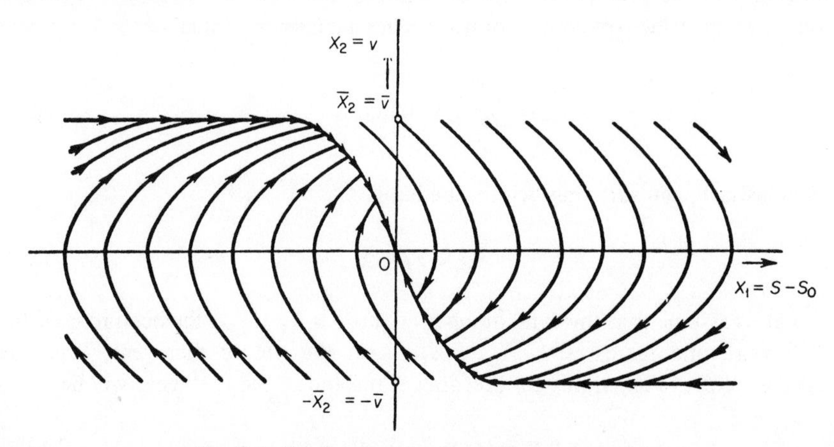


Fig. 8.4. *Family of optimal transition trajectories.*

Using the phase portraits Fig. 4.6, it is possible, on the basis of the above considerations, to construct on the phase plane $X_1 = S - S_0$, $X_2 = v$, a set of optimal transition trajectories from any initial state into any given state of the system considered (Fig. 8.4). One and the same diagram of optimal trajectories will be valid for any system in which the velocity and acceleration of the output quantity $X_1$ are limited and the effectiveness criterion of the control is the travelling time.

## 8.2. Optimal strategy

The problem of optimal control frequently turns out to be to select an optimal sequence of solutions for any *multi-stage process.* Such problems arise for instance when planning capital investments in a factory over a period of many years, when we try to select such a sequence of capital investments for each year so that over a certain period the economy as a whole should reap the maximum benefit. A similar problem occurs when selecting a sequence of solutions for modernising production, if we aim to achieve a maximum output during a certain period, taking into consideration the losses in production during the reorganisation period. Solution of problems of this type, working out an optimum strategy of decision-making, can be based on mathematical methods which have been worked out for this purpose: either the maximum principle of L. S. Pontryagin or the dynamic programming of R. Bellman.

Let us consider the problem of evolving an optimum strategy of the multi-stage decision-making on the example of running an animal farm.

Let us assume that the criterion of effectiveness $J$ is the income from the sale of livestock over a period of $N$ years. Designating by $Y_i$ the number of cattle (from a total of $X_i$) sold at the end of the $i^{-\text{th}}$ year, and by $C$ the cost per head, we find that the problem of control is to select a sequence $Y_1, Y_2, \ldots, N\, Y_N$ such that the effectiveness criterion $J$ reaches a maximum value

$$J = \sum_{i=1}^{i=N} CY_i = \max. \tag{8-1}$$

Obviously, $Y_i$ can vary only within the limits

$$0 \leqslant Y_i \leqslant X_i. \tag{8.2}$$

Let us assume that the total number of cattle is $X_{i-1}$ at the beginning of the $i^{-\text{th}}$ year, and increases to $X_i = k\, X_{i-1}$ at the end of that year. Then the number of cattle remaining for breeding at the end of the $i^{-\text{th}}$ year will be

$$X_i^* = k\, (X_{i-1} - Y_{i-1}). \tag{8.3}$$

The existence of an optimal strategy in this case follows from the fact that if the sale in the $i^{-th}$ year is increased, then the income during that year will increase but it will be lower in subsequent years because the number of livestock will grow more slowly. If, however, too few cattle are sold, each year's investment into the sum (8.1) will be excessively small and the optimal effectiveness will not be reached. Obviously, there exists a most favourable *optimal strategy which maximises the total income over N years*, and it is this strategy which must be found.

In accordance with the method of dynamic programming we will start from the last step, i.e. selecting the value $Y_N$. It is obvious that the maximum contribution to the sum (8.1) of the $N^{th}$ year will occur if we select the maximum possible value of $Y_N$ which, in accordance with the limitation (8.2), yields $Y_N = X_N$. For the preceding, $(N-1)^{th}$ year, we find that whatever the value $Y_{N-1}$, the contribution of the $N-1^{th}$ year would be $CY_{N-1}$ roubles (pounds, dollars), and a decrease in the contribution of the $N^{th}$ year (taking into consideration an increase in the number of cattle by $k$ times) caused by sale of cattle during the preceding year $Y_{N-1}$ would be $kCY_{N-1}$ roubles (pounds, dollars) which will have an adverse effect on the total value $J$.

Applying similar considerations for all the preceding years, we arrive at the conclusion that no cattle should be sold during any year except the last. Thus the optimal strategy for the example under consideration is:

$$\left.\begin{aligned} Y_1 &= 0,\\ Y_2 &= 0,\\ &\dots\\ Y_{N-1} &= 0,\\ Y_N &= X_N. \end{aligned}\right\} \qquad (8.4)$$

We will now show that the optimal strategy will change appreciably if we impose additional constraints on the system. Let us take into consideration, for instance, that the number of cattle on the farm is constrained to $\overline{X}$ head:

$$X_i \leqslant \overline{X}, \qquad (8.5)$$

determined by the available space, fodder etc. Then, applying calculations similar to the above and moving from the end of the process to the beginning, we arrive at the conclusion that the optimal strategy consists of building up the number of cattle to the limit value $\overline{X}$ in the shortest possible time and then retaining it at that level up to the last year, when the whole herd should be disposed of. The optimal strategy for the case $N$–5 is clearly shown in the diagram Fig. 8.5.

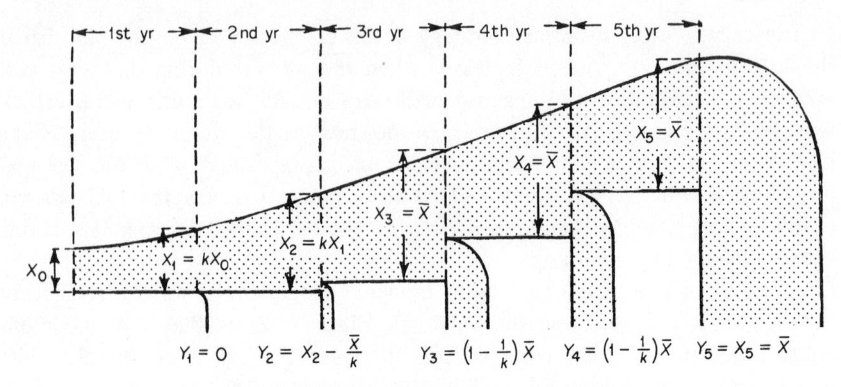


Fig. 8.5. *Optimal strategy for an animal farm.*

This strategy represents the following sequence of $Y$ values:

$$\left.\begin{aligned} Y &= 0, \\ Y_2 &= X_2 - \frac{\overline{X}}{k}, \\ Y_3 &= \left(1 - \frac{1}{k}\right)\overline{X} \\ Y_4 &= \left(1 - \frac{1}{k}\right)\overline{X} \\ Y_5 &= X_5 = \overline{X}. \end{aligned}\right\} \tag{8.6}$$

It can be seen from the quoted examples and also from consideration of optimal processes as described in Section 8.1, that at any instant of the optimal process the system should operate at its limit.

## 8.3. Iso-surfaces

In most cases the task of finding an optimal control is very laborious. Although methods of solution have been worked out for a wide variety of practical problems, the practical application of these methods usually requires so much computation that in many cases it is even too much for a modern computer. The calculations can, however, be considerably simplified if instead of seeking optimum trajectories of the transition of the system from the initial state into a given state, the problem is solved by determining the limits of the areas in the state space which are characterised by a certain value of the effectiveness criterion. It has been found that in the given case it is possible to obtain the necessary data also on optimal transitions.

Let us limit ourselves, for instance, to a class of problem where the effectiveness criterion is the transition time, i.e. a class of system optimal from the point of view of speed of response. We will consider the case where the constraints are on the speed and acceleration of the controlled quantity, as was done in the optimal control of a train considered in Section 8.1. The same limitations occur in the optimum control of lifts, servo systems, shaft winding machines, and many others.

We will try in the phase space to distinguish an area where from each point it is possible to pass into a definite point $a_0$ at the definite effectiveness criterion $J = \overline{J}$.

For this purpose we propose that at the initial instant $(t = 0)$ the representative point is in the position $a_0$, and *time runs backwards*, and in this way wᴖ follcw the 'history' of movement of the representative point into the given state. It can be imagined that we observe the movement of the representative point as recorded on a ciné film, but with the projector running backwards. Since we already know that the optimal movement occurs at the maximum permissible values, we will see to it that this condition is adhered to at all stages of movement of the system. Let the point $a_0$ be at the origin of the coordinates of the phase space (in the case under consideration the phase space is a plane with axes $S$ and $v$), as shown in Fig. 8.6. Since the optimal transition in this point could occur only when moving with the limit deceleration $\overline{u}$, see Fig. 8.4, we will consider that during the first stage of the 'reverse' movement the acceleration of the system equals $u$ in the time $T_1$. The representative point will approach the point 1 or the point 1* along the trajectory 0 – 1 or 0 – 1*, Fig. 8.6. However, the system may not have moved during the entire time $T_1$ with the deceleration $\overline{u}$, but only during a certain part of the time $t_3$. In the remaining time $T_1 - t_3$ the system could move with the limit acceleration $\overline{u}$.

Then, considering the movement as occurring in two intervals whose sum equals $T_1$, with the accelerations $+\overline{u}$ and $-\overline{u}$, we find the points 2, 2* 3, 3* . . . , where the representative point falls in the process. Plotting a sufficiently large number of such points, we can determine with the required accuracy the line $H_1$ on which these points lie, so that we can change over from any point on this line to the point $a_0$ along the *optimal path* during the time $T_1$. It has been proved that from any point inside the boundary $I_1$, it is possible to change over to the point $a_0$ during a time $T_1$ without disturbing the limitations imposed on the system, but no such transition is possible from any point beyond the limits of this area located inside the boundary $I_1$. The area inside a boundary $I_1$ is called the *isochronous area* from the time $T_i$, with respect to the *pole* $a_0$.

If the transition time $T_2 > T_1$ is given, we can then construct the boundary $I_2$ of the *isochronous area* with the transition time $T_2$. For sufficiently large values of $T$ a limitation as regards the velocity $\overline{v}$ begins to manifest itself and the boundary of the isochronous area is deformed, as can be seen from the shape of the $I_3$ isochrones in Fig. 8.6.

During the construction of the isochronous area boundaries a specific programme of control actions was given each time which brought the

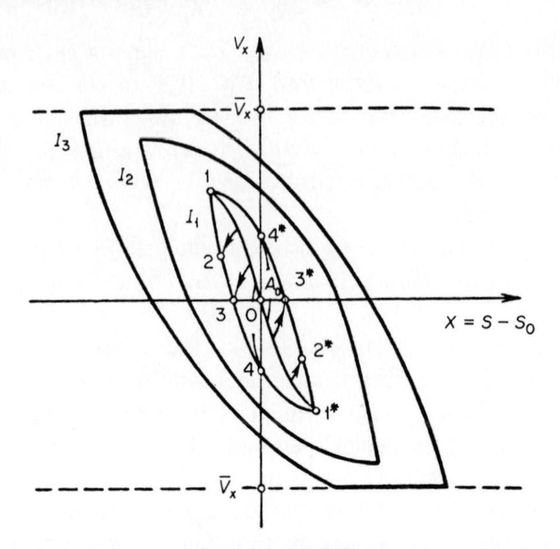


Fig. 8.6. *Isochrones.*

representative point to the boundary. It is obvious that for the time running in the opposite direction, the same programmes represent a programme of optimal transition of the system from this point of the boundary into the pole $a_0$. Therefore, by constructing the boundaries of the isochronous areas, we have also found the forms of the optimal control processes of the system.

Furthermore, construction of the boundaries of the isochronous areas allow the solution of other optimal control problems. For instance, for a given area of possible initial states of the system it is possible to determine the achievable speed of response $T_{in}$ along the areas $I_{in}$ into which we draw the area $R_{in}$ of the possible initial states, Fig. 8.7. It is obvious that if $R_{in}$ is wholly inside the $I_{in}$, then transition into the pole which is in the origin of the coordinate system can be achieved during the time $T_{in}$ from any point of the area $R_{in}$.

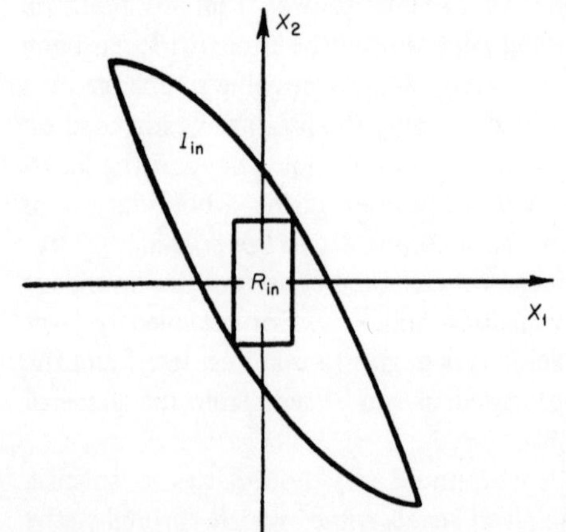


Fig. 8.7. *Determination of the speed of response of a system from the area of the initial states* $R_1$ *and the isochrone or isochronous area* $I_{in}$.

Obviously other isosurfaces can be constructed which in the state space of the system will separate areas characterised by specific values, not of the time $T$ but of another effectiveness criterion $J$, which may also prove useful in finding the optimal control or for determining the limit achievable $J$ for various transitions of the system from the initial state to a given state.

## 8.4. Optimal control system

Study of the laws of optimal control has created the conditions required for automatic realisation of optimal processes in engineering systems for improving their technical and economic indices. Optimal control systems are being used in fast servo systems, in systems for controlling electric drives, in the power industry, in transportation, and in rocket engineering.

In the realisation of optimal control systems the following problem arises: how should the structure and parameters of the control equipment be selected, to ensure that under real operating conditions of the system the movements should be optimal, or at least nearly optimal. Let us consider some methods of solving such problems.

If the system operates with fixed transitions, then the point which represents its state should move from fixed initial into fixed prescribed states, the optimal control programmes can be pre-selected and kept in the memory of the control equipment. When it becomes necessary to apply a certain transition, the appropriate programme is taken from the memory and realised in the form of an optimal sequence of control responses. Such control systems are extensively used in the control of freight and passenger lifts, excavators, electric drives of reversing rolling stands, for controlling the start-up and braking of electrically driven vehicles.

If the required transitions of a system are not known in advance, the problem of realising optimal systems cannot be solved on the basis of rigid control programmes, and the control responses must be generated during the activity of the system. Let us assume that it is necessary to realise a control system which is optimal from the point of view of speed of response for controlling a train under the conditions described in Section 8.1, or a hydraulic drive or any other system on which limitations are imposed on the rate of change and acceleration of the controlled quantity. The set of optimal processes in such systems is shown in Fig. 8.4, from which it can be seen that for realising an optimal transition of a system from any of its initial states to given equilibrium state it is necessary and sufficient that the control actions correspond to the diagram shown in Fig. 8.8. Here the phase space is subdivided into two regions: A and B. In any point of the region A the control response should be so chosen that the acceleration is $\bar{u}$, and in the range B it should equal $-\bar{u}$. The boundary separating the areas A and B is called the *switching line.*

Now the problem of optimal control reduces to determining in which region

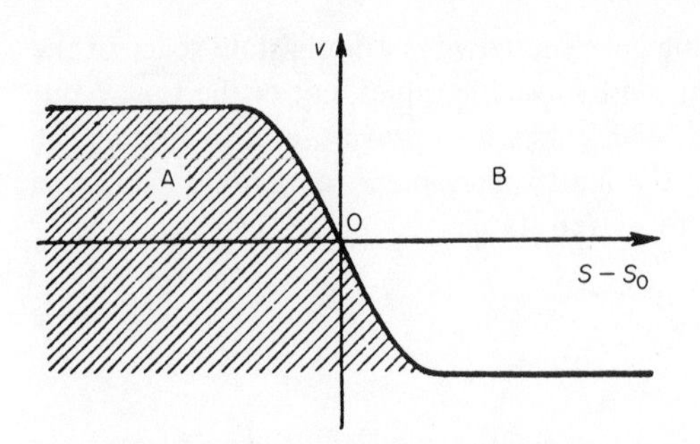


Fig. 8.8. *Switching line of control responses.*

(A or B) the point is located which represents the state of the controlled system, since the belongings of the representative point to one of these regions which is subdivided by switching lines, will determine unequivocally the control response values (values of the manipulated variable).

From the diagram of the set of optimal processes the position and form, and consequently also the equation of the switching line, can be determined. For the class of system under consideration the switching line in the range of $X_1$ values consists, as can be seen from Fig. 8.4, of sections of parabolas which pass through the origin of the coordinate system. Its equations can be written in the form

$$X_1 + cX_2^2 \operatorname{sgn} X_2 = 0 \tag{8.7}$$

where $c$ is the coefficient depending on the limit acceleration $\bar{u}$ of the controlled quantity $\bar{X}_1$, and the sign sgn $X_2$ designates the function of $X_2$ which will assume the values

$$\operatorname{sgn} X_2 = \begin{cases} +1 & \text{when} \quad X_2 > 0 \\ -1 & \text{when} \quad X_2 < 0. \end{cases}$$

If the values of $X_1$ and $X_2$ are such that the point which represents the state of the system does not lie on the switching line, then for the region A

$$X_1 + cX_2^2 \operatorname{sgn} X_2 < 0$$

and for the region B

$$X_1 + cX_2^2 \operatorname{sgn} X_2 > 0.$$

From what has been said above it follows that the control signal $u$ generated by the control equipment must assume the values

$$\left.\begin{aligned} u &= +\bar{u}, \quad \text{if} \quad X_1 + cX_2^2 \operatorname{sgn} X_2 < 0, \\ u &= -\bar{u}, \quad \text{if} \quad X_1 + cX_2^2 \operatorname{sgn} X_2 > 0, \end{aligned}\right\} \tag{8.8}$$

The conditions (8.8) express the control algorithm which realises the optimal transition in systems of the class under consideration.

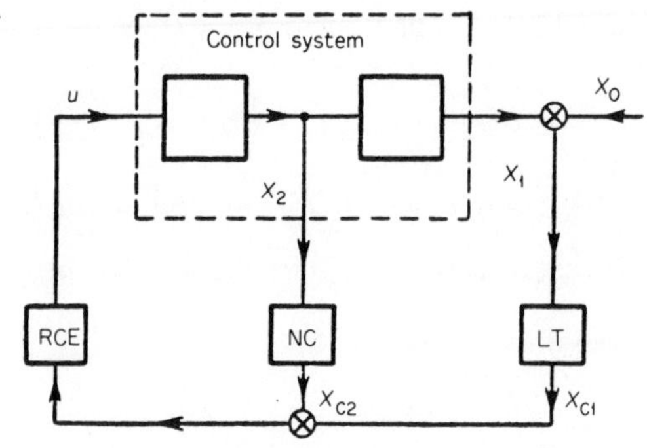


Fig. 8.9. *Diagram for realising the optimal control algorithm.*

This algorithm can be realised by means of a linear transducer LT, a nonlinear converter NC, and a relay correcting element RCE, which are connected into a circuit as shown in Fig. 8.9.

The components of this circuit realise the following transformations:

$$\left.\begin{aligned} X_{c1} &= kX_1, \\ X_{c2} &= kcX_2^2 \operatorname{sgn} X_2, \\ u &= -\bar{u} \operatorname{sgn}(X_{c1} + X_{c2}) \end{aligned}\right\} \tag{8.9}$$

It can easily be seen that for the characteristics of the transducers corresponding to (8.9) the control equipment of the control system shown in Fig. 8.9 will satisfy the control algorithm (8.8).

To realise systems for the optimal control of more complex equipment and for more complex effectiveness criteria, correspondingly more complex control equipment is necessary.

## Exercises

**1.** A freight car loaded with coal is unloaded at a station by means of a bridge crane. The distance from the freight car to the coal reception platform is given by $s$, where $s = 30$ metres. The maximum speed of movement of the crane $v = 1$ m/sec, the maximum acceleration at the beginning of the movement is $u_1 = 0.2$ m/sec$^2$, and the deceleration at the end of the movement $u_2 = 0.4$ m/sec$^2$. Construct the graph of the situation, and calculate the parameters of a control which is optimal from the point of view of the time of movement of the crane from the wagon to the unloading place.

*Solution.* Fig. 8.10 shows an optimal control of the crane on the path from $X_{\text{in}}$ (the wagon) to $X_{\text{k}}$ (reception point of the coal). To find $t_1$, $t_2$ and $t_3$, we write the equations of movement of the crane for the individual sections of travel:

$$S = \frac{u_1 t_1^2}{2} + v_{max}\, t_2 + \left(v_{max}\, t_2 - \frac{u_2 t_3^2}{2}\right);\ 20 = \frac{0.2 t_1^2}{2} + 2 t_2 - \frac{0.4 t_3^2}{2};$$

$$v_{max} = u_1 t_1 = 0.2 t_1 \text{ (for the section where acceleration is applied),}$$

$$v_{max} = u_2 t_3 = 0.4 t_3 \text{ (for the section where braking is applied).}$$

From these we obtain:

$$t_1 = \frac{v_{max}}{u_1} = 5 \text{ sec},\ t_3 = \frac{v_{max}}{u_2} = 2.5 \text{ sec},$$

$$t_2 = 16.25 \text{ sec}.$$

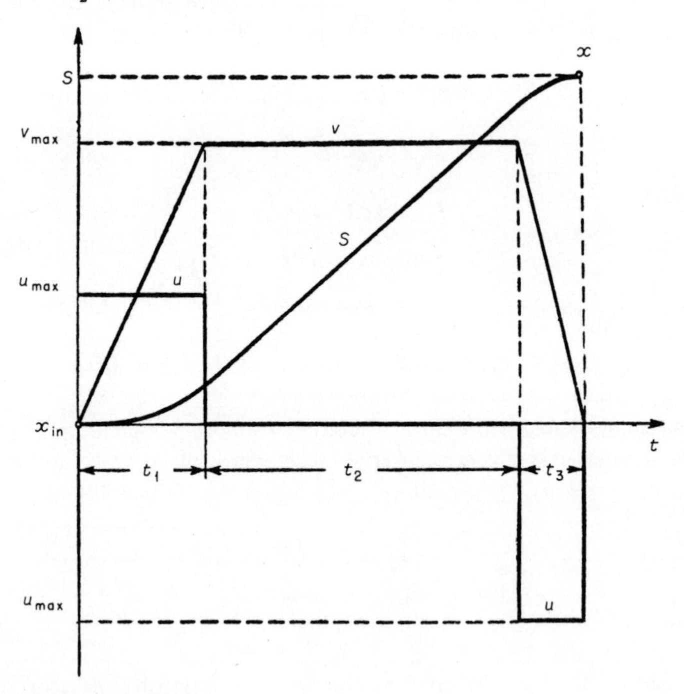


Fig. 8.10. *To Example No. 1.*

2. A motor car A (Fig. 8.11) moves out from a parking spot between two other cars. When the car moves out it must not cross the continuous centre line which is at a distance $H = 7$ metres from the edge of the road. The width of the car is $h = 2$ m, the length of the car A is $l = 5$ m, its turning radius $R = 6$ m. Determine the optimal sequence of manipulations by the driver if the car can only move forwards. Find the minimum distance between the cars B and C for which this manoeuvre is still possible, and the optimal sequence of movement of the right front wheel.

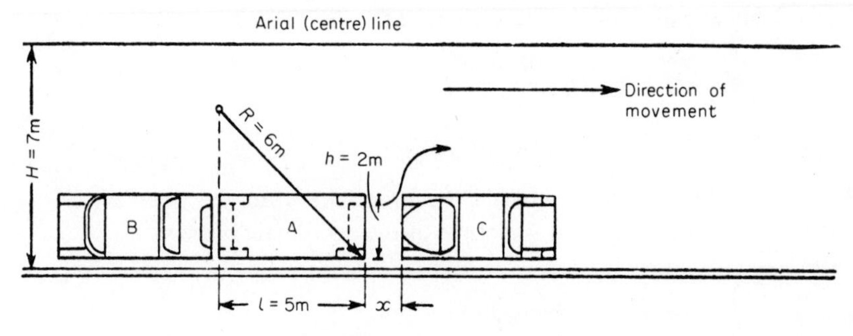


Fig. 8.11. *To Example No. 2.*

*Solution.* The sketch of the movement of the car is shown in Fig. 8.12. Optimal control is achieved as follows. From the initial position 1 the car moves into position 2 with the largest possible turning angle of the wheels to the left; then one side AB assumes the position A′B′, when the steering is moved fully to the right and the car will move into a position parallel to the basic movement (position 3). The minimum distance to the motor car C will be determined by the turning radius $O_1D$ and is about 1 m, which can easily be determined by geometrical construction. The solid lines in Fig. 8.12 show the optimal trajectory of the wheel.

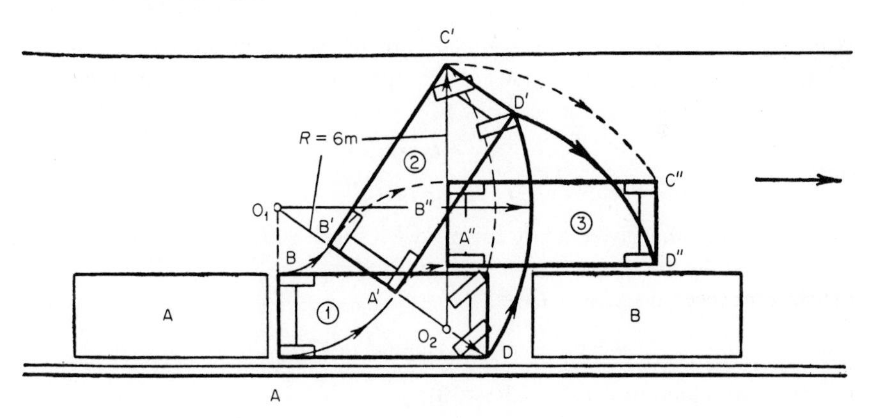


Fig. 8.12. *To Example No. 2.*

**3.** In Fig. 8.13 a block diagram is given of a two-stage hydraulic drive described in Section 4.2, which contains two astatic members with the transmission coefficients $K_1 = 1$ and $K_2 = 3$. The limitations on the coordinate movements are given by the inequalities $|x_1| \leqslant \bar{x}_1 = 2$ and $|x_2| \leqslant \bar{x}_2 = 30$.

The dependence of the path and the speed of the output coordinate on the duration of the intervals is:

(a) for regimes when the speed does not reach the limit values

$$x = Nt_1^2 - \tfrac{1}{2}Nt_n^2, \quad v_x = Nt_n - 2Nt_1;$$

(b) for regimes which include movement at the limit speed:

$$x = \tfrac{1}{2} N t_1^2 - M t_n + \tfrac{1}{2} \frac{M^2}{N}$$

where $V_x = M - Nt$; $N = k_1 k_2 \bar{x}_1$ and $M = k_2, \bar{x}_2$.

Construct the boundaries of the isochrone areas for the given system with respect to the pole located at the origin of the coordinate plane $x$, $v_x$.

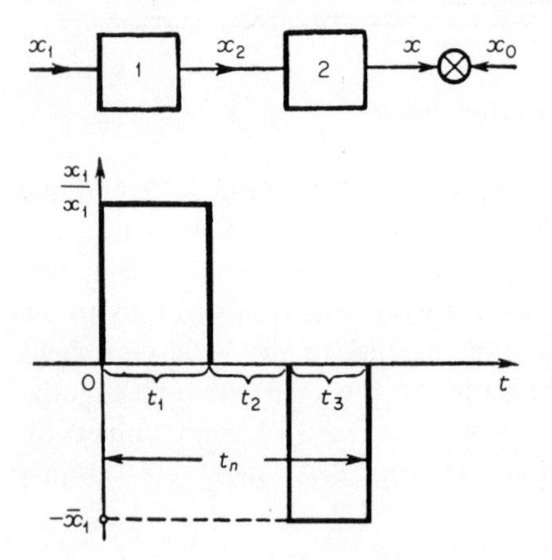


Fig. 8.13. *To Example No. 3.*

*Solution.* For constructing the boundaries of the isochrone areas $I_1$, $I_2$ and $I_3$, we will assume that the durations of the intervals of the process are such that their sum should be equal respectively to

$$t_1 + t_2 = t_i,$$

or

$$t_1 + t_2 + t_3 = t_i.$$

Then, using the given relationships, we obtain computation formulae for constructing the isochrones:

(1) without speed limitations ($N = 3.1.2 = 6$):

$$x = 6t^2 - 3t_i^2, \quad v_x = 6t_i - 12t_1;$$

(2) with speed limitation ($M = 1.30 = 30$):

$$x = 15t_i + 15t_1 - 3t_1^2 + 15t_2; \quad v_x = 30 - 6t_1.$$

From these formulae, curves shown in Fig. 8.6 were plotted.

**4.** The designers are faced with the problem of designing a multi-stage space rocket. The starting weight of the rocket $G$ and the weight of the space capsule $g_k$ are given. The rocket is to have $l$ stages. The starting weight comprises the weights of all the stages and the weight of the capsule, i.e. $G = \sum_{l=1}^{l} G_i + g_k$ where $G_i$ is the weight of the $i^{-\text{th}}$ stage. Each stage has an engine which requires a certain quantity of fuel. After some stage has consumed its fuel it is discarded and the next stage begins to operate. During the operation of the engine of the $i^{-th}$ stage the rocket gains the additional velocity $\Delta v_i$, which depends on the weight of this stage as well as on the weight of the load which it still has to carry.

The task is to find the optimal distribution of the weight of the rocket between the $l$ stages, for which the velocity after discarding all the stages will be maximal.

*Solution.* It is possible to imagine $l$ stages of the rocket as being $l$ stages of acceleration. Prior to each stage it is necessary to solve the problem of what part of the existing weight, not so far discarded, must be expended on the given stage, and what must be kept for the following ones.

We denote by $G_i$ the weight discarded on the $i^{th}$ stage, and by $G_i{}^* = \sum_{k=i+1}^{l} G_k$ the weight reserved for the remaining stages. In this case the increment in velocity after burning out of the $i^{th}$ stage can be written as a function of the weight of this stage and of the remaining weight, taking into consideration the weight of the capsule, i.e.

$$\Delta v_i = f(G_i, G_i^* + g_k).$$

The solution of the problem should be started from the last stage. Any weight $G_{l-1}^*$ which is retained after the previous stages have been discarded must be fully given to the $l^{th}$ stage. The maximum speed increment will be achieved corresponding to the given value of $G_{l-1}^*$

$$\Delta v_l = f(G_{l-1}^*, g_k).$$

We fix the weight $G_{l-2}^*$ remaining after the $(l-1)^{th}$ stage. Obviously

$$G_{l-1}^* = G_{l-2}^* - G_{l-1}.$$

The optimal control on the $(l-1)^{th}$ stage will be such as yields the maximum sum of the two velocity increments: $\Delta_{l-1}$, reached in the $(l-1)^{th}$ stage, and $\Delta v_l$, the maximal increment in the $l^{th}$ stage. This procedure for selecting the optimal control can be continued until the first step is reached.

After optimising the first step (selecting the weight of the first stage $G_1$) the sequence of stages is again calculated, starting this time from the beginning and proceeding to the end. As a result, a set of optimal weights of the individual stages is determined: $G_1, G_2, \ldots, G_l, \left(\sum_{i=1}^{l} G_i = G_0^*\right)$ which gives the payload (the capsule) the maximum velocity $\Delta v_{max}$.

# 9 Automata

The engineering term '*automaton*' is used to designate a system of mechanisms and equipment where the processes of obtaining, transformation, transmission and utilisation of energy (power), materials and information required for performing its functions are realised without direct participation of man. Examples of systems of this type are automatic machine tools, automatic packing machines, automatic machines for taking and printing photographs, vending machines, and many others.

In cybernetics, however, the term '*discrete automaton*' or simply 'automaton' has gained acceptance and is widely used for designating a much more abstract concept, namely a model which has the following features:

(a) at each of the discrete instants of time $t_1, t_2, \ldots m$ input values $x_1$, $x_2 \ldots, x_m$ each of which can assume a finite number of fixed values from the input alphabet $X$, are applied on the input of the model;

(b) on the output of the model $n$ output values $y_1, y_2, \ldots y_n$ can be observed, each of which can assume any finite number of fixed values from the output alphabet $Y$;

(c) at each instant of time the model can be in one of the states $z_1, z_2, \ldots, z_N$;

(d) the state of the model at each instant of time is determined by the input value $x$ at this time and the state $z$ in the preceding instant of time;

(e) the model realises the transformation of the situation at the input $x = \{x_1, x_2, \ldots, x_m\}$ into the situation at the output $y = \{y_1, y_2, \ldots, y_n\}$ depending on its state in the preceding instant of time.

Such a model, Fig. 9.1, is convenient for describing many cybernetic systems.

Automata in which the situation $y$ at the output is unequivocally determined by the situation $x$ at the input will be related to the class of *automata without a memory*. Automata in which $y$ depends not only on the value $x$ at a given instant of time but also on the state of the model $z$, which is determined by the value of $x$ in the previous instant of time, belong to the class of *automata with a finite memory*.

In Section 9.3 the term *automaton with an infinite memory* will be additionally introduced.

In the present chapter we will limit ourselves to considering only the simpler of the discrete automata, whose input and output alphabets consist only of the two symbols: 0 and 1. This is justified because, as is shown in the theory of automata, automata with such 'poor' alphabets are capable of solving the same problems as automata with any other alphabet, however complex.

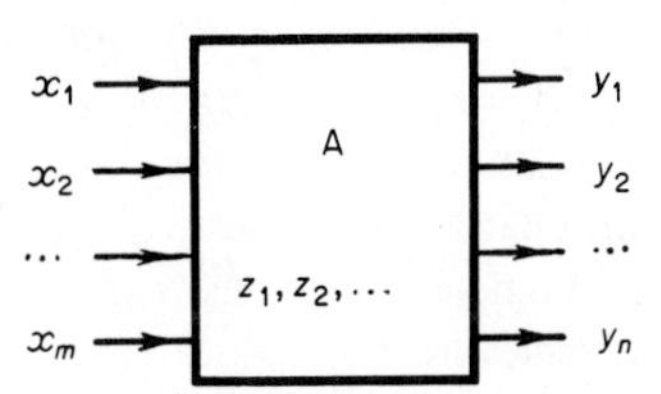


Fig. 9.1. *Discrete automaton.*

The theory of discrete automata, or, as it is sometimes called, the theory of relay circuit devices, is of great importance in solving some fundamental problems of cybernetics associated with the possibilities in principle of processing information in cybernetic systems and also for the analysis and synthesis of complex relay circuits and digital computers.

## 9.1. Logical automata

Transformation of input into output values realised by memoryless discrete automata operating with a two-letter alphabet is equivalent to transformations performed in formal logic. Therefore these will be called *logical automata* and the functions describing the transformations performed by logical automata will be called *logical functions*. The mathematical apparatus used for solving problems of analysis and design of logical automata is *the algebra of logic*. Historically, the first example of such an algebra was published in 1843 by the British mathematician George Boole, and as a result it is referred to as *Boolean algebra*.

Each output $y_i$ of a logical automaton can assume the value 0 or 1, depending on the values of the input variables $x$. Let us determine the number of all the possible logical functions of the $x$ to $y_i$ transformation, if the number of the input values equals $m$, each of which can assume either the value 0 or 1. For this purpose we expand all the input values into the series $x_1, x_2 \ldots, x_m$ and we will consider these as digits of a binary number. It is obvious that the number $r$ of various combinations of input values will equal the number of various binary numbers containing $r$ digits, and it thus follows that $r = 2^m$. But to each of the $r$ situations at the input there can correspond one of the two output values 0 or 1.

Therefore the total number $N$ of all the different logical functions for a logical automaton with $m$ binary inputs is given by

$$N = 2^r = 2^{(2^m)}. \qquad (9.1)$$

Logical functions are formed from some elementary logical functions. We will use three elementary logical functions:

(1) $\bar{x}$ = the negation of $x$ (read 'not $x$'). The negation function means that $\bar{x} = 0$, if $x = 1$, and $\bar{x} = 1$, if $x = 0$.

(2) $x_1$ & $x_2$: *logical multiplication* or conjunction (read '$x_1$ *and* $x_2$'). The function of logical multiplication means that its result equals unity only when both $x_1 = 1$ and $x_2 = 1$, and in all other cases it equals zero.

(3) $x_1 \vee x_2$: *logical addition* or disjunction (read '$x_1$ and $x_2$'). The logical addition means that the result will equal zero only if $x_1 = 0$ and $x_2 = 0$, and in all other cases will equal unity.

Logical functions can be given in the form of so-called 'truth tables' containing the values of the function $y$ (we will omit the index $(i)$ for all combinations of the arguments $x$. Table 9.1 gives values of two elementary logical functions of the two arguments $x_1$ and $x_2$. This table can be read along the lines: 'if $x_1 = \ldots$, and $x_2 = \ldots$, then $x_1$ and $x_2 = \ldots$, and $x_1$ or $x_2 = \ldots$'.

Table 9.1

| $x_1$ | $x_2$ | $x_1 x_2$ | $x_1 V x_2$ |
|---|---|---|---|
| 0 | 0 | 0 | 0 |
| 0 | 1 | 0 | 1 |
| 1 | 0 | 0 | 1 |
| 1 | 1 | 1 | 1 |

Logical functions are extensively used in the theory of neural networks and form part of the mathematical apparatus used in the study of information processing in the brain. These problems will be dealt with in Chapter 16.

From elementary logical functions, further logical functions can be built up which describe the properties of various logical automata including relay circuits. The components of relay circuits are electric relays whose input element is the winding 0, Fig. 9.2. and the output element – contacts of two types: normally

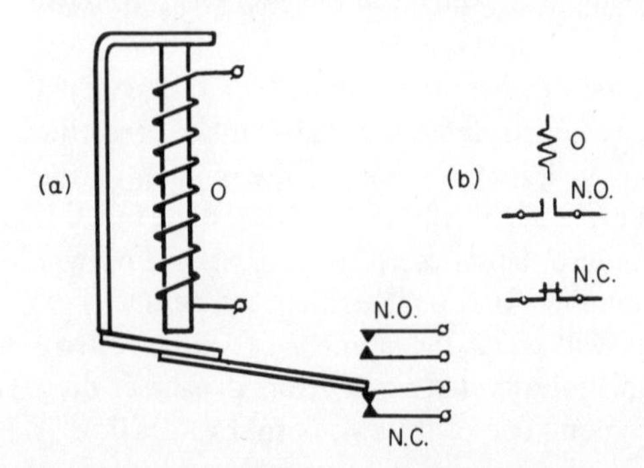


Fig. 9.2. *Components of relay circuits.*

open contacts (open when no current flows in the winding and closed when current does flow in the winding), and normally closed (closed when no current flows in the winding and open when there is a current flow). The conventional symbols of relay components in relay circuits are shown in Fig. 9.2(b).

We will assume that the current in the relay winding and the state of the contacts are logical variables which can have either the values 0 or 1. The logical variables characterising the state of one and the same relay will be designated by equal symbols and equal indices. Thus, if the state of the winding of the $i^{\text{th}}$ relay is designated by $x_i$, then the state of its normally open contact will also be designated by $x_i$ and the state of the normally closed contact by $\overline{x}_i$.

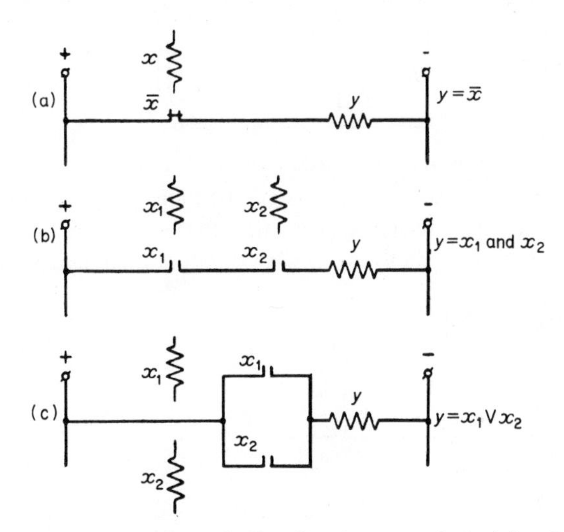


Fig. 9.3. *Relay circuits which realise elementary logical functions.*

Fig. 9.3 shows relay circuits which perform transformations corresponding to elementary logical functions of negation, Fig 9.3(a) logical multiplication, Fig. 9.3(b) and logical addition, Fig. 9.3(c).

Using methods of the theory of relays, relay circuits can be designed which will realise the required logical function.

For instance, it is necessary to construct a relay circuit whose transformation function $y = F(x_1, x_2, x_3)$ is given in Table 9.2.

Writing out the conditions for which $y = 1$, and joining these conditions together by logical addition, we obtain expressions for the function $F$ in the so-called normal disjunctive form

$$y = (x_1 \;\&\; \overline{x}_2 \;\&\; \overline{x}_3) \vee (x_1 \;\&\; x_2 \;\&\; \overline{x}_3) \vee (\overline{x}_1 \;\&\; x_2 \;\&\; x_3) \vee (x_1 \;\&\; x_2 \;\&\; x_3). \tag{9.2}$$

The expression (9.2) means that $y$ should equal 1, if $x_1 = 1$ and $x_2 = 0$ and $x_3 = 0$ or $x_1 = 1$ and $x_2 = 1$ and $x_3 = 0$ or . . ., as can be read off from Table 9.2. From the expression (9.2) it is possible to construct a circuit (Fig. 9.4) in which each bracket corresponds to a loop of successively joined contacts (normally open for symbols without negation and normally closed for symbols with negation), and for every expression there is a parallel connection of all the four loops.

Table 9.2

| $x_1$ | $x_2$ | $x_3$ | $y$ |
|---|---|---|---|
| 0 | 0 | 0 | 0 |
| 1 | 0 | 0 | 1 |
| 0 | 1 | 0 | 0 |
| 0 | 0 | 1 | 0 |
| 1 | 1 | 0 | 1 |
| 1 | 0 | 1 | 0 |
| 0 | 1 | 1 | 1 |
| 1 | 1 | 1 | 1 |

It can easily be seen that such a circuit will realise the transformation given in Table 9.2. However, in addition to this circuit there are an infinite number of other circuits which also realise a given transformation function. It is obvious that the circuit obtained directly, according to the normal disjunctive form, may

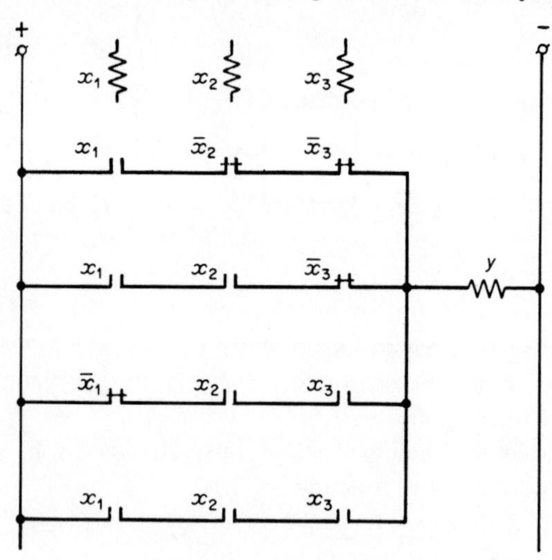


Fig. 9.4. *Circuit realising the logical function (9.2).*

not prove to be the best as regards the number of elements or the reliability of faultless operation. In the given case the circuit can be simplified without changing the function it realises, as can be seen from Fig. 9.5. As a result, the number of contacts used will decrease from 12 to 7.

A logical automaton can also be built from contactless elements, in particular in the form of a diode circuit.

Fig. 9.6(a) and (b) show diode circuits which realise the logical functions

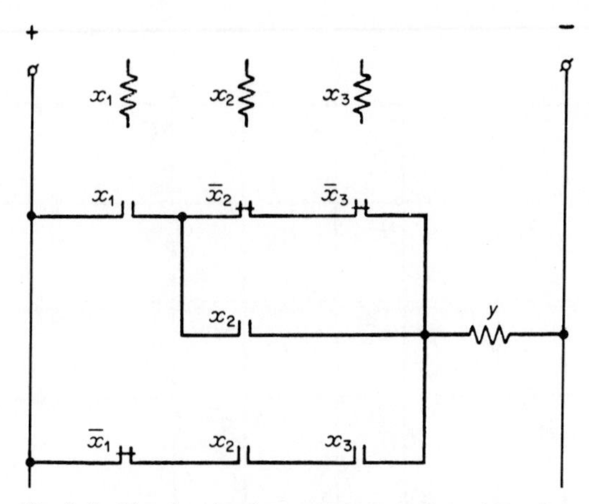


Fig. 9.5. *Simpler circuit, equivalent to that of Fig. 9.4.*

AND and OR. The operation of these circuits is based on the properties of a diode to vary its resistance, depending on the polarity of the voltage on its terminals. The diode resistance is low if the potential fed to the terminal, designated by a triangle, is higher than the potential fed to the terminal, designated by a stroke (line), and is high if the polarity is reversed. In the circuits 9.6, the zero signal is the relatively low positive potential $V_0$ and the unity signal is a relatively large positive potential $V_1$. It is obvious that in the circuit 9.6 a the potential of the output terminal $y$ will be near to $V_1$ (i.e. designated as unity) only when the potentials on both terminals $x_1$ and $x_2$ equal $V_1$. If this is not the case, $y$ will prove to be connected to the higher potential $V_1$ through a high resistance and to the lower potential $V_0$ through a low resistance, and will assume the (lower) potential of the latter. In Fig. 9.6 b the potential of the

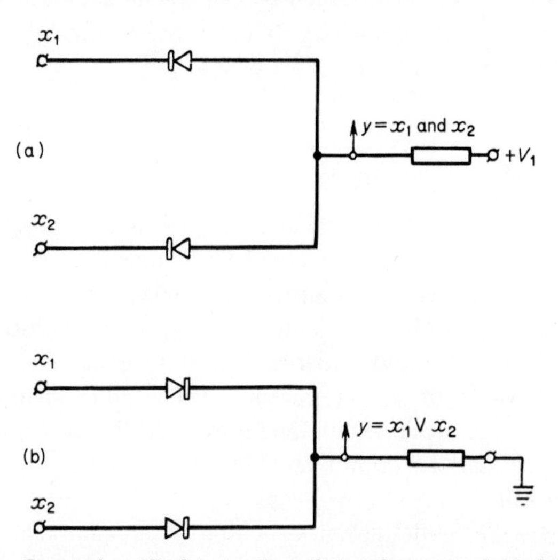


Fig. 9.6. *Contactless (diode) circuit, realising elementary logical functions.*

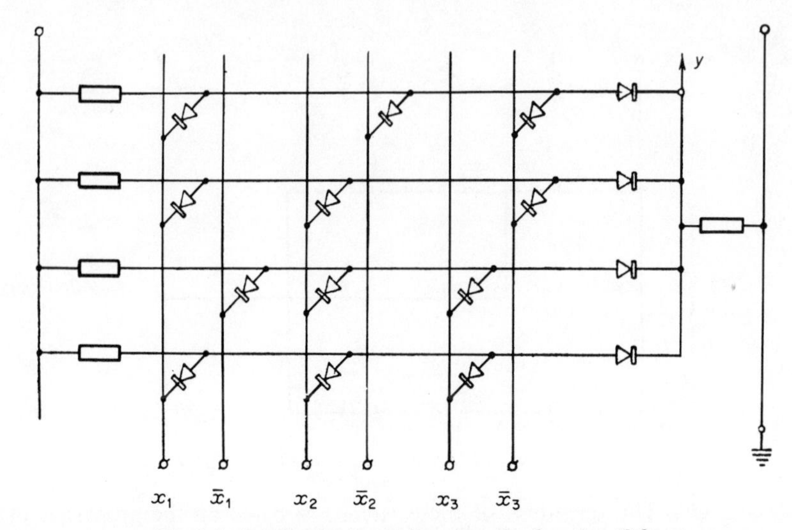


Fig. 9.7. *Diode circuit which realises the function (9.2).*

terminal $y$ will be near to that of $V_1$ if at least one of the terminals has the potential $V_1$ because it is particularly the terminal with the high potential which will be connected through a low resistance to the output.

Logical automata in the form of diode circuits are usually built in the form of matrices (i.e. in rows and columns). Fig. 9.7 shows a diode matrix which realises the function (9.2). The circuit of this matrix follows directly from the normal disjunctive form, since the number of rows here is determined by the number of expressions in the brackets. In each row the diodes are connected to the appropriate input (or its negation* depending on how this input figures in the expression. All the rows are connected by the OR circuit, which generates the output value $y$ in the form of a voltage between the terminal $y$ and the negative busbar.

## 9.2. Automaton with a finite memory

When studying automata with a finite memory, we are usually interested only in the steady state which is established a sufficiently long time after changes in the input variable have occurred. The transition of the system from one steady state into another is assumed to be sufficiently rapid compared with the time interval between changes in the input variable. For this reason it is convenient to consider the behaviour of an automaton with a finite memory during the discrete times $t_1$, $t_2$ . . ., separated by the intervals $\Delta t$. The additional assumption is made that in this case the output variables can only change at the instants $t_1$, $t_2$, . . ., which are called 'cycles'.

In accordance with the definition the output of the automaton with a finite

* Elements which realise the negate operation in the circuit are not shown.

memory during the $j-th$ cycle depends on the state of the automaton in the $(j-1)-th$ cycle and on the state of the inputs in the $j-th$ cycle. Therefore the transitions of such an automaton from one state to another can generally be described by the expression

$$\left.\begin{aligned} y^j &= F(z^{j-1}, x^j), \\ z^j &= G(z^{j-1}, x^j), \end{aligned}\right\} \tag{9.3}$$

where $y^j$ is the output of the automaton during the $j-th$ cycle, $z^{j-1}$ is the state of the automaton in the $(j-1)-th$ cycle

$$x^j = \{x_1^j, x_2^j \ldots\ldots, x_m^j\}$$

is the input to the automaton during the $j-th$ cycle $F$ and $G$ are a logical state and input function.

In order that an automaton should realise the transformation 9.3 it is necessary that in addition to elements realising logical functions it should also contain *delay elements*, whose outputs are determined by the value of its state during the preceding cycle, i.e. elements whose output $y$ is related to the input $x$ in accordance with the expression

$$y^j = f(z^{j-1})$$

or, in particular,

$$y^j = z^{j-1}.$$

The delay element should contain a memory for storage of the sequence of the previous state, otherwise its state could not depend on the previous state. One widely used discrete element with a memory is the flip-flop, a circuit with two stable states and the ability to change over from one of these states to the other under the effect of a control signal. The flip-flop can be made of various components, in particular of electronic components, as shown in Fig. 9.8. Here,

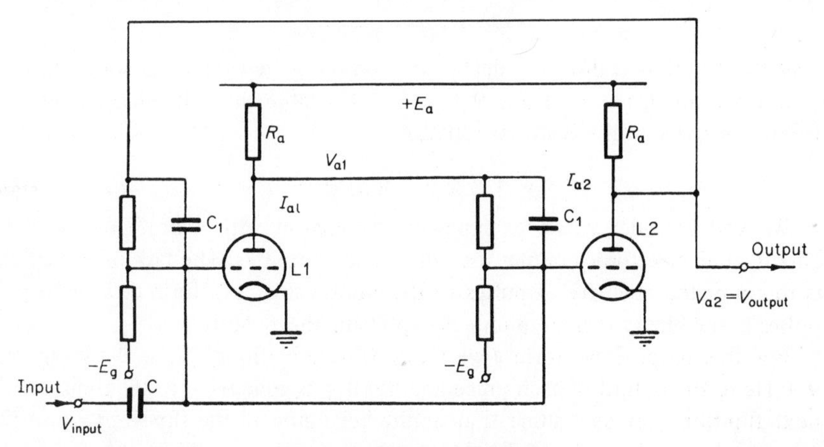


Fig. 9.8. *Trigger with triodes.*

due to positive feedback (connection between the anode current of one triode and the grid voltage of the other), the circuit can be in one of the following three conditions:

(1) the state $z_1$: the triode L1 open-circuited (grid potential positive); anode current $I_{a1}$ large; voltage $V_{a1}$ low; triode L2 closed (negative grid potential); anode current $I_{a2}$ small; output voltage $V_{out}$ large, $V_{out} = V_1$.

(2) the state $z_2$: triode L1 closed (potential on the circuit negative); anode current $I_{a1}$ small; voltage $V_{a1}$ high; triode L2 open (grid potential positive); anode current $I_{a2}$ large; output voltage low, $V_{out} = V_0$. This circuit can be changed from the state $z_1$ into the state $z_2$ by acting on its input. If a voltage pulse is capacitor fed to the input, it passes the capacitor c and becomes transformed into two pulses: a positive and a negative. The circuit parameters are so chosen that the positive pulse has no effect on the operation, but the negative pulse cuts off the open-circuited triode which in turn triggers the triode which was previously cut off, i.e. leads to the circuit changing over into the second stable state. As a result, each control impulse changes the state of the circuit and each even pulse returns it to its initial state.

We will call a low voltage on the flip-flop output zero, and a high voltage unity. It is obvious that the output of the flip-flop at a specific instant of time (the cycle) $y^j$ is unequivocally determined by its state $z^j$. Therefore the dependence of the output of the flip-flop on the input at this moment and on the state during the preceding moment can be expressed as the dependence on the input at this moment and on the output at the *preceding* moment. In this

Table 9.3

| $y^{j-1}$ | $x^j$ | $t^j$ |
|---|---|---|
| 0 | 0 | 0 |
| 0 | 1 | 1 |
| 1 | 0 | 1 |
| 1 | 1 | 1 |

way the flip-flop realises the delay function, i.e. it remembers a state. Bistable transitions are listed in Table 9.3. From this table the following expressions follow for the logical transitions function:

$$y^j = x^j \,\&\, \bar{y}^{j-1} \vee \bar{x}^j \,\&\, y^{j-1}. \tag{9.4}$$

We will consider as an example an automaton with a finite memory the circuit of an electronic counter used in digital computers. The task of this circuit is to count the quantity of pulses to the input, i.e. to transform the quantity of pulses in the binary code of numbers expressing this quantity.

For this purpose we form a circuit consisting of flip-flops, as shown in Fig. 9.9. Here, the output of each preceding flip-flop is connected to the input of the next flip-flop. Let us assume that at the beginning all the flip-flops are in the zero state, i.e. the voltage on their outputs equals $V_0$. When the first pulse

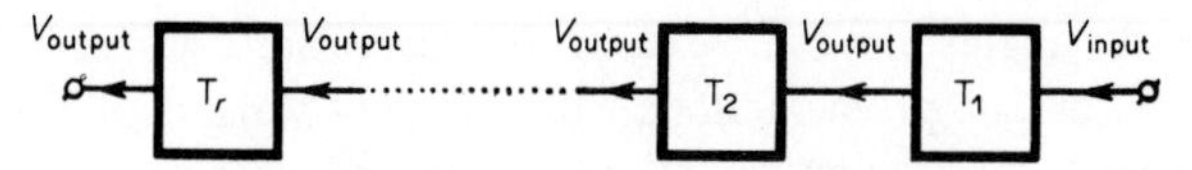


Fig. 9.9. *Counting of pulses on the flip-flop.*

arrives at the input of flip-flop $T_1$ there will be the voltage $V_1$ on its output, and at the input of flip-flop $T_2$ there will be a positive voltage pulse to which it does not react. The second pulse causes $T_1$ to revert to the zero state, as a result of which the voltage on its output will change from a value of $V_1$ to $V_0$, producing a negative pulse on the input of $T_2$ and its transition into the unity state. In this way $T_1$ will change its state after each input pulse, $T_2$ after each second pulse, $T_3$ after each fourth pulse, etc., $T_k$ after each $2^{k\text{-}th}$ pulse at the input of the circuit. If we consider the state of each flip-flop as the value of the appropriate (corresponding) discharge of the binary number, then the state of the entire chain of $r$ flip-flops will represent the number (in the binary system of counting) of pulses which have been fed to the input of the circuit.

The capacity of this circuit, the maximum number, $R$, of pulses which can be counted, is determined by the number $r$ and equals the maximum binary number consisting of $r$ digits, namely

$$R = 2^r - 1.$$

Automata with finite memories are widely used in both automation and computing. The theory of these automata also helps in the understanding of some relationships governing information processing in biological systems.

## 9.3. The Turing machine

Purposeful information processing operations in control and communication systems, solution of problems of various types by computer or manually, can be considered as an ordered sequence of operations. *The instructions which determine the content and sequence of operations for transforming the initial data into the sought result are called an algorithm.*

As examples of simpler algorithms we can consider the sequences of operations which lead to arithmetic problems, the solution of algebraic equations, calculation of the area of shapes. The method of designing a logical automaton on the basis of a given logical function, described in Section 9.1, which the automaton is required to realise, can also be considered as the *algorithm* for the design of the circuit for a logical automaton. It was mentioned in Chapter 6 that information processing in the control device which generates control signals is also realised by means of a specific algorithm – the control algorithm.

Any algorithm should satisfy the following requirements: '*definiteness*', '*generality*' and '*resultativeness*'. By 'definiteness' we understand here its accuracy and uniqueness without leaving room for arbitrariness. 'Generality' means that the algorithm can be used for an entire class of problems and therefore not for one problem alone with different variants of the initial data. The requirement of 'resultativeness' is satisfied by an algorithm which, after performing a *finite number* of operations, leads to the sought results; we sometimes call this 'effectiveness'.

If the set of all possible initial data processed by any type of algorithm A is represented as the sequence

$$\alpha_1, \alpha_2, \ldots, \alpha_i,$$

and all the possible results in the form of the sequence

$$\beta_1, \beta_2, \ldots, \beta_j,$$

then any algorithm $A_k$ which transforms the initial data $\alpha_i$ into the result $\beta_j$ can be reduced to calculating the function $\phi_k$, which indicates the number of the result $j$ according to the number of the set of initial data $i$

$$j = \phi_k{}^{(i)} \tag{9.5}$$

The indices $i$ and $j$ are integers and can always be written in the binary system using a finite set of zeros and ones. In this case the function $\phi_k$ can be considered as a logical function which transforms the situation $i$ at the input of the automaton into the situation $j$ at its output.

Consequently, in principle, any algorithm can be realised by means of an appropriate discrete automaton.

The English mathematician A. M. Turing proposed an abstract circuit of an automaton which in principle is suitable for realising any algorithm. This automaton, which is called the '*Turing machine*', is an *automaton with an infinite memory*.

The memory in the Turing machine is a tape which is marked off in the cells No. 1, No. 2, . . . so that it has a beginning (No. 1) but no end (it extends to infinity as a sequence of natural numbers).

In each of these cells it is possible to record either zero or one. Above the

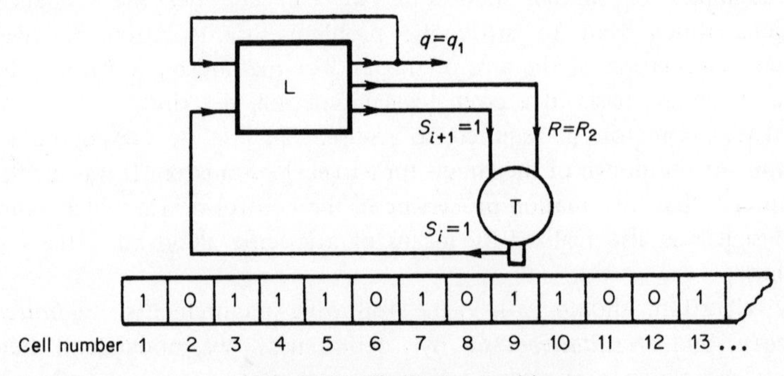


Fig. 9.10. *The Turing machine.*

tape moves a head T, which is controlled by the automaton L, possessing a finite memory (Fig. 9.10). The automaton L operates in cycles. Information on the symbol (0 or 1) read off by the head from the tape is fed to its input. Acting on commands obtained during each cycle from the automaton L, the head may either remain still or move by one cell to the left or to the right. Simultaneously the head receives commands from the automaton L and as a result of these it can substitute the symbol recorded in the cell above which it is located.

The operation of the Turing machine is determined uniquely by the initial filling of the cells of the tape and the transformation operator of the control automaton, which can be given in the form of a table of transitions. We denote by $S_i$ ($S_0 = 0, S_1 = 1$) the symbol read off the head; by $R_j$($R_0$ (stop), $R_1$ (left), $R_2$ (right)) the command for displacement of the head; by $g_k$ ($k = 1, 2, \ldots, n$) the state of the controlling automaton. Then its table of transitions can be represented as follows:

Table 9.4

| | *State* | |
|---|---|---|
| *Input* | $S_0 = 0$ | $S_1 = 1$ |
| $q_1$ | $S_0, R_2, q_k$ | $S_1, R_1, q_m$ |
| $q_2$ | $S_1, R_0, q_s$ | $S_0, R_1, q_l$ |
| $q_3$ | $S_1, R_1, q_p$ | $S_0, R_2, q_2$ |

As can be seen from Table 9.4, the action of the automaton L depends on the input $S$ and on its state $q$. To definite values $S_i$ and $q_i$ there will correspond a certain set of values of the three quantities: $S$, $R$ and $q$, which designate respectively what symbol $S$ the head will record on the tape, what will be the command $R$ for the displacement of the head, and into what new state $q$ the automaton L will change. We must bear in mind that amongst the states $q$ of the automaton L there should be at least one state $q^*$ for which the head does not change the symbol $S$, the command $R = R_0$ (stop) and the automaton L remains in the standstill position $q^*$. Arriving at the state $q^*$, the automaton terminates the execution of the algorithm, and the further operation of the Turing machine is stopped.

Let us assume, for instance, that the table of transitions of the automaton L is of the following sequence:

Table 9.5

| | $S$ | |
|---|---|---|
| $q$ | 0 | 1 |
| $q^*$ | $0, R_0, q^*$ | $1, R_0, q^*$ |
| $q^1$ | $1, R_0, q^*$ | $1, R_2, q_1$ |

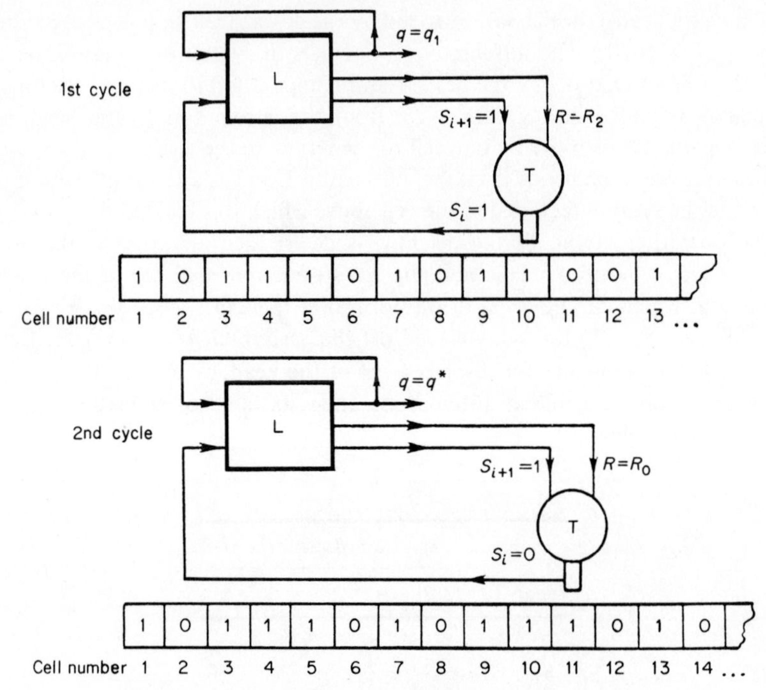


Fig. 9.11. *Example of the operation of a Turing machine.*

If at the initial instant of time the automaton is in the state $q_1$ and the head is above the cell in which the symbol 1 is recorded, then the head will move to the right until a cell with the symbol zero is detected, will substitute it by the symbol 1, and then stop. If the initial state of the system (state during the zero cycle) and the filling up of the tape correspond to what is shown in Fig. 9.10, then according to Table 9.5 the system will change over during the subsequent two cycles into the state shown in Fig. 9.11 and will stop during the second cycle.

The given example shows only one of the simplest problems solved by the Turing machine. If the table of transitions is appropriately expanded, the Turing machine can be used for solving problems however complicated, and in any case for a problem that can be solved by any other machine, hence the title of 'Universal Turing Machine'. We must, however, bear in mind that in practice the use of automata built according to the Turing principle is impractical because the number of steps required for solving very complicated problems is extremely large, which means that a very long solution time is required.

In spite of this, the theory of Turing machines is of great importance for solving such important problems as the existence of an algorithm for the solution of one or another class of problem, which also means for describing what functions can and what functions cannot be performed automatically, in particular by such a universal automaton as the digital computer.

## 9.4. The probabilistic automaton

In the previous paragraphs consideration was given only to automata whose state was a single-valued function of the input variable and the state at the instant preceding it. The behaviour of such automata can be accurately predicted if the transfer operator is known and given in the form of a table or a logical function, the initial state and the input sequence also being known. For this reason such automata are called *deterministic*. However, systems often have to be dealt with whose future state is not uniquely determined by the initial state and the input values. From one and the same state with the same input the system may change into different states, and consequently the output values may also prove different and only the probability of transition into one state or another is known. Such a system is called a *probabilistic automaton*. Obviously a deterministic automaton is a particular case of a probabilistic one: for the latter the probability of one definite transition equals unity and for all other transitions it equals zero, whatever its initial situation.

It is pointed out that the difference between a deterministic and a probabilistic automaton is not absolute. On the one hand we can never distinguish how accurate the various physical states of a system are, because due to random fluctuations and noise which cannot be taken into consideration, a state which we may consider as being definite will only be the homomorphic image (see Chapter 3) of an entire set of various states. Because of this and also due to greater perturbations (the system being out of order) its behaviour during its operation can generally speaking not be predicted with absolute certainty, although it may be adequate for the given purpose.

On the other hand, a system may have to be considered as a probabilistic automaton, although if its states could be distinguished in more detail, it would have to be considered as a deterministic one.

From the point of view of a telephone subscriber the network is a probabilistic automaton. On dialling, it will respond with the probabilities $p$ and $1 - p$, depending on the average loading of the network, either the signal $y_1$ 'engaged' or $y_2$ 'ringing'. However, if we knew the entire set of calls and connections in the network we would be able to predict exactly what the answer to dialling will be – the system would be a deterministic automaton.

Let us assume that a probabilistic automaton may be in any of the states $z_1$, $z_2, \ldots, z_n$, the input value $x$ may assume any value $x_1, x_2, \ldots, x_m$ and the output value $y$ the values $y_1, y_2, \ldots, y_l$. Then such an automaton can describe for given conditions of the probability of the transition into the state $z^i = z_k$, with the condition that the preceding state was $z^{i-1} = z_k$ and the input signal $x^i = x_r$:

$$p(z_k^i / z_j^{i-1}, x_r^i) = p_{jk}(r), \tag{9.6}$$

Furthermore, the output value would have to be given as a function of the state of the automaton:

$$y^i = f(z^i). \tag{9.7}$$

Here by the $i^{\text{th}}$ cycle, we mean the number is the instant of time $t_i$. Equations (9.6) and (9.7) fully describe the probabilistic automaton. Equation (9.6) can be graphically represented in the form of a set of matrices of the transition probabilities, each of which corresponds to a fixed set of input values $x_r$. One of these matrices is given in Table 9.6. Altogether there are as many of them as there are different $x$ values, i.e. $m$ matrices.

Table 9.6

| | $z^i$ | | | | |
|---|---|---|---|---|---|
| $z^{i-1}$ | $z_1$ | $z_2$ | ... | $z_n$ | $\sum_k P_{jk}$ |
| $z_1$ | $p_{11}$ | $p_{12}$ | ... | $p_{1n}$ | 1 |
| $z_2$ | $p_{21}$ | $p_{22}$ | ... | $p_{2n}$ | 1 |
| . | . | . | | | |
| . | . | . | | | |
| . | . | . | | | |
| $z_n$ | $p_{n1}$ | $p_{n2}$ | | $p_{nn}$ | 1 |

It is pointed out that the sum of the probabilities of all possible transitions of the automaton will of necessity équal unity. Therefore the sum of the probability values in each row always equals unity, as is shown in the additional right-hand row of the matrix.

The concept of the probabilistic automaton is also important for solving numerous problems in cybernetics and, in particular, those concerned with adaptive systems, which will be considered in Chapter 11. Probabilistic automata are of particular importance in studying very complicated systems encountered in biology, psychology and sociology. Such systems usually represent a 'black box' (see Chapter 3), and their state can be judged only from the output signals. For instance we can consider as a black box (probabilistic automaton) a chess player who, in answer to the input signal (his opponent's first move) (1 e 4 or 1 d 4, etc.) – will select one or another move (1 . . . e 5; 1 . . . c 6, etc.) with probabilities that can be evaluated on the basis of his practical training. The laws of heredity, the Brownian movements of microscopic particles, economic processes, military operations – all this multitude of phenomena can be described by the concept of the probabilistic automaton.

## Exercises

1. Compile a table for the logical function given by means of the following expression:

$$f(x_1, x_2, x_3) = \bar{x}_1 x_2 x_3 \vee x_1 \bar{x}_2 x_3 \vee x_1 x_2 \bar{x}_3.$$

*Solution.* See Table 9.7.

Table 9.7

| $x_1$ | $x_2$ | $x_3$ | $f(x_1, x_2, x_3)$ | $x_1$ | $x_2$ | $x_3$ | $f(x_1, x_2, x_3)$ |
|---|---|---|---|---|---|---|---|
| 0 | 0 | 0 | 0 | 0 | 1 | 1 | 1 |
| 0 | 0 | 1 | 0 | 1 | 0 | 1 | 1 |
| 0 | 1 | 0 | 0 | 1 | 1 | 0 | 1 |
| 1 | 0 | 0 | 0 | 1 | 1 | 1 | 0 |

2. Let $f(x_1, x_2, x_3)$ be as given in Table 9.8.

Table 9.8

| $x_1$ | $x_2$ | $x_3$ | $f(x_1, x_2,, x_3)$ | $x_1$ | $x_2$ | $x_3$ | $f(x_1, x_2, x_3)$ |
|---|---|---|---|---|---|---|---|
| 0 | 0 | 0 | 0 | 0 | 1 | 1 | 1 |
| 0 | 0 | 1 | 1 | 1 | 1 | 0 | 1 |
| 0 | 1 | 0 | 1 | 1 | 0 | 1 | 1 |
| 1 | 0 | 0 | 1 | 1 | 1 | 1 | 0 |

Write for this function the normal disjunctive form. Try to simplify it. Draw a diode circuit which would realise this function.

*Solution.*

$$f(x_1x_2x_3) = \bar{x}_1\bar{x}_2\bar{x}_3 \vee \bar{x}_1x_2\bar{x}_3 \vee x_1\bar{x}_2\bar{x}_3 \vee$$

$$\vee \bar{x}_1x_2x_3 \vee x_1x_2\bar{x}_3 \vee x_1\bar{x}_2x_3.$$

After simplification the function assumes the form

$$f(x_1, x_2, x_3) = \bar{x}_1x_3 \vee x_2\bar{x}_3 \vee x_1x_2.$$

The diode circuit for realising this function is given in Fig. 9.12.

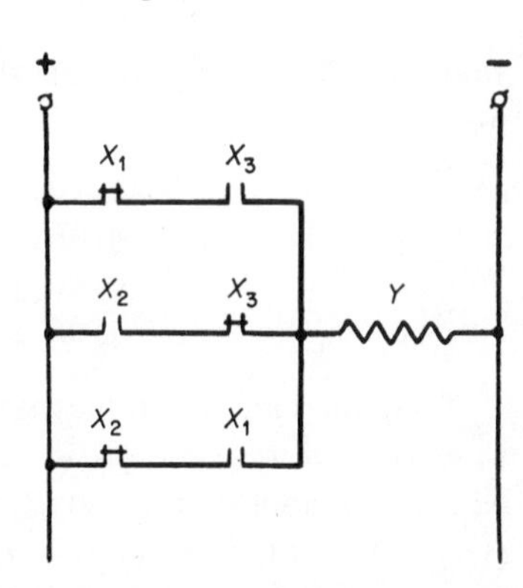


Fig. 9.12. *To Example No. 2.*

3. The function $x_1 \rightarrow x_2$ (read: 'from $x_1$ follows $x_2$') is given in Table 9.9. Write the disjunctive normal form for this function and simplify it.

Table 9.9

| $x_1$ | $x_2$ | $f(x_1, x_2)$ | $x_1$ | $x_2$ | $f(x_1, x_2)$ |
|---|---|---|---|---|---|
| 0 | 0 | 1 | 1 | 0 | 0 |
| 0 | 1 | 1 | 1 | 1 | 1 |

*Solution.* After simplification the function assumes the form

$$f(x_1, x_2) = \bar{x}_1 \vee x_2.$$

**4.** Write the logical expression in the variables $p$, $q$ and $r$ for the following statement: 'If it is going to rain in the evening we will not go to the cinema and we will study'. (Use the logical function from Example No. 3).

*Solution.* We will denote by $p$ the event 'it rains in the evening', by $q$ 'we go to the cinema', by $r$ 'we will study'. Then $p - \bar{q}br$.

**5.** An explorer fell into the hands of cannibals. They decide to make him make a statement and to impose the condition that if his statement is true they will boil him, and if it is false they will roast him. What statement must he make so that he does not perish?

*Solution.* The explorer must say: 'You will roast me.'

**6.** Compile a table of the transitions of a Turing machine consisting of any two natural numbers $n_1$ and $n_2$, which are represented on a tape with two series of $n_1 + 1$ and $n_2 + 1$ units, separated by a cell on which is written the symbol 0. (The result of the calculation is considered as the number of units in the expression written on the belt after the calculation is terminated). In the initial state the reading head is over the symbol 1.

*Solution.* The Turing machine should operate in accordance with the following rules:

1. $q_1 1 0 q_1$,
2. $q_1 0 R_2 q_2$,
3. $q_2 1 0 q_2$,
4. $q_2 0 R_2 q_3$,
5. $q_3 1 0 q_3$,
6. $q_3 0 R_0 q^*$.

These rules mean that the machine in the state $q_1$ substitutes 1 by the symbol 0, remains in this state $q_1$, in the state $q_1$ it shifts to the right if it scans a cell with the symbol 0, in the state $q_2$ it shifts to the right if it perceives (scans) the symbol 1, passes into the state $q_3$ and moves to the right by one cell if it scans the symbol 0, in the state $q_3$ it substitutes the symbol 1 by the symbol 0, then it stops. The machine erases two units, after which $n_1 + n_2$ units remain on the tape.

7. In his classical works for elucidating the laws of heredity, Gregor Mendel (1865) used white and purple peas. He established that the colour gene has two modifications – purple A and white a, with the modification A being the dominant one; the gene type Aa gives a purple colour, as does the gene type AA. We will consider the sequence of gene types of offspring of one plant as the states of a probabilistic automaton, the colour of the flowers as the output signal, and the genotype of the second parent as the input signal. Draw the matrices of the transition probabilities of the automaton and the dependences of the output signal on the state. All the possible combinations of allelae during cross-breeding should be considered as equally likely.

*Solution.* The matrices of the transition probabilities can be written as follows (Table 9.10):

Table 9.10

$x^j$ = AA

| | $z^j$ | | |
|---|---|---|---|
| $z^{j-1}$ | *AA* | *Aa* | *aa* |
| AA | 1 | 0 | 0 |
| Aa | ½ | ½ | 0 |
| aa | 0 | 1 | 0 |

$x^j$ = Aa

| | $z^j$ | | |
|---|---|---|---|
| $z^{j-1}$ | *AA* | *Aa* | *aa* |
| AA | ½ | ½ | 0 |
| Aa | ¼ | ½ | ¼ |
| aa | 0 | ½ | ½ |

$x^j$ = aa

| | $z^j$ | | |
|---|---|---|---|
| $z^{j-1}$ | *AA* | *Aa* | *aa* |
| AA | 0 | 1 | 0 |
| Aa | 0 | ½ | ½ |
| aa | 0 | 1 | 0 |

$y_1$ denotes purple, $y_2$ denotes white. The dependence of the output signal on the state is shown in Table 9.11.

Table 9.11

| $z^j$ | *AA* | *Aa* | *aa* |
|---|---|---|---|
| $y^j$ | $y_1$ | $y_1$ | $y_2$ |

# 10 The Computer

Using a suitable code, any data can be converted into a number or a set of numbers, and any data transformation can be realised by appropriate operations of these numbers. Thus any text can be transmitted telegraphically by means of a series of electric pulses which are equivalent to some number represented in the binary system as was shown in the previous chapters. Any picture can be transmitted by television, using radio signals, which can also be represented in the form of a sequence of numbers. Transformation of data contained in the initial data of any problem, for example a chess problem, into information on the results of its solution can also be reduced to coding the data into numbers and operations on these numbers.

Any number represents a set of discrete digits such as zeros and ones, and therefore development of the necessary equipment for operations with these numbers is of fundamental importance for mechanising data processing, which means also for solving a wide range of cybernetics problems.

Digital computers produced initially for carrying out purely calculating operations, also proved to be suitable for applications in various branches of science and engineering and their use has extended far beyond the mechanisation of computing operations. In addition to carrying out cumbersome calculations, digital computers function as control devices for production plants, and are used for modelling complicated engineering, economic and biological systems, and they also assist in the planning of complex operations to select a strategy of behaviour in competitive situations, for example for playing chess or draughts, for developing plans of military operations, for diagnosing disease, or the design of engineering equipment including new digital computers.

A high level of development has been reached in programme-controlled digital computers. However, for solving individual classes of problem specialised digital computers are being developed which are intended for controlling production processes (control machinery), for modelling dynamic systems (digital models), etc.

## 10.1. The digital computer

The modern universal digital computer with programmed control consists of five basic functional devices: input, control, arithmetic, memory and output units which are interconnected by communication channels, as shown in Fig. 10.1.

The input unit transforms the input data and the method of its solution into electrical signals, and presents this data to the machine. The control unit ensures coordinated operation of all the units in the machine, determines the data circulation path in the machine, realises the stopping of the machine after it has processed the data. The arithmetic unit carries out arithmetical operations on the numbers introduced into the computer. The memory is used to receive, store and retrieve numbers which contain the initial data, command sequences which determine the order of carrying out computations, and the intermediate results. The output unit is intended for presenting the results in a form suitable for further use and this usually involves punched cards, punched tape or a printout on paper.

The solution of any problem on a computer reduces to performing a specific set of operations on numbers. As only a fixed set of arithmetic operations are performed in the computer, for instance addition, subtraction, multiplication

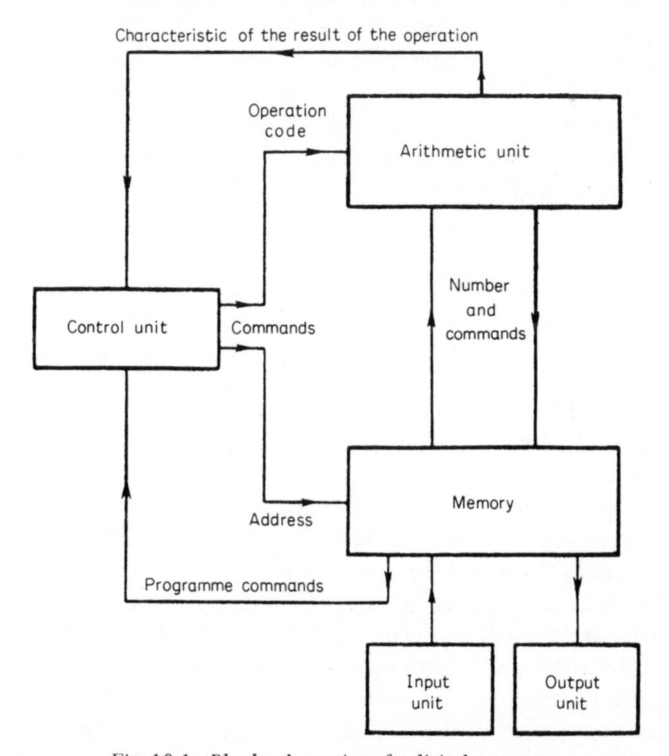


Fig. 10.1. *Block schematics of a digital computer.*

and division, the algorithm for solving problems must be in the form of a sequence of operations from this set of operations as well as some logical operations which the machine can perform.

Each operation is performed by the machine on a specific control signal – a command. The sequence of commands forms the *programme* for solving the problem by the computer. The programme instructions are coded by means of appropriate numbers, and introduced into the memory of the computer before the solution of the problem is started. The memory consists of individual cells with fixed numbers (addresses), in each of which one number can be stored. A number characterising an instruction consists of several parts. One part, the operation code, indicates the type of operation to be performed by the computer, the other part, the address, determines the numbers of the cells where the numbers on which the indicated operation is to be performed are stored, and the number of the cell into which the result should be fed. The control unit ensures the sequence of execution of the programme instructions and thus also realises the given algorithm for solving the problem.

During the calculations it is often necessary to change the order of the computation, depending on the intermediate results obtained. For instance the procedure of calculating the roots of a quadratic equation depends on the sign of the expression under the square root sign. In such cases the computer should automatically select the length of the computations. This is achieved by introducing into the programme particular *conditional* instructions so that on the basis of a certain characteristic the machine will select one of several commands which determine the further progress of the computation. An example of such a branching programme is the programme for calculating the roots of the quadratic equation

$$x^2 + px + q = 0$$

in accordance with the formula

$$x_{1.2} = -\frac{p}{2} \pm \sqrt{\frac{p^2}{4} - q},$$

the scheme for which is given in Fig. 10.2.

The sequence of the instructions forming the programme is presented to the machine in the form of numbers, and therefore it is possible to perform operations also on the numbers which represent the instructions. This permits changes in the programme during the computations and multiple execution of individual sections (cycles) of the programme.

The computer is a discrete automaton operating in cycles. Most frequently the numbers are presented in the machine in a binary form, by means of electric signals. Arithmetical and logical operations in the machine are performed by means of logical elements, as described in Chapter 9.

The basic operations in the computer are realised by means of adders and shifters. Addition of two one-digit binary numbers A and B can be effected with

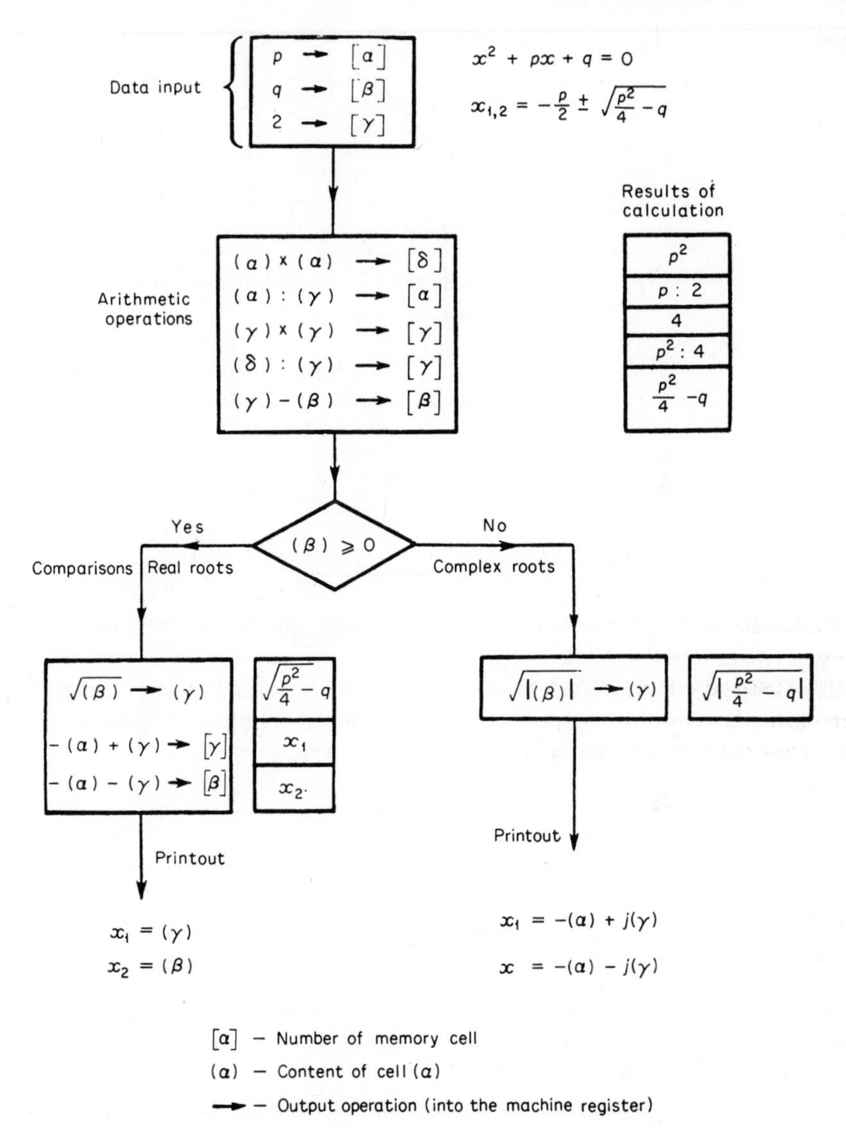


Fig. 10.2. *Scheme of the programme for calculating the roots of the quadratic equation.*

the use of a *one-digit adder* circuit, Fig. 10.3, assembled from simple logical elements. Such a circuit is realised in Table 10.1, where C denotes the result of adding A and B, which is placed in the correct digit, and D gives the transfer to the next digit.

A somewhat more complex circuit than that shown in Fig. 10.3 will enable adding three binary single-digit numbers. Such single-digit adders SA with three inputs are necessary for adding multi-digit numbers, when in each digit three components are to be added: $i^{\text{th}}$ digit $a_i$ of the number A, $i^{\text{th}}$ digit $b_i$ of the

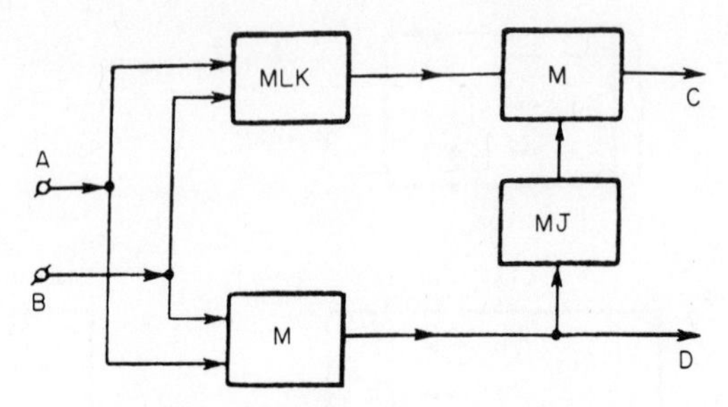


Fig. 10.3. *Diagram of a single-digit adder.*

Table 10.1

| *A* | *B* | *C* | *D* |
|---|---|---|---|
| 0 | 0 | 0 | 0 |
| 1 | 0 | 1 | 0 |
| 0 | 1 | 1 | 0 |
| 1 | 1 | 0 | 1 |

number B and transfer $d_i$ from the $(i-1)^{\text{th}}$ digit. The diagram of a multi-digit adder is shown in Fig. 10.4.

Multiplication is realised in the arithmetic unit by adding and shiftng the numbers, as is done in manual computations. For this purpose a multiplication table is used, which is very simple for the binary system:

$$0 \times 0 = 0, \quad 1 \times 0 = 0$$

$$0 \times 1 = 0, \quad 1 \times 1 = 1$$

It can easily be seen that this table corresponds to the operation of logical multiplication, realised by means of the AND element.

The shift operation which is required for multiplying multi-digit numbers is realised in the machine by means of the *shift register.* One of the possible methods of designing a shift register is based on the use of two electronic counters, as described in Section 9.2.

On a command from the control circuit the number in the counter A is

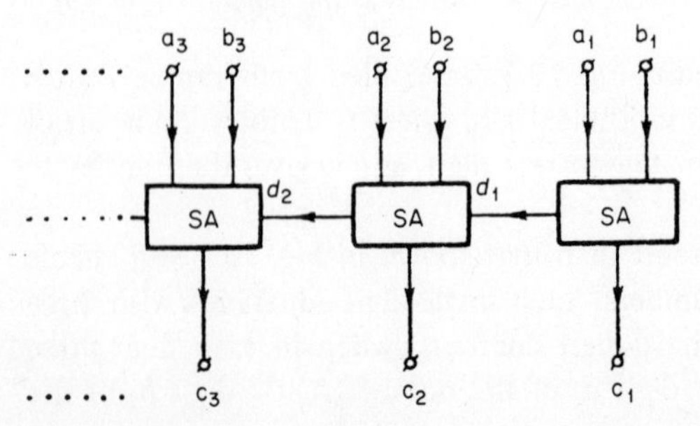


Fig. 10.4. *Diagram of a multi-digit adder.*

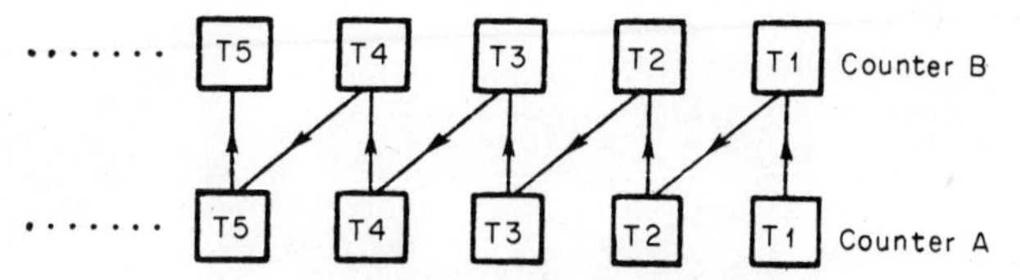


Fig. 10.5. *Diagram of a shift register.*

transmitted to the counter B and then returned to the counter A, but with a shift of one digit so that the unit in the $i^{th}$ digit returns to the $(i + 1)^{th}$ digit, as shown in Fig. 10.5. As a result of such an operation the initial number will double after each command. If, for instance, the initial number was

0001100010110,

then after the first shift we obtain

0011000101100,

and after the second shift

0110001011000,

and so on.

Subtraction and division are performed in much the same way as adding and multiplication, and we will not deal with these in greater detail.

Arithmetic and logical operations in modern computers are performed at a very high speed. Addition time can be as little as 0.2 μs, and multiplication time 1.2 μs. Due to progress in computer engineering and the application of new, compact and low-inertia components, it is possible to ensure not only fast operation of the machines but also to increase their reliability, reduce their dimensions and increase their memory capacity, automate the detection and correction of errors. Entirely new methods for the improvement of computers are made available by lasers and molecular electronics, which achieve extremely compact computer elements of several million elements per cubic centimetre, therefore are comparable to the neuron density in the human brain.

Fig. 10.6 shows a photograph of a modern digital computer. Such computers are widely used for carrying out automatically complicated and cumbersome computations in every branch of science, engineering and economics. Computers are being used for determining the trajectories of space ships, for weather forecasting, for working out the most favourable profile of the wing of an aircraft and optimal relationships of the dimensions of electrical machinery, planning the construction of railways, planning the production of fertilisers, statistical processing of data on the increase in population and on the turnover in the retail trade. Computers also intervene decisively in various branches of creative activity which were previously considered to be exclusive to human thought: computer translation from one language to another, the scores of

Fig. 10.6. *A system 4-70 digital computer supplied by International Computers Ltd at the heart of the British Giro system. (Photograph by courtesy of International Computers Ltd.)*

orchestral music, deciphering ancient manuscripts, proving mathematical theorems, and many others.

There is no area of modern life which has remained unaffected by the computer and its most powerful weapon, the electronic digital computer.

## 10.2. The control machine

During recent years specialised computers have been developed which are adapted for direct control of production processes, in addition to the great strides made in the development of general purpose computers. Dispensing with general purpose designs enabled considerable simplification and increase in the reliability of operation, as well as increasing the speed of solution of a limited number of problems which are specific to the controlling given processes, with a speed of response increased to correspond to the speed of the processes in the controlled plant.

The main distinction of process-control machines as compared to computers is that their connection with the outside world is made directly via measuring instruments at the input and final actuating elements at the output. This dispenses with the need for human participation in the work of the system: controlled plant + control computer.

In such control computers the input and output units must be matched for working with the measuring and actuating members of the controlled system. The actuation of the control computer on the process can be performed either directly on the control organs, CO, of the plant or through local control devices, CD, by replacing with local systems. Both these alternatives in the structure of

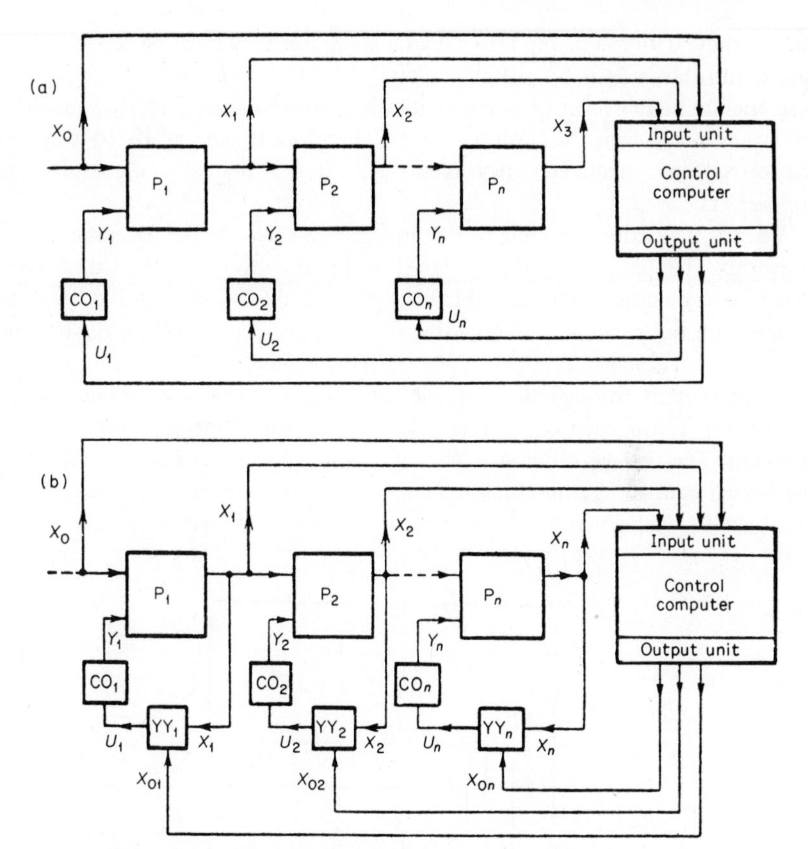


Fig. 10.7. *Actions of the control computer on the process: (a) Direct action on the control elements CO; (b) Through local control devices CD.*

the control system using control computers are illustrated in Fig. 10.7(a) and (b) on the example of a plant consisting of series-connected processing elements P.

The circuit Fig. 10.7(a) is simpler but it requires faster computers, and is less reliable because if there is a fault in the control computer the whole control system is put out of operation. The advantage of the circuit Fig. 10.7(b) is that if the computer is out of operation the normal working regime of the system will be maintained by the local control systems. Although in the absence of the control computer, the coordination of the operation of the individual elements of the system is not ensured, the process can continue for a certain time which allows the fault to be corrected and the control computer to be put back into operation. In addition, rapidly varying disturbances can be compensated in such systems by means of local control devices and therefore there is the possibility that it is not necessary to impose excessively rigid requirements on the speed of response of the control computer.

The signals obtained from the measuring instruments, which carry the data on the state of the controlled plant, are usually represented in analogue form.

Before these data can be fed into a digital computer they must be converted into digits, and this is done by *analogue-digital converters*. The output signals of the machine are in the form of numbers which cannot be used directly in analogue actuation mechanisms or control devices. Therefore it is necessary to re-convert the digital into analogue information and this is done by *digital-analogue converters.*

We will now consider some of the principles involved in the designing of such converters. Let us assume that it is necessary to convert electrical d.c. voltage into a binary number. One possible way of doing this is shown in Fig. 10.8, and is based on the principle of equalising the voltage $E_x$ to a voltage $E_y$ obtained from a resistance box $R_1 - R_4$. The current regulator CR maintains a strictly constant current through the reference resistances of the box, so selected that the voltage across each of them is twice the voltage of the adjacent resistor to the right. The voltage difference, $E_x - E_y$, amplified by the amplifier A, acts on the logical control circuit which selects a combination of excited relays $y_1 - y_4$ at which the imbalance of the metering circuit $E_x - E_y$ is a minimum. At the

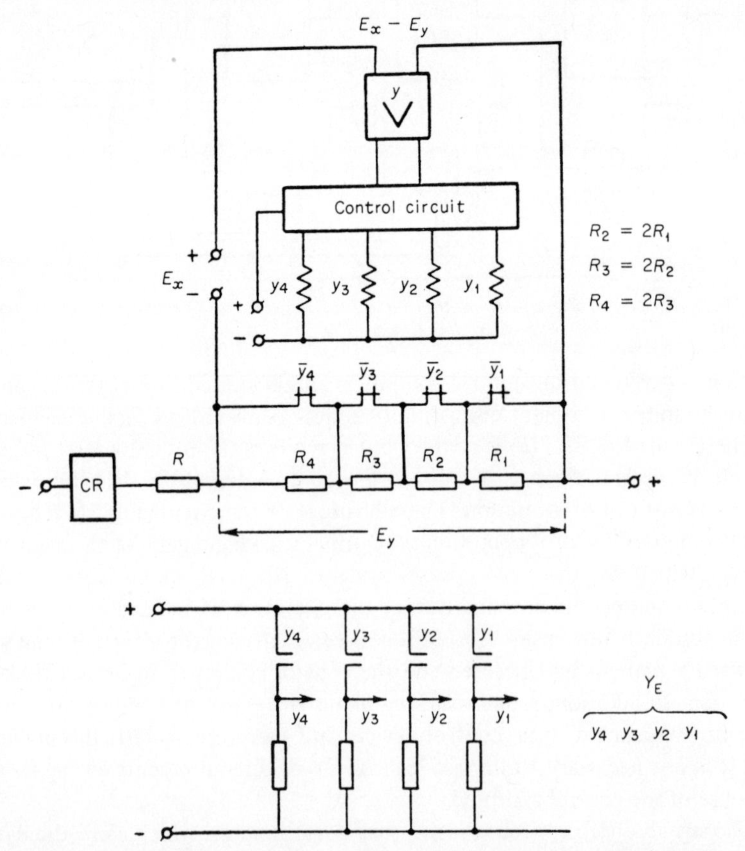


Fig. 10.8. *Example of conversion of electrical d.c. voltage into a binary number.*

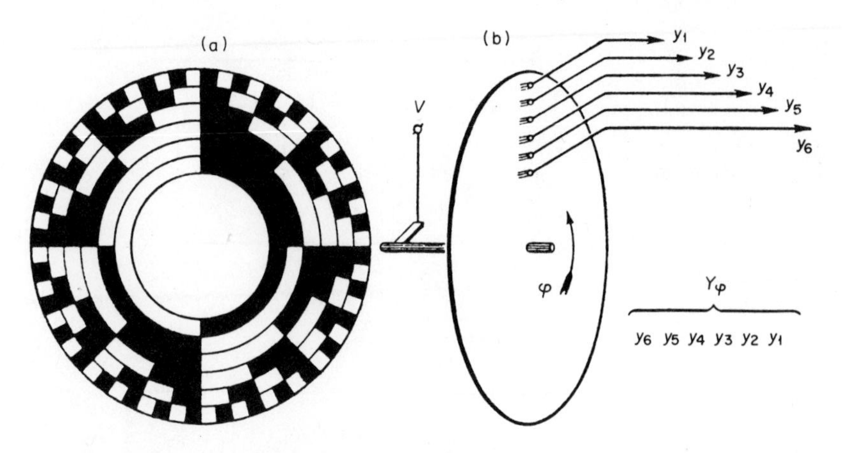


Fig. 10.9. *Scheme for converting the turning angle $\varphi$ into a binary code by means of an analogue converter: (a) coding disc; (b) electrical circuit.*

output we obtain a set of signals $Y_E = \{y_1 - y_4\}$ which represents in the binary code the voltage $E_x$. The number of digits which is equal to the number of relays, are chosen in accordance with the possible and necessary accuracy of conversion.

Frequently the analogue quantity which is to be introduced into the machine represents an angle of rotation $\varphi$ of a shaft (the shaft of the mobile system of an instrument, the output shaft of a servo system, etc.). In such a case conversion of the angle into a binary digital code $Y_\varphi$ can be realised by means of a coding disc, as shown in Fig. 10.9. The disc contains a certain number of rings, an even number of bits of the code $Y_\varphi$. The outer ring is divided into segments whose angular dimension determines the value of the most recent bit $y_1$. The angular dimension of the segments in all subsequent rings is doubled. The black segments represent the conducting area of the disc, whilst the bright ones represent the area covered with an insulating layer. If contact brushes are placed along the radius of the disc and a voltage $V$ is fed in, as shown in Fig. 10.9(b), then the voltage on the brushes will be the binary number $Y_\varphi$ corresponding to the rotation angle $\varphi$ of the disc.

For converting the binary digital code into an analogue value a digital-analogue converter is used. One of the possible circuits of such a converter, intended for conversion of the number $Y_E$ into the voltage $E_y$, is shown in Fig. 10.10. Here, in the same way as in the above described analogue-digital converter, a resistance box $R_1 - R_4$ is used, which is fed by a stabilised current source CS.

Many other input and output devices for converting information between the machines and the outside world have been developed and are in use. There is equipment for introducing into the machine moving and stationary images, spoken commands, data on the state of controlled systems. The output units of

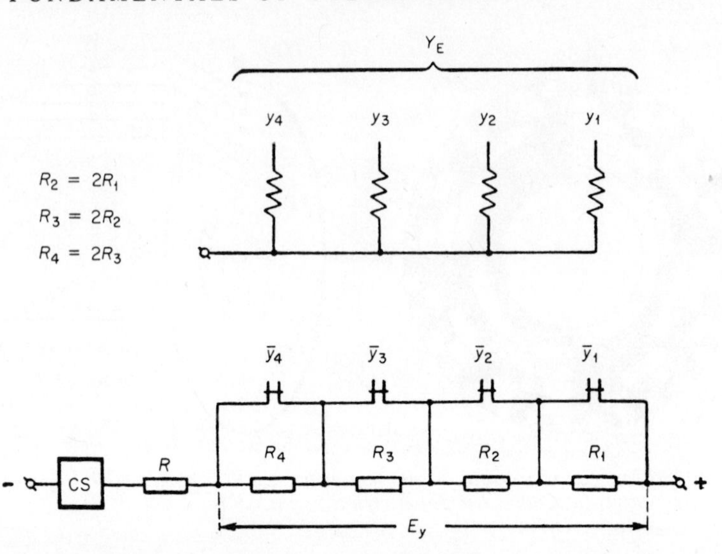


Fig. 10.10. *Example of conversion of a binary digital code into an analogue quantity, using a digital-analogue converter. CS: stabilised current source.*

the machines are adapted for presenting the output in the form of electrical signals, texts, drawings, and speech.

A digital computer which is adapted for solving problems of process control in a plant of a specific class, fitted with the above described input and output units, is capable of direct interchange of signals with the controlled systems. Indeed the control computer is one of the most remarkable achievements of modern engineering.

## 10.3. Applications of control computers

The use of digital control systems in industry began about 150 years ago. The French inventor, J. M. Jacquard, produced a weaving machine which was fitted with a simple programming device, in which holes in cardboard contained the code of the pattern of the fabric and ensured the required inter-weaving of the yarn; this was later widely used in the textile industry. However, the importance of digital control systems, and their general applicability, especially to a great variety of production processes, has only been realised during the last few years; this has opened up a whole new technology.

At present, digital control machines are being used for direct control of production processes in power generation, engineering, metallurgy, chemical and food industries and railroad transportation.

Two typical examples of the use of digital systems for direct control of production processes are described.

*Digital Programmed Control of Machine Tools.* The control computer permits realising a principally new method of manufacturing components in engineering: automation of the production process; avoidance of inaccuracies in the shape and dimensions of components, caused by errors of the operator controlling the operation of the machine tools; reduction to a minimum of the time necessary for mastering the production of new types of component and expenditure on the manufacture of technological equipment (for example dies, jigs, copying devices).

Using specialised control computers, it has become possible to realise *digital programmed control of machine tools*, where all the movements of the tools which are required to produce the necessary shape of the blank are performed automatically according to a programme stored in the memory of the machine. This programme contains, in coded form, all the data which during manual production are written on the drawing. Fig. 10.11 shows a widely used circuit of the digital programmed control of a milling tool. The memory of the machine 1 is a magnetic or punched tape or a ciné film, on which is recorded a sequence of digits characterising the required displacement of the tool, the milling tool 7 in this case, during the process of manufacture of the component 8. The reading device 2 which operates in conjunction with the drive for displacement of the blank, receives the information recorded on the programme belt and generates signals that characterise the necessary direction of movement of the tool, the number of steps and the size of each step. The control device 3 controls the movement of the stepping motor 4 which ensures realisation of the programme of displacement of the milling tool. The stepping motor is so constructed that each current pulse in its windings causes a turn of the motor shaft by a strictly predetermined angle. The movement of the stepping motor displaces the drive 6

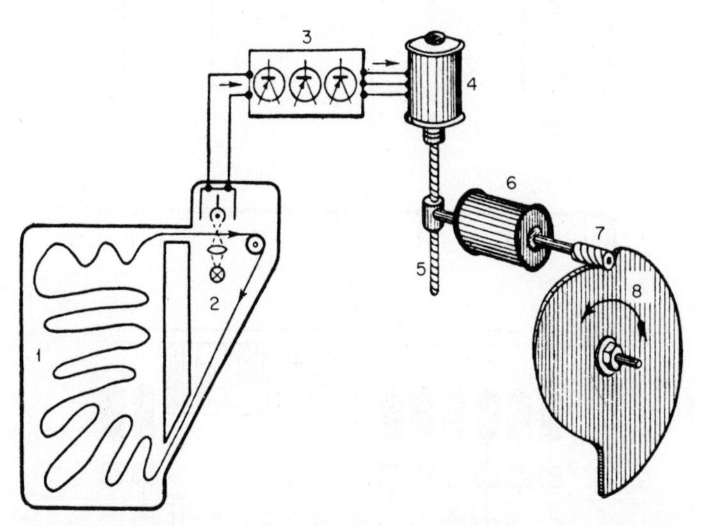


Fig. 10.11. *Diagram of a digital programme-controlled milling machine.*

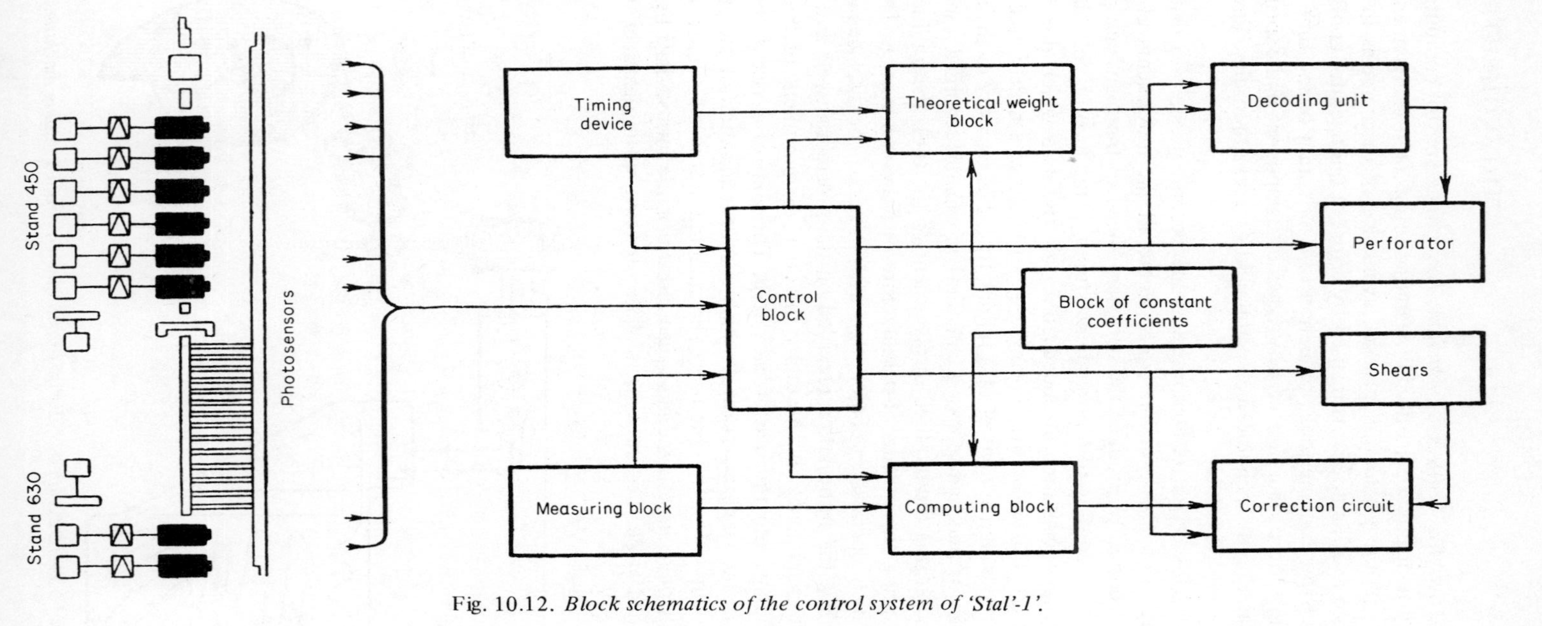


Fig. 10.12. *Block schematics of the control system of 'Stal'-1'.*

and the milling tool 7 through the threaded screw 5 by an amount which is proportional to the number of pulses sent from the stepping motor winding to the control device. In this way, to each position of the revolving blank 8 there corresponds a specific position of the tool which machines the blank according to the given programme of the required contour.

This simple system is called a single-coordinate system because the movement of the milling tool is in one direction only. If it is necessary to extend the potentialities of machining to components of complicated contour and to large components, two- or even three-coordinate systems may be used. In this case two or three programmes will be written on the tape, these would be processed respectively by two or three stepping motors which move the tool in mutually perpendicular directions.

*Digital Control of Metal-Cutting.* A specialised digital computer 'Stal'-1' has been built for optimal control of the in-line cutting of hot-rolled strip.

The 'Stal'-1' control system consists of a control block which determines the programme of optimum cutting-up of the strip, shears which realise this programme, and a number of sensors fitted along the rolling mill, as shown in Fig. 10.12.

The general formulation of the optimal control is as follows.

The rolling mill produces strip of length $L_i$ (i is the serial number of the rolled strip), which is cut by special shears into blanks of length $a_j$. It is permissible to reduce this length to $b_j$· furthermore the length $q_j$ of the last blank should not be less than $c_j$. The length $L_i$ fluctuates within wide limits and therefore the last blank will often be shorter than $c_j$ and will have to be scrapped. The task is to ensure that all the blanks are within the limits of the specified tolerances.

The length $L_i$ cannot be directly measured because the shears are positioned side by side with the rolling stand and cut the strip as it emerges from the stand.

The constants $a_j$, $b_j$, $c_j$ are agreed on when the order for the blanks is given.

The present-day manufacturing technology does not ensure obtaining constant lengths $L_i$, due to variations in ingot weight and burn-off during heating, and variations in the amount of cropping of defects in the bottom and top parts of the ingot. The length of the strip after rolling also depends on the grade of steel, the metal temperature, on its cross-section, and on a number of other factors which are difficult to assess. Therefore the control computer is required to compute not only the programme for optimal cutting and to control the shears in accordance with this programme, but also to forecast the length $L_i$.

The problem is additionally complicated by the difficulties involved in measuring the parameters of the hot metal.

As the initial value for forecasting $L_i$ the length $l_i$ of the strip is used, this being measured before the strip enters the rolling stand. The length $l_i$ can be conveniently represented as $l_i = a + h_i$, where $a$ is some constant component and $h_i$ is a variable component which can be measured before the cutting-up starts. Since the speed of rolling of each strip is sufficiently constant, measurement of the length is substituted by measurement of the rolling time. The value $h_i$ is generated in the register, a shifter with a single digit adder, on whose second input pulses of a fixed frequency are fed in from a timing device, which is in turn controlled by signals from the sensors A and B (spaced at distance $a$). The control circuit connects the timing device (time monitor) to the adder of the register when rolling begins (triggered off by the signal from the sensor A) and the counting of pulses stops on a signal from the sensor B, when the end of the strip passes beyond this sensor. Similarly, the code is generated which corresponds to the length $a_j$. The value of the respective $L_i$ is forecast from the results of measuring and calculating $l_i$, based on the law of constancy of the flow of metal at the entry and exit of the rolling stand.

The optimal cutting plan is determined in two stages according to a programme recorded in the computing block. At first the number $z_i$ of blanks which will be obtained on cutting up $L_i$ to the length specified in the order, is calculated; then the length is obtained of the last blank $q_i$ and the length $x_i$ which has to be added to the last one in order that it should be within the tolerance limits. The minimum length by which each of the $z_i$ blanks must be shortened in order that $q_i$ of the last blank should increase by $x_i$ is then calculated. Obviously, if the length of each blank is reduced by one 'quantum' (for instance by $m$ mm, if this is the unit according to which the machine operates), then on $z_i$ blanks $z_i$ 'quanta' are saved, with a corresponding increase in the length of the last blank. Therefore the optimal cutting can be found by using the following procedure of successive calculation of the $r_k$ values:

$$-x_i + z_i = r_0, -x_i + 2z_i = r_1, \ldots, -x_i + nz_i = r_{n-1}$$

If some $r_{n-2} > 0$ and $r_{n-1} < 0$, then, shortening the first $r_{n-1}$ blanks by $(n-1)z_i$ and all the other blanks by $nz_i$, we obtain cutting without scrap and a minimum shortening of the lengths of the individual blanks compared with the initial length. This can easily be verified.

On the basis of the calculated values of $n$ and $r_{n-1}$ the computer will feed the necessary control data into the circuit controlling the shears.

To compensate inaccurate operation of the shears, feedback was built in: the calculated and the real lengths of the blanks are continuously compared and the required corrections made. The machine contains a blocking circuit which in the case of any defects in the system will change the operation of the shears so that lengths $a_j$ are cut. At the same time a signal that there is a fault is generated.

Use of a digital computer for controlling the cutting of the rolled strip permits easy combination of the control function with the function of counting the production. The computer will count the number of rolled ingots and of the cut-up blanks, will convert the measured length into weight of the accumulated total for all the ingots in the batch (heat) and then these will be recorded, together with other necessary data, for subsequent processing on the computers of the plant computing centre.

The use of a 'Stal'-1' control computer has increased the output of prepared rolled material by 20,000 to 25,000 tons per annum.

The control computer can be amortised over less than one month's operation, and this gives some idea of the great savings which can be achieved by using similar control computers.

### 10.4. The potentialities of digital computers

The fantastic speed of operation of digital computers, the fact that they are all-purpose machines, and the large capacity of their memories, have made an enormous impression on people all over the world. Specialists in cybernetics, mathematics and other exact sciences are on the whole unanimous in their appraisal of the potentialities and prospects of digital computers. On the other hand, less knowledgeable people have very different opinions, based on conjecture and prejudice.

On the one hand there is great misunderstanding as to the difference between the theoretical potentialities of cybernetic systems (including artificially created ones) and the real potentialities of modern computers. The sensational reporting relating to some examples of the use of digital computers makes it difficult for people to realise how long and laborious is the development required before it is possible to produce a machine which will 'think' in a manner similar to intelligent beings. On the other hand, attempts are being made to erect an impassable barrier between the potentialities of artificial machines and those of human thought. It must be pointed out straightaway that the latter point of view has no scientific basis and is purely a psychological, and 'bad' psychological, explanation: the idea that 'the machine can think as man does' has a disturbing effect on people who cannot think scientifically, in the same way as their ancestors were disturbed by the idea that 'man originated from monkeys'.

The champions of the privacy of human thought do not notice that they themselves belittle its potentialities by stating that man will never be capable of developing artificially a machine which is superior in intelligence.

Both these extreme points of view interfere with scientific judgement and the solution of serious problems associated with the present and future developments of computers, both with the advantages to be gained, and also to an understanding the dangers to mankind of the increasing role of computers in all branches of the life of society. To understand correctly the forces at the disposal of mankind as a result of the advent of digital computers, one must take into consideration not only their real potential but also their limitations, both in principle and in practice.

The effectiveness of digital computers is based on their ability to overcome a number of difficulties associated with the solution of many problems posed by modern science, engineering and economics. The most important of these difficulties are the following:

(a) *The enormous number of operations* required for solving many important problems, due to the very large number of steps into which the algorithm for solving the problems branches out.

Such difficulties are met, for instance, in astronomical calculations, when determining the trajectories of space ships and satellites, in the preparation of maps from data of geodesic photographs of a locality.

(b) *The very time-consuming nature of the calculations*, associated with considering a large number of alternative solutions of a problem so as to select from them the optimum one. Such problems arise when choosing the most favourable design of a bridge, the most favourable shapes of the contour of a ship, the profile of an airfoil or the shape of a turbine blade.

(c) *The necessity of processing huge masses of data* in such problems as for instance the processing of statistical data, financial calculations, processing of meteorological data for weather forecasting.

(d) *The pressing need to perform the calculations rapidly*, arising in cases when the solutions are used to control fast processes, such as the rolling of metal, chemical reactors, manoeuvering of an aircraft or a ship in combat, and space flights.

(e) *The indivisibility of the solution of a problem between several actuating units.* Situations in which complex problems have to be solved in a single centre within a limited time; taking into consideration the large amount of data which characterise the situation arising during the control of extensive power systems, when making decisions in a defence system against aircraft, in operative production control. Obviously, problems of this type cannot be solved by increasing the number of human operators, because in this case the decision of any one of these would be taken without adequate consideration of the situation as a whole and it would be impossible in practice to coordinate their activities.

The above mentioned circumstances, which make manual solution of problems difficult and in some cases impossible, stimulated the use of computing techniques which in a number of cases ensures effective solution due to the high speed of response, flexibility of the programme, the facility of using and storing a very large volume of information. The potentialities of digital computers have increased even more during recent years, when the prerequisites were made for building *computer systems*, capable of exchanging information and of 'collectively' solving complex problems.

However, even under these conditions the potentialities of computers are not without limit. There are a number of limitations which in principle are insurmountable for any machine. This means, for any cybernetic system, artificial or natural, and however perfect its design, there are limitations. These limitations comprise:

(a) *The impossibility of producing an effect before its cause.* The law of cause-effect relations is the basis of Ashby's postulate.

'Any system which realises a purposeful selection at a level higher than a random selection, achieves this on the basis of information obtained.' This means that an answer to any problem posed to the machine cannot be obtained before the problem is formulated and information fed into the machine which is adequate for working out a justified answer.

(b) *The algorithmic insolubility of some classes of problem.* It follows from the theorems formulated by the American mathematicians Gödel and Church, and the work of Turing, that classes of problem exist for which it is impossible in principle to work out an effective algorithm. The Soviet mathematicians A. A. Markov and P. S. Novikov have worked out concrete examples of algorithmically insoluble problems. These results were used by some scientists and were the basis of excessively pessimistic evaluations of the potentialities of computers. In reality, the proof of the algorithmic insolubility of some classes of problem does not mean that amongst individual problems of this class there are those that are insoluble; it means rather that there exists only a very general and wide class of problem whose solution cannot be achieved by a *single algorithm*.

(c) *Limitation of the speed of a computer.* Since the operation of computers involves exchanging signals between its individual parts, the speed of operation is limited by the time of travel of a signal from one part of the machine to the other. The speed of transmission cannot exceed the speed of light, and the distance between the components cannot be smaller than the dimensions determined by the sizes of the molecules from which these elements are made. Therefore the time taken to execute each operation in the machine cannot be shorter than a certain, although very short, finite value. It can be argued that this limitation is of no importance, but in reality even now it considerably impedes increasing the speed of operation of computers.

Obviously these basic limitations of the potentialities of solving problems by computers *also apply fully to the potentialities of man*, who cannot go beyond the limits which follow from the law of causality, algorithmic insolubility and limitations on speed of response. Therefore, forecasts that a certain type of problem can never be solved by machines are either not substantiated, or mean that these problems are also insoluble by man.

In addition to the theoretical limitations of computers, practical limitations associated with the natural requirement of *effectiveness* in solving problems are of great importance. By effectiveness we understand, in the first instance, completion of a solution within an acceptable period of time, and secondly, obtaining the result in a form suitable for use. Many problems which in principle are insoluble require an astronomical number of computing operations. Even such an apparently simple problem as the selection by complete scanning of the optimal regime of a system in the presence of hundreds of control actions, which must be selected with an accuracy of 1%, may require for its solution millions of years of operation of even the fastest computer, and is obviously inapplicable. It may transpire that a computer gives the sought solution in an acceptable time, but the output is in such a form that its conversion to a form convenient for use cannot be realised in practice, as for instance in the case when the solution consists of an enormous mass of data. It is obvious that to obtain ineffective solutions is equivalent in practice to insolubility.

The difficulties involved in the practical application of machines for solving complex problems are frequently associated with the inability to formulate the problem correctly and clearly. It is very often difficult to determine what in particular has to be clarified, i.e. what we want the machine to do.

### 10.5 Analogue principles of computer design

In the previous chapters we have dealt almost exclusively with digital computers. There is also another type of computer – the analogue computer. Their operation is not based on processing discrete signals representing numerical data but on designing a physical model of the process being investigated. This has already been mentioned in Chapter 3. For instance, since the current distribution in electrical circuits of a certain type obeys the same equations as the temperature distribution in a blast furnace, the pressure distribution in air streams which flow past aircraft and a number of other processes of interest, it is possible to build an analogue machine which gives a numerical solution of such problems in the form of certain values of current density at the machine output. However, the potentialities of modern analogue computers lag behind those of digital computers owing to two reasons: The first is the extremely narrow 'specialisation' of such machines which usually solve an extremely limited class of problem of a single type. It is very difficult to eliminate this limitation due to the 'attachment' of the functions of the machine to the properties of the physical process in it whilst computers which contain this information in digital form are free of this drawback. The second reason is the low accuracy of solutions obtained by analogue computers (usually not better than tenths of a percent).

There are very weighty arguments that in future this situation may change and analogue principles will play a more important role in the development of computers. Obviously they will be used very differently from present-day analogue computers, and in the first instance they will be used in close connection with digital methods of information processing.

This conclusion is based in the first instance on a comparison of the operation of a computer with that of the human brain. The analogy between a digital computer and the brain has proved very incomplete. It has been found that in the brain, analogue processes play an important role and the information changes its form many times from digital to analogue and vice versa. The accuracy of presentation of numerical data in the brain is low: of the order of $10^{-2}$ to $10^{-3}$. Since the magnitude of the error increases strongly when operations are performed successively, the maximum number of successive operations in the brain – the so-called *logical depth* of information processing – cannot be very large. It is assumed that the logical depth is of the order of 10. In modern digital computers almost all the operations are carried out successively, requiring an accuracy millions of times higher than in the brain, and in addition it leads to a low loading of the individual components of the machine. The collossal capability of the brain, the high accuracy and reliability of its operation, are not achieved by speed of response, accuracy and reliability of carrying out each operation, but as a result of the extremely complicated mechanism of *parallel* processing of information and the peculiar forms of presentation of this information, which combine digital and analogue principles. For instance, the information is not in the accurate form of a sequence of pulses, but in the statistical properties of this sequence. All these considerations were summed up by J. von Neumann in the aphorism: 'The language of the brain is not the language of mathematics.' Cybernetics has still to learn to speak the language of the brain.

Another range of problems is associated with the following considerations. It is known that solution of complex problems as for example in finding the maximum of a function of many variables, in the digital form will lead partly to highly unfulfillable requirements as regards the number of operations and the memory capacity of the computer. For instance, if a computer is required to find one of all the possible Boolean functions of $n$ binary variables, it will be found that even for $n = 100$ the necessary number of memory cells exceeds the number of atoms in the Pacific Ocean, and the duration of calculation of the computer with a speed of $10^{12}$ operations per second (a million times higher than the best existing modern computers) will exceed the age of the Earth. On the other hand the simplest physical systems can solve much more 'laborious' problems more easily and faster at every step, for instance, using a beam of light to find the shortest path in an optical nonuniform medium, or using a gas in a vessel which, on changing from the nonequilibrium to the equilibrium state, will find the maximum entropy – the complex function of an astronomical number of variables. Obviously, this is possible only in view of the 'analogue' solution of the problem, where all the molecules of the system virtually play the part of 'computing elements' working in parallel. The idea of using similar processes in cybernetic devices is very tempting. The example of living organisms where strictly ordered processes occur in every cell indicates that in principle this is possible.

Among the first attempts in this field was the work of the British scientists Gordon Pask and Stafford Beer. They produced a colloidal solution containing green vitriol, into which they submerged amplifiers and platinum electrodes which served for feeding in electrical input pulses. As a result, metallic fibres grew in the solution, the configuration of which depended on the input signals, and also (due to feedback) on the 'previous history' of the system. This simple system revealed a number of extremely interesting and unexpected properties. It was able to 'solve' very cumbersome systems of linear and uniform equations containing dozens of equations. It proved capable of learning and adapting itself to the environment and to input actions. In particular, it could be taught to react in a certain way to changes in the acidity of the individual sections of the medium, to various configurations of the magnetic field and to vibrations. What is more, this sytem could also be taught relatively quickly to react to sound and to distinguish between sounds of various frequencies (50 and 100 Hz). In the system an 'ear' grew containing fibres which resonated at particular frequencies. The prospects of similar investigations are fascinating.

**Exercises**

1. Subtraction in computers is most frequently carried out using an artificial method which can also be used in the decimal system. According to this method the subtraction is substituted by addition with the so-called supplementary form of the subtrahend. If, for instance, it is necessary to calculate the difference 1340 – 864 = 476, then we find the supplement of the subtrahend (to one thousand) equalling 136, and perform the addition:

$$\begin{array}{r} 1340 \\ +\ \ 136 \\ \hline 1476 \end{array}$$

It can be seen that the result so obtained differs from the previously obtained difference by a unit of a higher order, which is discarded when reading the result. In the binary form the supplementing can be determined by the simple rule: the number should be divided into two parts in front of the last unit to the right, on the right-hand side nothing is changed, whilst on the left-hand side the zero is replaced by one and the one by zero. For instance the number 1001101110 is divided into the two parts: 10011011/10, and then we obtain the supplementary form: 0110010010.

Calculate the following differences, using the supplementary form:

(a) 101101 – 10110 (b) 1001110111 – 0001111100
(c) 111100000101 – 101110100010 (d) 100011110 – 10000

Verity your results by taking the decimal equivalents of the binary numbers.

*Solution.*

(a) $\begin{array}{r} 101101 \\ +001010 \\ \hline 110111 \end{array}$ or $\begin{array}{r} 45 \\ -22 \\ \hline 23 \end{array}$ (b) $\begin{array}{r} 1001110111 \\ +1110000100 \\ \hline 10111111011 \end{array}$ or $\begin{array}{r} 631 \\ -124 \\ \hline 507 \end{array}$

(c) 111100000101 or 3845 (d) 100011110 or 286

| | | | |
|---|---|---|---|
| (c) 111100000101 | or 3845 | (d) 100011110 | or 286 |
| +010001011110 | −2978 | +111100000 | − 32 |
| 1001101100011 | 867 | 1011111110 | 254 |

2. The coding disc shown in Fig. 10.9 may yield large errors at some turning angles relative to the contact brushes. For instance, if the turning angle $\varphi$ is such that at the output of the converter ones appear in all the six digits. Any additional turn to the right by a small angle may produce a large error because as brushes cannot be arranged in an ideal manner they do not move off simultaneously from the sectors. Therefore the coding is frequently made to differ from the binary one. For instance, it is possible to use a code, given for the first ten numbers, Table 10.2.

Table 10.2

| *Decimal number* | *Binary number* | *Code* | *Decimal number* | *Binary number* | *Code* |
|---|---|---|---|---|---|
| 0 | 0000 | 0000 | 6 | 0110 | 0101 |
| 1 | 001 | 001 | 7 | 0111 | 0100 |
| 2 | 0010 | 0011 | 8 | 1000 | 1100 |
| 3 | 0011 | 0010 | 9 | 1001 | 1101 |
| 4 | 0100 | 0110 | 10 | 1010 | 1111 |
| 5 | 0101 | 0111 | | | |

Draw a coding disc (or drum) which would realise this code.

Estimate the maximum possible error caused by the non-simultaneous meeting of the brushes when reading off the code.

*Solution.* Fig. 10.13 shows the coding disc which realises the Grey code (this is the name of the code written in Table 10.1). Changeover from one number to the next in the Grey code involves a change of the number only in one digit, as

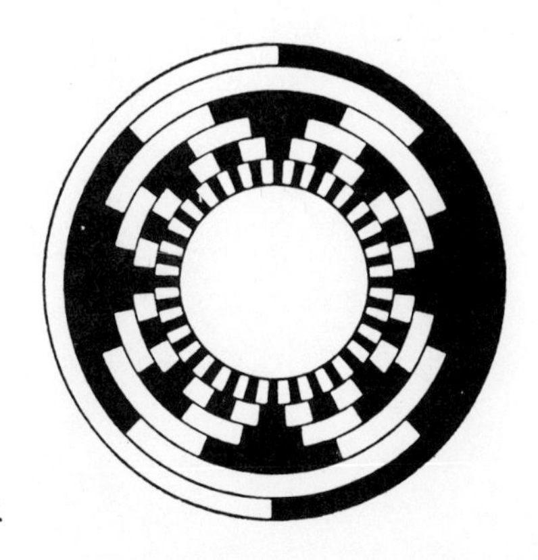

Fig. 10.13. *To Example No. 2.*

can be seen from Table 10.1, and therefore the maximum read-off error does not exceed ± 1 digit.

**3.** It is necessary to design a digital-analogue converter of the type shown in Fig. 10.10. How many relays are required to ensure a conversion accuracy of 0.2%.

*Solution.* 9 relays. In this case the conversion accuracy will equal $2^{-9} < 0.2\%$.

# 11 Adaptation

Observation of the multitude of highly organised systems existing in nature, their adaptability to changes in their surroundings, their development and self-reproduction, leads us to believe that the basis of all these remarkable phenomena is some cognitive mechanism which gives these systems the property not only of not losing their order and organised state, but also of increasing it with the progress of time. The formation and development of such systems is undoubtedly one of the most grandiose manifestations of the creative force of Nature.

The origins of highly organised systems can be explained by the fact that amongst an enormous number of random combinations of elements, sooner or later combinations will arise which are adapted to the surrounding medium to an extent at least sufficient for survival, i.e. this means that they have the ability to retain in a stable manner their organised structure. However, it is impossible to visualise the development of a system which from the very beginning would be so perfect as to exhibit optimal or at least acceptable behaviour when the conditions change in one way or another. To maintain the viability of a system when conditions arise in which the organisms are not capable of choosing the required reaction, it is necessary to change the structure of the system and the form of its behaviour, to change the nature of the reaction of the system to perturbations.

*The process of changing the properties of a system, which allows it to achieve the best, or at least an acceptable, functioning under varying environmental conditions is called adaptation.*

The essential features of natural adaptive processes have so far been little studied, nevertheless there has been some success in constructing models that achieve adaptation although in a very crude and simplified manner. These models of *self-organising systems* give an idea of the possibilities of spontaneous structure formations, which are capable of retaining their organised state, and also have the ability to generate a behaviour that enables them to adapt to changing conditions.

The property of adaptation manifests itself clearly in the *homeostatic* mechanism, namely that living organisms have the ability to maintain their most important variables within acceptable physiological limits in spite of considerable variation in the conditions in which the organism has to exist. The adaptability of the homeostatic mechanism can be illustrated by the example of changing the reaction of the body of warm-blooded animals to temperature changes. At relatively low temperatures the thermoregulation in the body is achieved by changes in the flow of the blood in the surface tissue so as to ensure optimal heat transfer conditions between the body and the medium. On the other hand, if the temperature of the surrounding medium is high, temperature control is by means of sweating and breathing to ensure intensive release of surplus heat. In this way the body adapts itself to changes in its surrounding conditions – it adapts and changes its behaviour, tending to ensure homeostasis and in the given example, striving to maintain the body temperature within permissible limits.

Artificial control systems usually operate under conditions which vary considerably, and to ensure that they function effectively it is tempting to use a mechanism similar to the adaptation mechanism developed by nature. The adaptability of technical or economic control systems to changes in the conditions can be achieved by using various methods of selection with respect to the conditions of this operation. The changes in the conditions of operation may consist of changes in the surrounding medium or changes of the properties and characteristics of individual parts of the control system. Choice of the most favourable conditions for the operation of the system can be, and frequently are, realised by the *search* method.

## 11.1. Selection of the most suitable regime

For any system to function normally it is necessary that the quantities which determine its conditions of operation should not exceed the limits of the permissible values. Thus, to retain the viability of an animal, quantities such as body temperature, content of oxygen and glucose in the blood, must be within the permissible limits; for the normal operation of a steam boiler it is necessary to maintain the steam pressure and temperature, the water level in the drum and the feedwater temperature, within permissible limits.

Let us denote the values which characterise the operating regime of the system and are available by $R_1, R_2, \ldots, R_n$.

In the simplest case, when the limit values of the system's operating regime do not depend on the operating conditions of the system, these limitations can be written as:

$$R_1' \leqslant R_1 \leqslant R_1'', \; R_2' \leqslant R_2 \leqslant R_2'', \; R_n' \leqslant R_n \leqslant R_n'', \tag{11.1}$$

where the prime symbol denotes the minimum permissible and the double prime denotes the maximum permissible operating values.

In the *space of the operating conditions* of the system, along the axes of which the operating values are plotted a D-area ($n$-dimensional rectangular parallelepiped) can be singled out for the inequalities (11.1), inside which there should always be a point which represents the conditions of operation of the system. Obviously, the control system should always ensure maintenance of the operating values within the limits which ensure that the representative point is inside the D-area. However, this does not exhaust the control problems, because points which are inside the D-area are not equivalent in all respects.

To each regime of the system there corresponds a specific value of the effectiveness criterion $\alpha$, which characterises the degree of correspondence of each regime with the aims of the control. For living organisms, the criterion $\alpha$ can determine the probability of their survival or the intensity of development; for a steam boiler this may be the efficiency or the productivity expressed in terms of steam; for a transportation system, the freight revenue or the ton-kilometres of transported freight. We will mark the $\alpha$ values in each point of the D-area and connect the points characterising equal $\alpha$ values, as has been done for instance in Fig. 11.1 for the case of a two-dimensional space of operating conditions.

If the most favourable value of $\alpha$ is its maximum value, then the most favourable conditions of operation of the system will be that represented by the point $\alpha^*$ and the control system may have the task of establishing the operating values $R_1^*$, $R_2^*$, ..., $R_n^*$, which ensure the most favourable effectiveness criterion $\alpha = \alpha^*$. If changes in the condition of the system do not lead to important changes of the coordinates of the point $\alpha^*$, this problem is solved by an appropriate system of stabilisation, using the methods described in Chapter 7. However, if the position of the point $\alpha^*$ changes during

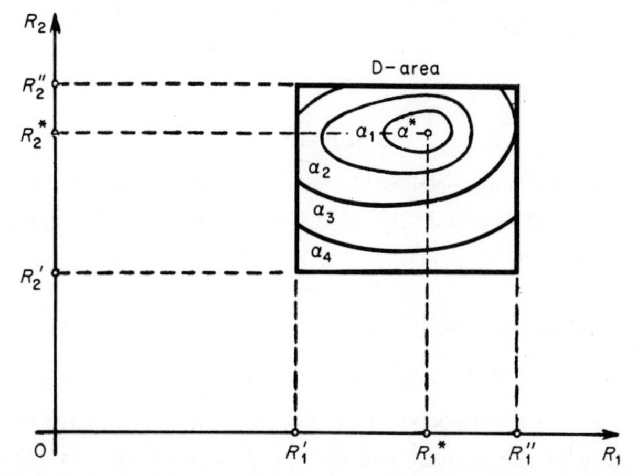


Fig. 11.1. *Curves of equal α values in the space of the regime of the system.*

operation of the system and the change is in a form which cannot be forseen, the control problem becomes more complicated.

The control system should always establish such operating values $R_1^*$, $R_2^*$, ..., $R_n^*$ for which $\alpha$ will have the most favourable value $\alpha^*$, compatible with the limitations imposed on the operating values, whatever the change in the position of the point $\alpha^*$.

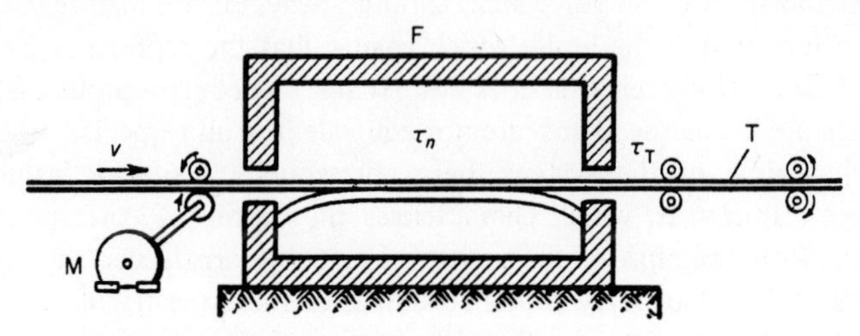


Fig. 11.2. *A continuous furnace for heat treatment of steel tubes: M – motor for the forward movement of the tubes, F – furnace, T – tubes.*

Let, for instance, the controlled plant be a continuous furnace F for the heat treatment of steel tubes T, as shown in Fig. 11.2. The operating regime of this unit is determined by the temperature $\tau_n$ in the operating space of the furnace and the speed of movement $v$ of the tubes. As the effectiveness criterion of the unit, we will consider the fuel consumption $q$ in the furnace per ton of heat-treated material

$$\alpha = \frac{q}{rv} = \min_{v,\tau_n} \tag{11.2}$$

where $r$ is the tube weight per metre. In the given case the value $\alpha$ must be minimised with respect to $v$ and $\tau_n$, i.e. the permissible $v$ and $\tau_n$ values must be so chosen that $\alpha$ will assume the lowest possible value.

The boundaries of the D-area in the given case are determined by the following considerations:

*Speed limitations*

$$v' \leqslant v \leqslant v'', \tag{11.3}$$

where $v'$ is the minimum permissible speed of movement of the tubes which ensures the required productivity of the unit, and $v''$ is the maximum speed developed by the drive.

*Temperature limitations*

$$k'v \leqslant \tau_n < k''v, \tag{11.4}$$

where $k'$ and $k''$ are coefficients which determine the permissible ratios between the temperature in the furnace and the speed of movement of the tube which will ensure respectively a minimum permissible and a maximum permissible

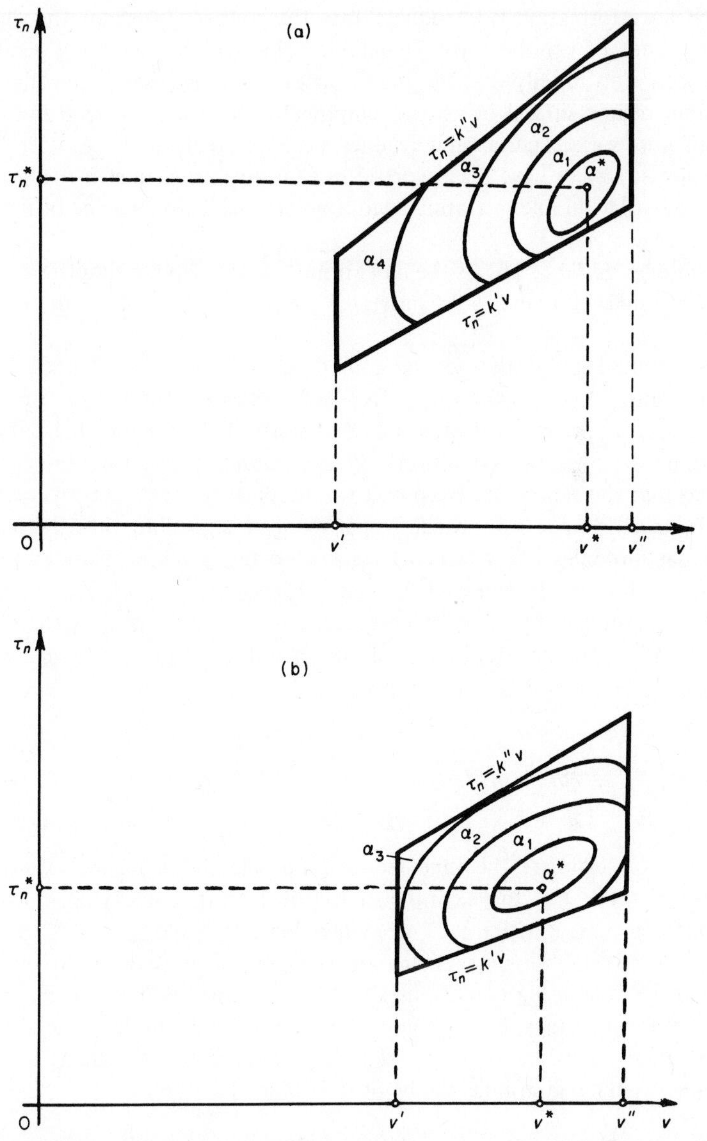


Fig. 11.3. *Lines of equal values of the criterion $\alpha$: (a) for thick-walled tubes; (b) for thin-walled tubes.*

temperature $\tau_T$ of the tube at the exit from the furnace. The D-area in the space of the operating conditions $v$, $\tau_n^*$ determined in this way is plotted in Fig. 11.3, which also contains lines of $\alpha$ = const. The most favourable $\alpha = \alpha^*$ corresponds here to the point with the coordinates $v^*$ and $\tau_n^*$. The diagrams show the boundaries of the D-areas the distribution of the lines of the level $\alpha$ for thick-walled tubes (Fig. 11.3(a)) and thin-walled tubes (Fig. 11.3(b)).

Since no information is introduced into the control system on the shape of the level lines or on the most favourable values of the operating conditions (which are not usually known in advance) it is necessary that the most favourable regime should be chosen empirically, by trial-and-error and by the comparison of various permissible regimes, i.e. by searching.

Definite strategies have been worked out for organising an effective search to find the most favourable operating conditions. Let us consider some of these.

*The Scanning Method.* This is the most primitive but also the simplest and most reliable method of search. It consists of successive trials of all the different regimes.

When searching by the scanning method the D-area is divided into $N$ compartments whose dimensions after each measurement equal respectively $\Delta R_1$, $\Delta R_2$, ..., $\Delta R_n$, selected on the basis of the required accuracy of establishing the operating conditions. By determining successively the operating values so that the representative points would remain in each compartment, and remembering the value $\alpha$ for each compartment, it is possible when the scanning process has terminated to select such a set of operating values as will correspond to the most favourable values of the effectiveness criterion $\alpha^*$. The number of cells $N$ is determined by the numbers $m_1, m_2, \ldots, m_n$ of the discrete operating values, which ensure establishment of the operating regime with the required accuracy:

$$N = m_1 m_2 \ldots m_n.$$

If $m_1 = m_2 = \ldots = m_n = m$, then

$$N = m^n. \tag{11.5}$$

Let us now estimate the time required for selecting a regime by using the scanning method. For the example of the heat treatment of tubes, assuming $m = 20$ (5% accuracy, with $n = 2$), we obtain $N = 20^2 = 400$.

To achieve establishment of the operating value it is necessary that each change in the operating conditions should be retained without further change (constant) for a certain time $T_\omega$ and only then to estimate the value $\alpha$. For the process under consideration the time $T_\omega$ cannot be less than 1 minute. Consequently the search time will be of the order

$$T_n = N T_\omega = 400 \text{ min.}$$

which is obviously inapplicable if the tubes to be treated change every 2 to 3 hours, as is normally the case.

If we wish to choose the most favourable regime by scanning, for a process characterised by 10 operating quantities, with an accuracy of 1%, then even for a time expenditure of 1 second per step we would require (according to Eq. (11.5))

$$N = 100^{10} \text{ steps}$$

and the search time would be

$$T_n = 100^{10} \text{ sec}, \approx 3 \cdot 10^{12} \text{ years},$$

i.e. a period of time in excess of the age of the Earth. Obviously the method of scanning is unsuitable for searches in systems with any degree of complication. The search method which will now be described is considerably more effective.

*The Gauss-Seidel Method.* Let us assume that it is necessary to find $\nu$ and $\tau$ which minimise $\alpha$ in the D-area, shown in Fig. 11.4, where the point $a_{in}$ represents the initial condition. Using the Gauss-Seidel method, we increase $\nu_{\text{in}}$ by $\Delta\nu$, verify in which direction $\alpha$ changed, and then select the direction for reducing the value of $\alpha$. We add to $\nu$ values of $\Delta\nu$ at each step, verifying each time that $\alpha$ does in fact decrease. In the point $a_1$ where changes in $\nu$ no longer produce a decrease in $\alpha$, movement is started along the coordinate $\tau$ in the same way until the point $a_2$ is reached where $\alpha$ no longer decreases. The process is then repeated along the $\nu$ axis until a regime $a^*$ is established such that changes in $\nu$ by $\pm\ \Delta\nu$ or changes in $\tau$ by $\pm\ \Delta\tau$ produce no further reduction of $\alpha$. The number of steps which have to be made in this case in order to arrive in the neighbourhood of $\alpha^*$ is determined by the nature of the dependence of $\alpha$ on the operating values and the initial conditions of the system, but the order of magnitude of the number of steps can be roughly estimated at

$$\mathrm{N} = m \cdot n \tag{11.6}$$

For a furnace with $m = 20$, $n = 2$, $T_\omega = 1$ minute, we find the search time to be

$$T_s = m \cdot n\, \mathrm{T}_\omega = 20 \cdot 2 \cdot 1 = 40 \text{ min.}$$

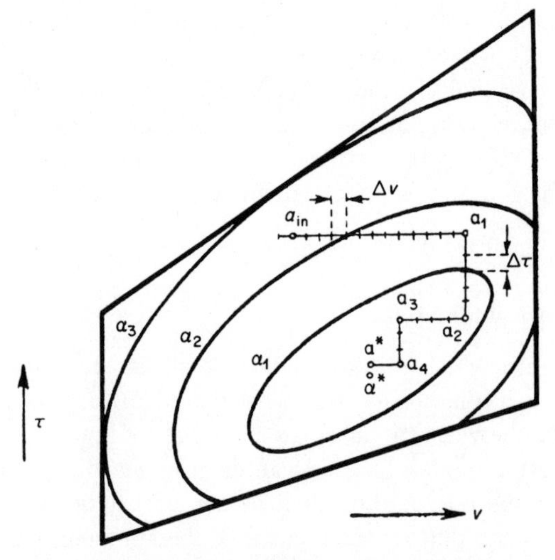


Fig. 11.4. *Search using Gauss-Seidel method.*

a time shorter by one order of magnitude than scanning, and for the quoted example $m = 100, n = 10, T_\omega = 1$ sec,

$$T_s = 100 \cdot 10 \cdot 1 = 1000 \text{ sec}; \quad \text{i.e. 3 hours}$$

instead of $T_s = 3 \cdot 10^{12}$ years for search by scanning.

An even greater reduction in the search time can be achieved by using methods based on the simultaneous variation of several operating conditions in the required direction (the method of the steepest descent, Fig. 11.5, the gradient method, the ravine method).

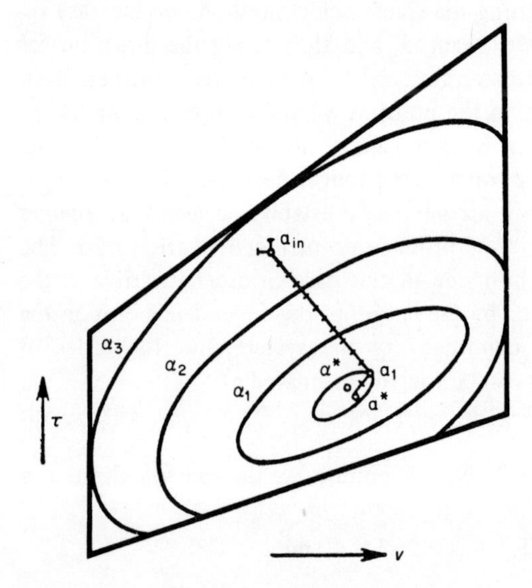


Fig. 11.5. *Search according to the method of the steepest descent.*

Not only the problem of selecting the most suitable conditions but also many other problems, for instance determination of the dimensions of a transformer, a motor or generator; choosing the optimal configuration of a water, electricity or gas supply system, working out an optimal production plan for a plant, reduce to the problem of finding the extremum of a function of many variables, which is solved using the above mentioned search strategy.

## 11.2. The adaptive automaton

An automaton into which no information at all has been introduced on the properties of the medium may also have the adaptive property if there is a certain interaction between the automaton and the medium.

As the simplest example of such an automaton we consider an automaton A with one input $x$ and one output $y$, which can exist in one of the two states $z_0$ and $z_1$. Let us assume that on the input $x$ of the automaton the reactions of a random medium, which depend on the output $y$ are fed in. Let us assume that $x$

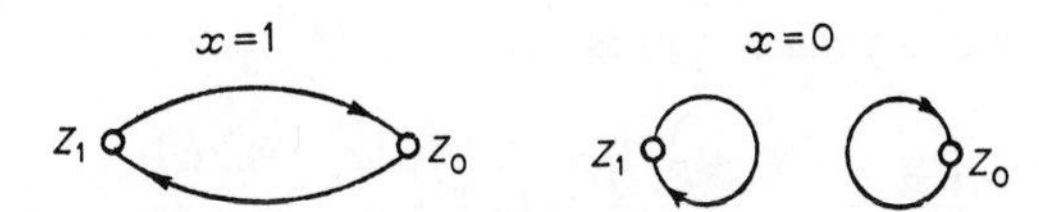


Fig. 11.6. *Diagrams of the transitions of the automaton A.*

and $y$ may only be either 0 or 1, where $x = 1$ means a 'penalty' (punishment) to the automaton and $x = 0$ means 'no penalty' (reward). Fig. 11.6 shows the diagrams of changeover of the automaton A for $x = 1$ and $x = 0$, from which it can be seen that the automaton is so designed that in the case of 'penalty' it changes its state. The output of the automaton is unequivocally determined by its state. Let us assume that in the state $z_0$, $y = 0$ and in the state $z_1$, $y = 1$. The properties of the medium are characterised by the 'penalty' probability values for various output values, given in Table 11.1.

Table 11.1

| | $y = 1$ | $y = 0$ |
|---|---|---|
| $x = 1$ | $p_1$ | $p_0$ |
| $x = 0$ | $1 - p_1$ | $1 - p_0$ |

If the automaton with equal probabilities would change over into each of the states regardless of the reaction of the medium, then the average number of 'penalties' over a number of cycles would be the mean arithmetic value of the probabilities of obtaining penalties in each of the states of the automaton. The probability of a penalty (or, in other words, the average number of penalties in one cycle) would in this case equal

$$p_{\text{pen}} = \frac{p_0 + p_1}{2} \tag{11.7}$$

If the transition of the automaton from one state into the other depends on the reaction of the medium, then the probability of penalties can be lowered. This reduction in the probability of 'penalties' is explained by the fact that an automaton with such a diagram of transitions, shown in Fig. 11.6, will more often be in the state in which there is less likelihood that it will be 'penalised' or punished. It can be proved that final (steady-state) probabilities exist for the automaton A to find itself in each of the states, regardless of its initial state and determined only by the structure of the automaton and the properties of the medium.

Using final probability states, the final probability of punishment can be determined.

For the automaton shown in Fig. 11.6 the final probabilities of existing in the states $z_0$ and $z_1$ are respectively

$$r_0 = \frac{p_1}{p_0 + p_1} \quad \text{and} \quad r_1 = \frac{p_0}{p_0 + p_1}$$

and the probability of a penalty equals

$$p^*_{\text{pen}} = p_0 r_0 + p_1 r_1 = \frac{2p_0 p_1}{p_0+p_1} \leqslant \frac{p_0 + p_1}{2} . \tag{11.8}$$

For $p_0 = 0.9$ and $p_1 = 0.1$ the probability of a penalty according to the formula (11.7) will equal $p_{\text{pen}} = 0.5$, and according to the formula (11.8) $p_{\text{pen}} = 0.18$.

It can be seen that even this primitive automaton is capable of behaving expediently in a certain sense, in a random medium, a property which was not taken into consideration when it was designed.

It is pointed out that the described automaton can be realised by means of an ordinary trigger, for which the 'penalty' is that it receives a pulse on the input, which causes a switching of the trigger.

By complicating the design of the automaton, for instance by increasing the number of states, a high degree of adaptability to the medium can be achieved even if the properties of the medium change with time.

## 11.3. The Homeostat

The first model of an artificial device which simulates the adaptive properties of living organisms, their adaptability to changing conditions of the medium, was made by the English scientist Ross Ashby. Since the behaviour of this model is to a certain extent similar to the behaviour of a living organism, expressed in the phenomenon of homeostasis, it was called a *Homeostat.* In a manner similar to living organisms who react to changes in the surrounding medium in order to maintain their essential variables within permissible limits determined by the physiological properties of the body, so a Homeostat also reacts to every change in its operating conditions by actions such that its basic coordinates remain within permissible limits.

The Ashby Homeostat consists of four main blocks, each of which contains a magneto-electric instrument with four metering coils. The moving system of each instrument shifts the slide of a potentiometer connected into the winding circuits of all the four instruments. As a result the position of the moving system of each instrument depends on the positions of the moving systems of the other three instruments. The connected instruments I1 – I4 form a dynamic system S (Fig. 11.7), which can be stable relative to any combination of intermediate positions of the moving system, or unstable, depending on the values of its parameters (resistances in the metering coil circuits). A control device Q containing a relay R and four step switches s1 – s4 are connected to the system S. The relay is energised if one of the moving systems reaches its extreme position. In this case the switches s1 - s4 begin to move and change the parameters of the system S by changing the resistances in the windings. If amongst the combination of the parameters there exists a combination for which the system S is stable, then the equipment Q will change its parameters until one

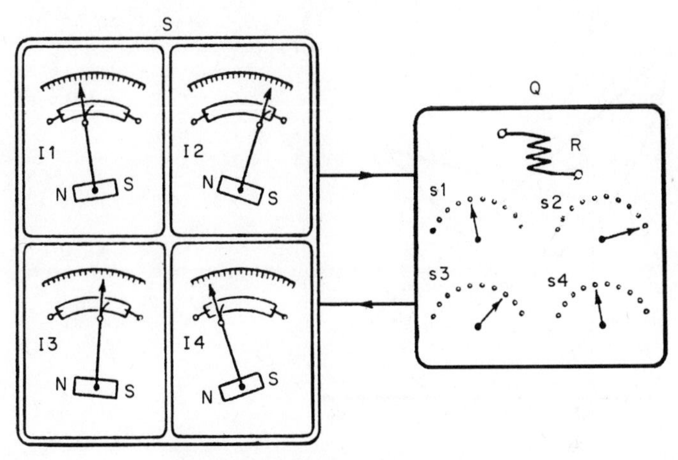


Fig. 11.7. *The Ashby Homeostat model.*

such combination is found. Then not one of the moving systems of instruments will reach its extreme position and the relay R will not be energised, and consequently the step switches will not move.

If the stable state found by the system is disturbed by any change in its parameters, the moving systems of the instrument will reach their extreme positions and the control device Q will again start the search for a new combination of parameters which ensures stability of the system S.

One possible trajectory of movement of a system of this type (but in fact a simpler one (second order system)) is shown in Fig. 11.8.

Here the position $X$ of the system can change within the limits

$$X' \leqslant X \leqslant X'',$$

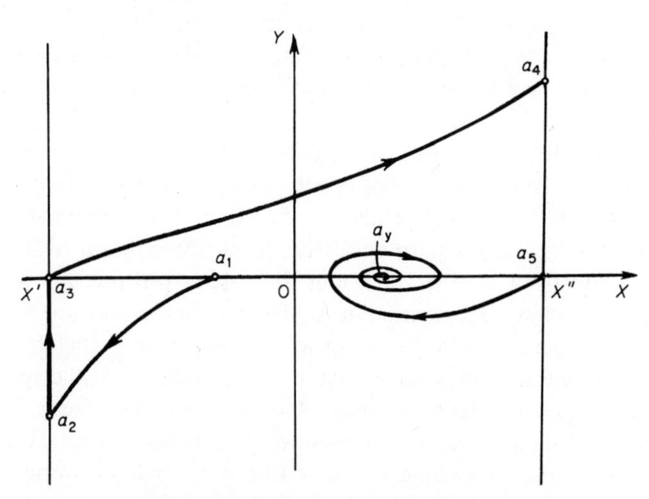


Fig. 11.8. *Trajectory of a Homeostat.*

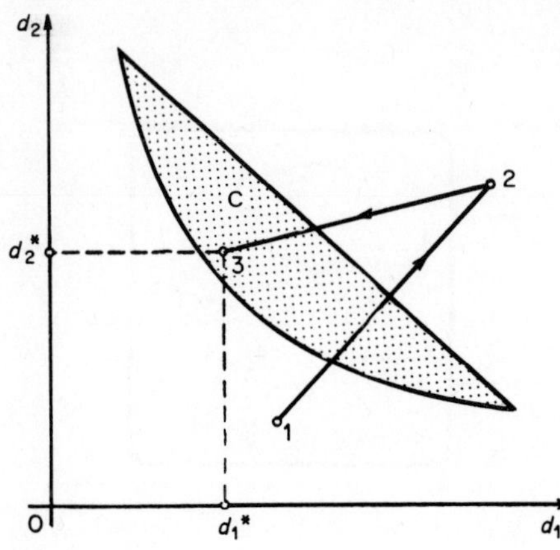


Fig. 11.9. *Range of stability of a Homeostat in the state space.*

determined by the stops. When hitting the stop the position of the system ceases to change – its velocity $Y$ reverts to zero. In the extreme positions of the system the control devices respond, changing its parameters. If the movement of the system starts from the position $a_1$, then due to its instability it reaches one of the stops, moving successively along sections of the trajectories which reach the stops $(a_1 - a_5)$.

The parameters $d_1$ and $d_2$ of the system will in this case change in a random manner, so that the point representing the set $d_1, d_2$ *in the parameter space* (in the given case on the plane $d_1, d_2$) will shift as shown in Fig. 11.9. If in the position 3 the point which represents the set of parameters $d_1^*, d_2^*$ of the system is in the region of stability (region C), then owing to displacement along the trajectory $(a_5 - a_y)$ on Fig. 11.9 the point representing the state of the system will move into the stable position $a_y$.

From what has been said above, it can be seen that however the operating conditions of the system change, whatever the external effects on it, as long as a set of parameters exists which corresponds to the stable regime a system like the Ashby Homeostat will sooner or later find a set of parameters and as a result it will re-establish its stability.

In conclusion it is mentioned that all the methods of adaptation described in this chapter are based on the idea of *search*: search for preferential conditions of operation, search for a preferential state, search for a preferential set of parameters. For realising a purposeful search the system must *obtain information* on the effectiveness of its behaviour. In systems which search for the most favourable condition of operation it is the effectiveness of the steady-state conditions of operation; in an adaptive automaton it is the frequency of obtaining 'instructions'; in a homeostat it is the stability of the system. Thus each adaptive system makes a series of experiments and draws from these experiments the data necessary for improving its behaviour, i.e. it acts in the same way as a conscious being which studies itself and the world around it, learning from it lessons which determine its behaviour.

**Exercises**

1. The dependence of the flame temperature in an open-hearth furnace on the air-fuel ratio $k$ can be represented by curves as shown in Fig. 11.10. If the optimal ratio $h_{opt}$ is constant for the entire heat, then it would be possible to choose the parameters of a controller to maintain the air-fuel ratio $k$ in such a way that its value is optimal. However, there are a number of additional factors due to which the optimum 0 will continuously move in the plane $t - k$ as shown by the dotted lines in Fig. 11.10. Therefore the controller is made adaptive so

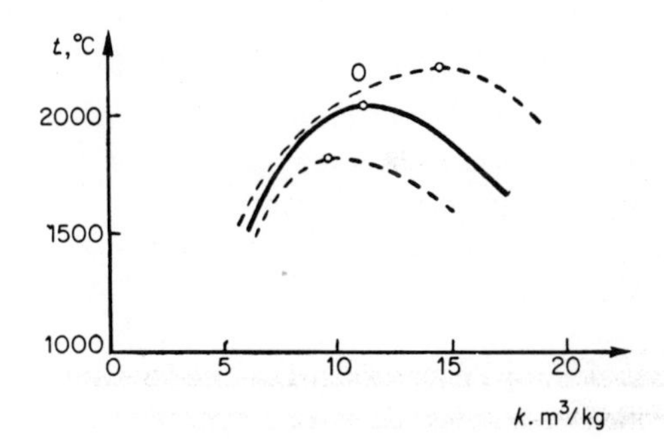


Fig. 11.10. *Dependence of the flame temperature on the air-fuel ratio.*

that it searches for a shifting optimum. Such a controller is called extremal. In the simplest case the controller operates as follows: $k$ changes by the value $\Delta k$ in any direction and the temperature $t$ is measured. If it increases, then the ratio $k$ is changed by a further step $\Delta k$ in the same direction. If it decreases, a step is made in the opposite direction, until the system reaches its maximum point. It will be assumed that the control system has found its maximum and maintains it, if $k$ differs from $k_{opt}$ by no more than one step. The maximum speed of shifting of the point 0 along the $k$ axis (Fig. 11.10) is 3 $m^3$/kg min. With what frequency did the controller make the steps so as to maintain an optimum, if the step $\Delta h = 0.25$ $m^3$/kg? (It is assumed that initially the system found the maximum).

*Solution.* 12 steps per minute.

2. For the automaton described in Section 11.2, calculate the final probability of 'penalty' $p^*_{pen}$ for the case when $p_0 = 0.5$ and $p_1 = 0.5$. What does the result indicate?

*Solution.* $p_{pen} = 0.5$. In this case the probability of penalty is the same as in an automaton whose behaviour does not depend on the reaction of the medium ($p^*_{pen} = p_{pen}$).

**3.** Which of the systems enumerated below is homeostatic: (a) an electric drive of a machine tool; (b) the human eye; (c) a clock mechanism; (d) an arithmometer; (e) the root system of a plant; (f) the system for controlling the surface body temperature of a sheep.

*Solution.* The systems (b), (e) and (f) are homeostatic.

# 12 Games

Up to now we have considered the behaviour of systems whose controls operate in such a way that the system exhibits the most favourable behaviour in the case of randomly varying disturbances. However, situations exist when some factors which have a great influence on the operation of the system $S_1$ with the controls $Q_1$, depend on the action of the control device $Q_2$ of the system $S_2$ (Fig. 12.1). It may prove that the 'interests' of the systems $S_1$ and $S_2$ are in opposition to each other in the sense that any improvement in the functioning of the system $S_1$ is associated with a deterioration in the functioning of the system $S_2$, and vice versa. Then the control devices which strive to improve the functioning of 'their' system will generate control responses which should be as harmful as possible to the 'foreign' system. Such situations are called *conflicting situations*. Conflicting situations arise for instance in the struggle of living organisms for survival, for example, in economic competition, in military combat, and in many cases when the control problems must be solved, taking into consideration the purposeful counteraction of the improved functioning of the control system of the adversary. The existence of an intelligent adversary radically changes the nature of the problem forming purposeful control since whatever control is chosen, the adversary will take this into consideration and will still try to put us into the worst possible conditions.

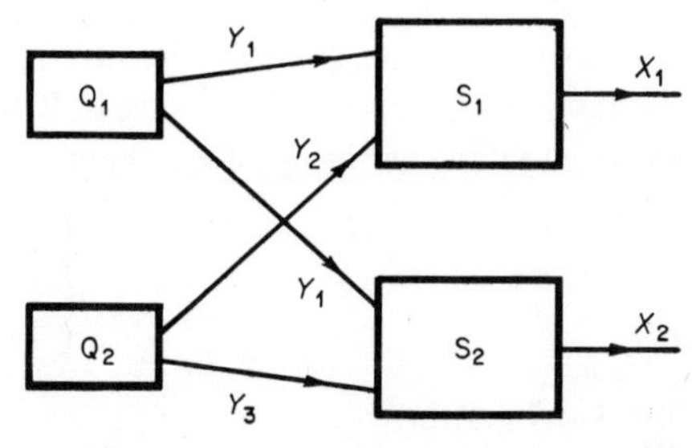


Fig. 12.1. *Cross-coupling effect of two control systems on the controlled plant.*

In a conflict situation, the control problem consists of generating reactions to a situation arising during combat, and a *strategy* for which the controlled system would be in the best possible position even in the case of the most unfavourable actions of the adversary. *Game theory* is concerned with developing methods for solving this type of problem.

## 12.1. The pair game. The minimax strategy

A simplified formalised model of a conflicting situation is called a *game*. A game characterises a set of rules which determine the permissible action of the sides participating in the game and the methods of determining their gains or losses. If two sides participate in the game (two players), it is referred to as a *pair game*, or a 2-person game. A zero-sum game is a game in which the algebraic sum of the losses and gains of all sides equals zero.

Let us consider a simple class of 2-person games with a zero sum. In such games the interests of the sides are directly opposed to each other because the gain of one player is equal to the loss of the other. Therefore we can limit ourselves to a thorough consideration of the game of one player, where the one player is striving to maximise his payoff, and his adversary to minimise it. The game consists in each adversary making one of a number of possible moves. It is assumed that the choice (of move) by each player is not known to his adversary. However, both adversaries know all the possible variants of their own actions and those of the other, and the size of the win for any of the combination pairs. Such games (referred to as *finite games*) can conveniently be represented as a matrix M, each element $a_{ij}$ of which indicates the amount of gain of the player A if he chooses the action $A_i$, and his adversary B chooses the action $B_j$. If the player A performs $m$ different actions and the player B performs $n$ different actions, then the game matrix will contain $m$ rows and n columns. Table 12.1.

Table 12.1 *Game m × n*

| | $B_1$ | $B_2$ | ... | $B_n$ |
|---|---|---|---|---|
| $A_1$ | $a_{11}$ | $a_{12}$ | ... | $a_{1n}$ |
| $A_2$ | $a_{21}$ | $a_{22}$ | ... | $a_{2n}$ |
| ... | ... | ... | ... | ... |
| $A_m$ | $a_{m1}$ | $a_{m2}$ | ... | $a_{mn}$ |

The matrix of the game is also the payoff matrix (its elements represent the payment to be made by the player B to the player A). By a matrix we simply mean a rectangular array of numbers, so that

$$\begin{bmatrix} 1 & 2 & 3 \\ 4 & 5 & 6 \\ 7 & 8 & 9 \end{bmatrix}$$

is a 3 × 3 matrix.

Solution of the game $m \times n$ means finding for each player such a strategy, i.e. a method of operation such that his average game over a large number of games is a maximum value. *The game theory recommends each player to select those actions for which the maximum possible gain can be obtained in the case of least favourable action of his adversary.* This is called the *minimax strategy.*

The most favourable (optimal) strategy $A_k$ of the player A can be determined from the game matrix by finding an element which satisfies the conditions

$$a_{kp} = \max_i \min_j a_{ij},$$

i.e. an element which can be selected as the maximal (along the rows) from minimal ones in each row (along the columns).

The most favourable strategy $B_l$ for the player B is determined from the element

$$a_{ql} = \min_j \max_i a_{ij}$$

i.e. an element which is minimal (along the columns) from the maximal ones (along the rows) in each column.

Let us consider the following example of a conflicting situation. A plant (player A) has to acquire a stock of black dyes for fabrics, the consumption of which depends on the market conditions (player B) during the coming season. The requirement may amount to 10 tons if the demand is low, 15 tons if the demand is normal, and 20 tons if the demand for black fabric is large. The cost of black dye when the decision is made is £100 per ton, which corresponds to a low rate of demand. If the demand increases to normal, the price of the dye will rise to £150 per ton and if the demand is high to £250 per ton.

How should the player A select one of the three available strategies: to buy 10, 15 or 20 tons of dye. Player B can ensure one of the three levels of demand: low, normal or high. The game matrix in this case has the dimensions 3 x 3. Its elements are the cost of purchasing the dye, which can easily be calculated from the data available for the nine possible strategy combinations of the two players. For instance, the element $a_{12}$ corresponding to a purchase of 10 tons of dye (strategy $A_1$) and normal market demand in the forthcoming season (strategy $B_2$), is calculated as follows: purchase of 10 tons of dye at £100 per ton will cost £1000. In addition, if the demand is normal it will be necessary to buy during the season an additional 5 tons at £150 per ton, so that the cost will be £1750, i.e. $a_{12} = -$ £1750. The complete matrix of the purchasing costs is as shown in Table 12.2.

It can be seen from this table that the minima of the rows are respectively –£3000, –£2500 and –£2000, and the maxima of the columns are –£1000, –£1500 and –£2000 and the maximal minimum of the rows (along the columns) coincides with the minimal maximum of the columns along the rows (both these equal $a_{22}$). Consequently, a reasonable solution is to buy 20 tons of dye at £100 a ton.

Table 12.2

| *Stocks of dye (tons)* | *Demand* | | |
|---|---|---|---|
| | $B_1$ *reduced* | $B_2$ *normal* | $B_3$ *high* |
| $A_1 = 10$ | $a_{11} = -1000$ | $a_{12} = -1750$ | $a_{13} = -3000$ |
| $A_2 = 15$ | $a_{21} = -1500$ | $a_{22} = -1500$ | $a_{23} = -2500$ |
| $A_3 = 20$ | $a_{31} = -2000$ | $a_{32} = -2000$ | $a_{33} = -2000$ |

In the given example $a_{kp} = a_{ql} = a_{kl} = a_{33}$, evidencing the presence in the game matrix of a point which is called the *saddle point*, which is simultaneously the maximum for player A and the minimax for player B. Such a coincidence does not always occur and in the general case the game matrix may not contain a saddle point.

The saddle point $a_{kl}$ has the following important property: if one of the players, for instance A, selects a strategy $A_k$ corresponding to the saddle point, then for the other player B any deviation from the strategy $B_l$ would be unfavourable because he can only maintain or increase the profit of A but he cannot reduce it. In the given example the player A can apply the strategy $A_3$ and his game will not be smaller than in the saddle point whatever strategy the player B uses. The optimal strategies $A_k$ and $B_l$ corresponding to the saddle point $A_{kl}$ are referred to as *pure strategies*. It is pointed out that if a saddle point exists, it is not necessary for it to be kept secret. Both players can openly show their selected strategies.

## 12.2. Mixed strategy. The value of a game. Domination.

If the game matrix does not contain a saddle point, then no purely optimal strategies exist for each of the players. Such games have a more complex solution.

Let us consider the example of a game 2 x 2, which has no saddle point. Let us assume that the player A is commander of a military unit which is to attack one of two defence positions of the enemy (player B). The enemy can successfully defend only one of these positions but not both at the same time. It is known that the first position is three times more important than the second. How should the two protagonists proceed?

We will denote the strategies of attack and defence of the first position by the index 1, of the second by the index 2. Then the payoff matrix will have the form

| | $B_1$ | $B_2$ |
|---|---|---|
| $A_1$ | 0 | 3 |
| $A_2$ | 1 | 0 |

(If B defends the structure which is attacked by A, then the game of A will equal zero). It can easily be seen that this matrix has no saddle point:

$$\max_i \min_j a_{ij} = 0, \min_j \max_i a_{ij} = 1.$$

The use of pure strategies in the given case is unfavourable for the players. Indeed, if A always adheres to one type of action (for instance, attacks the more important position), then, knowing this, B will reduce the gain of A to zero. In the same way, if B will, for instance, always defend his most important position (which one would assume is dictated by 'common sense') then A may be successful in winning 1 – capture of the second position. Therefore the one who applies pure strategy will be in a poorer position than his more 'flexible' adversary. It is obviously more favourable to select, secretly, sometimes one and sometimes another pure strategy, not according to any previously known relationship but at random, for example using dice or a table of random numbers. For instance, if the player A attacks the positions with an equal probability, then whatever B may do, the average gain of A will be less than 0.5, i.e. it will in any event be larger than zero. The question arises, what are the optimal probabilities of using pure strategies? In the case of 2 x 2 games the game theory yields a relatively simple answer: the probabilities of pure strategies should be calculated according to the formulae

$$p(A_1) = \frac{a_{22} - a_{21}}{(a_{11} + a_{22}) - (a_{12} + a_{21})}, p(B_1) = \frac{a_{22} - a_{12}}{(a_{11} + a_{22}) - (a_{12} + a_{21})}$$

$$p(A_2) = \frac{a_{11} - a_{12}}{(a_{12} + a_{22}) - (a_{12} + a_{21})}, \quad p(B_2) = \frac{a_{11} - a_{21}}{(a_{11} + a_{22})(a_{12} + a_{21})} \qquad (12.1)$$

For our problem we obtain the following probability:

$$p(A_1) = 1/4, \quad p(A_2) = 3/4$$
$$p(B_1) = 3/4, \quad p(B_2) = 1/4.$$

The average gain of A (i.e. the loss of B) will equal 0.75 in this case. *The strategy which consists of random application, with certain probabilities, of some pure or other strategies is called mixed strategy.* Mixed strategy is given with indications of probabilities of using the pure strategies forming part of it. In our case the optimal mixed strategy of the players $S_A$ and $S_B$ can be written in the form

$$S_A = \begin{vmatrix} 1/4 \\ 3/4 \end{vmatrix}, \quad S_B = \begin{vmatrix} 3/4 \\ 1/4 \end{vmatrix}$$

where the pure strategies are numbered from the top downwards.

In contrast to the case of games with a saddle point, the selection of a specific action in the case of a mixed strategy must remain unknown to the adversary, though it is also true that the optimal mixed strategy of the player can always be calculated by his adversary on the basis of the game matrix.

An important property of the optimal mixed strategy is that for any strategy of his adversary (pure or mixed) the player is ensured an average gain not less than if his adversary uses his optimal mixed strategy. This average gain, which can be achieved by a 'good' player from a 'good' player is called the *game value*. In the given example the game value is 0.75 (in favour of A). In a game with a saddle point the game value equals the payoff in the saddle point. If the game value equals zero, the game can be considered 'just'. If this is not the case, one of the players (for whom the game value is negative) would be better advised to stay away from the game, if this is possible. This, however, would also represent a pure strategy which would change the game to a game with a saddle point and a value equalling zero.

It is pointed out that the optimal strategy of the players does not change if all the payoff values are multiplied by a constant number (if, figuratively speaking. they are expressed in a different currency) or if a constant number is added to them. Obviously the *game value* will change appropriately.

The solution of games with a number of pure strategies in excess of 2 x 2 is complicated. Von Neumann has, however, proved that each $m \times n$ game has at least one solution in the form of optimal (pure or mixed) strategies for both players.

We will show the property of the payoff matrix, which in some cases permits reduction of the game to one with a smaller number of strategies. Consider for instance the game with a 3-person matrix as given in Table 12.3

Table 12.3

| | $B_1$ | $B_2$ | $B_3$ |
|---|---|---|---|
| $A_1$ | 2 | -1 | 2 |
| $A_2$ | 3 | 4 | 0 |
| $A_3$ | 1 | 2 | -2 |

We will compare the payoff values corresponding to strategies $A_2$ and $A_3$. It can easily be seen that whatever strategy the player B chooses, the strategy $A_2$ gives the player A a larger gain than $A_3$. In this case we say that the strategy $A_2$ *dominates* over the strategy $A_3$. It is obviously inadvisable for the player A to use the strategy $A_3$ and therefore this need not be considered. In the same way, comparing the strategies $B_2$ and $B_3$, it is obvious that $B_1$ is less favourable for B.

Table 12.4

| | $B_2$ | $B_3$ |
|---|---|---|
| $A_1$ | -1 | 2 |
| $A_2$ | 4 | 0 |

To be more accurate, it is *no better*, because in the case ot the strategy $A_1$ the loss to B is the same. Therefore it is sufficient to consider the matrix shown in Table 12.4.

Domination has enabled the 3 x 3 game to be reduced to a 2 x 2 game. Using the formulae (12.1) and taking domination into consideration, we find the optimal mixed strategy for the initial game:

$$S_A = \begin{vmatrix} 4/7 \\ 3/7 \\ 0 \end{vmatrix}, \quad S_B^{\cdot} = \begin{vmatrix} 0 \\ 2/7 \\ 5/7 \end{vmatrix}$$

The game value = 8/7. A more complicated case of domination may occur if any pure strategy is less favourable than a mixed strategy composed of the other pure strategies.

The simplest types of zero-sum finite pair games have been described. Games with a larger number of strategies are much more complicated, in particular the so-called *infinite games* (with an infinite number of pure strategies) and also games between many people, in particular *coalition games* where the players form a coalition for the purpose of increasing the overall gain. The achievements of the game theory are being extensively applied in a great variety of fields, such as in the analysis of the behaviour of people and social groups in society, selection of an optimal form of action in the case of incomplete information on an object (the so-called 'game against nature'), and some biological problems.

## 12.3. Machine games

In order to take decisions in conflicting situations as follows from game theory, it is necessary to carry out a number of calculations and logical operations. These operations consist in calculating the elements of the game matrix, separating the maximum and minimum numbers from some set of numbers, calculation of the probabilities of using pure strategies which form part of mixed strategies. Obviously, this type of work is well within the potentialities of a digital computer, which can solve game problems.

To realise mixed strategies the computer must also have available a source of random signals with certain statistical characteristics, in particular those which give a random sequence of strategy numbers with given values of the frequency of each number.

A random physical process, for instance radioactive decomposition or thermal noise of a cathode of an electron tube, can serve as a source of random signals. Random signals can also be obtained from random value tables stored in the computer memory or generated by means of a specific algorithm (the algorithm for generating *pseudo-random* numbers).

Due to their high speed of operation, digital computers are capable of finding and realising optimal strategies in complicated cases (for large values of $m \times n$) much more rationally than can be done by man. Another advantage of the computer is that it can easily achieve 'unpredictability,' i.e. absolute randomness of the successive steps when applying mixed strategies, whilst man usually supplies sequences of steps in which a certain regularity can always be detected, even if he tries to combine the steps in a random manner. It is for this reason, in particular, that many competitions of man against a computer have proved the evident superiority of computers, at least in single move games.

The position is very different when the computer is required to carry out a multi-move strategic game, where the gain is determined as a result of a relatively long sequence of moves by the adversary, as is the case in chess, for instance. Here the computer is inferior to man, who uses accumulated knowledge, inventiveness and an economic selection of alternative moves based on heuristic methods, which generalise great experience in solving tactical and strategic problems.

Success in choosing a strategy for multi-move games depends largely on the possibility of sorting out the alternatives by the machine, which is determined by speed of operation and memory capacity. In playing chess, the necessary number of operations and the required memory capacity increase very rapidly with increasing number of moves for which the calculation has to be made. To determine the results of a game four moves in advance, the computer must examine a number of alternatives of the order of $10^{12}$, which involves great difficulties even for powerful modern computers.

The successful operation of computers in such situations also depends to a considerable extent on the selection of the *estimation function*

$$y = F_0(x)$$

where $y$ is a number characterising the value of the intermediate position from the point of view of getting nearer to achieving the aim, and $x = \{x_1, x_2, \ldots, x_n\}$ – factors which have to be taken into consideration when estimating the position. In a game of chess these factors are the forces of both sides and their positions on the chessboard. In military operations the factors which characterise the position are the manpower, the equipment, and their deployment. In game models of military operations the economic potential, morale and other factors can also be taken into consideration.

Investigations on writing programmes for games of chess, draughts and other strategic games are in progress in scientific establishments in various countries all over the world. The interest in developing and improving game machines is associated with the prospects of using game methods for solving important problems in the distribution of capital investment in industry, transportation and agriculture, utilisation of natural resources, choosing strategies in military operations and with many other topical problems.

**Exercises**

**1.** Find $\max_i \min_j a_{ij}$ and $\min_i \max_j a_{ij}$ for the following matrices:

(a) $\begin{vmatrix} 1 & 3 & 4 \\ 2 & 1 & 3 \\ 5 & 2 & 1 \end{vmatrix}$ (b) $\begin{vmatrix} 1 & 8 & 4 & 6 \\ 2 & 6 & 17 & 81 \\ 9 & 15 & 0 & 5 \\ -2 & 1 & 7 & 3 \end{vmatrix}$

*Solution.*

(a) $\max_i = 5, 3, 4$ $\min_f \max_i = 3$ (b) $\max_i = 9, 15, 17, 81$ $\min_j \max_i = g$

$\min_j = 1, 1, 1$ $\max_i \min_j = 1;$ $\min_j = 1, 2, 0, -2$ $\max_i \min_j = 2$

**2.** In the game matrices described below, find the saddle points:

(a) $\begin{vmatrix} 2 & 4 \\ -1 & 8 \end{vmatrix}$ (b) $\begin{vmatrix} 2 & 6 & 1 & -1 \\ 3 & 5 & 8 & 4 \end{vmatrix}$ (c) $\begin{vmatrix} 5 & 11 & 5 & 5 \\ 6 & 3 & 4 & 4 \\ 1 & 2 & 3 & 8 \end{vmatrix}$

*Solution.*

(a) $a_{11} = 2$ (b) $a_{21} = 3$ (c) $a_{13} = 5$

**3.** The two players A and B write simultaneously and independently of each other one of the three numbers: 1, 2 and 3. If the sum of the written numbers is even, then B pays to A the same sum in dollars (pounds); if it is odd, then A pays this sum to B. Write the game matrix.

*Solution.* The player A has three strategies: $A_1$ – to write 1, $A_2$ – to write 2, and $A_3$ – to write 3. The player B also has three similar strategies. The game matrix 3 x 3 is as written in Table 12.5.

Table 12.5

| | *B1* | $B_2$ | $B_3$ |
|---|---|---|---|
| $A_1$ | 2 | −3 | 4 |
| $A_2$ | −3 | 4 | −5 |
| $A_3$ | 4 | −5 | 6 |

**4.** When programming computers for games of draughts or chess it is necessary to analyse and evaluate the moves to be taken after each move of the adversary.

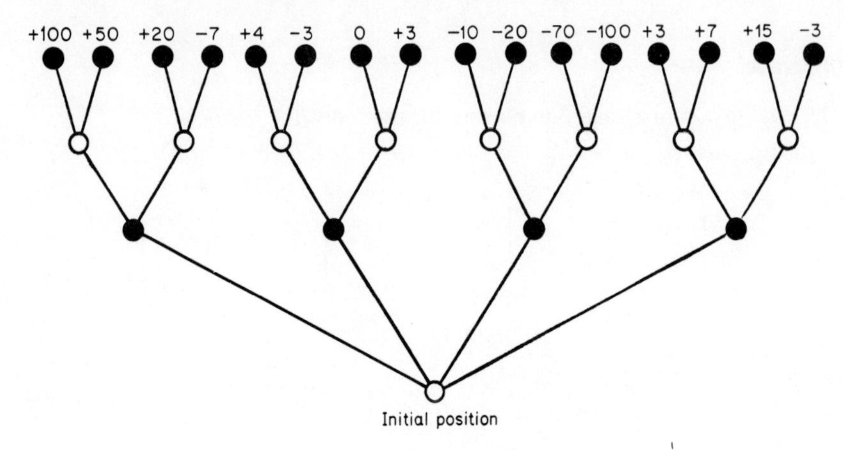


Fig. 12.2. *To Example No. 4.*

The task of analysis for each of them consists in following all the variants of subsequent moves from a given position on the board, several moves in advance, and to select the optimum variant. For this position, which arises after each move, an evaluation by some method or another and a selection is made of the minimum move, i.e. of the move which produces a position with a minimal of the maximum possible evaluations. Fig. 12.2 shows the so-called analysis tree of continuations for the first move of the white chessmen. For positions which occur as a result of analysing three moves in advance, evaluations are given which have been obtained by some method. Using the minimax strategy, determine the best moves for the white chessmen from the initial position.

*Solution.* If we consider the positions which arise on analysing three moves in advance, the best position of the white chessmen will be evaluated at + 100. However, this position can only be reached if the black chessmen make a move which is unfavourable for them. Applying similar considerations for the black

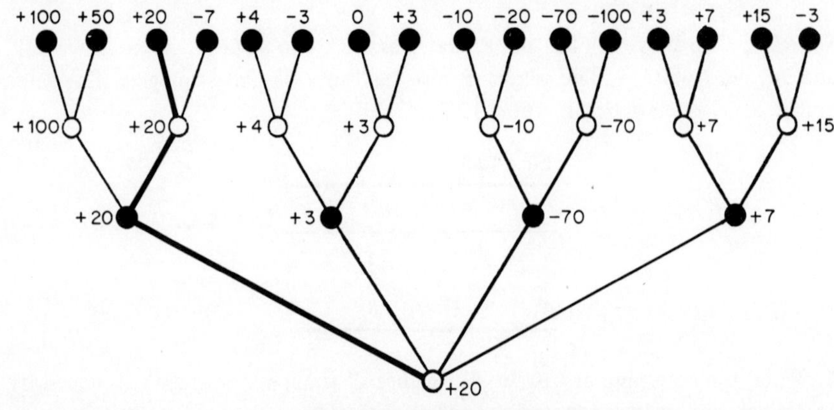


Fig. 12.3. *To Example No. 4.*

ones, it is possible to find the optimal path of movement along the tree from the initial to the final position. In the given case the optimal path terminates in the position with the estimated value + 20. The optimal path and intermediate evaluations are shown in Fig. 12.3.

5. A patient is suffering from a dangerous illness which can have three different forms. In this case it is not possible to establish the form. The relevant statistics are also not known. The doctors can use three different methods of treatment: X-ray therapy, surgery, and a new chemical preparation. X-ray therapy cures 30% of patients suffering from forms I and III and 10% suffering from form II of the disease. Surgery is successful for 20% of the cases of form I and 50% of forms II and III. Chemical therapy helps 25% of patients suffering from form I and form II, and 50% suffering from form III of the disease. It is not possible to give all three treatments at the same time. What is the optimal strategy of the doctors? What are the patient's chances of recovery if the optimal strategy is applied?
*Hint.* Games of the 2x $m$ type have an optimal strategy which is based on no more than two pure strategies, i.e. they can be reduced to 2 x 2 games.

*Solution.* The matrix of this 'game with nature' can be written as follows:

| | $B_1$ | $B_2$ | $B_3$ |
|---|---|---|---|
| $A_1$ | 30 | 10 | 30 |
| $A_2$ | 20 | 50 | 50 |
| $A_3$ | 25 | 25 | 50 |

where $A_1$ = X-ray therapy, $A_2$ = surgery, $A_3$ = chemotherapy, $B_1$ = form I of the disease, $B_2$ = form II of the disease, $B_3$ = form III of the disease.

Obviously, $B_1$ dominates over $B_3$. Excluding the strategy $B_3$, we obtain a 2 x 3 game. In this game the strategy $A_3$ may prove poorer than the mixed strategy $A_1$ and $A_2$. Therefore we consider the 2 x 2 game with the matrix

| | $B_1$ | $B_2$ |
|---|---|---|
| $A_1$ | 30 | 10 |
| $A_2$ | 20 | 50 |

This matrix has the solution $S_A = \begin{vmatrix} 3/5 \\ 2/5 \end{vmatrix}$, $S_B = \begin{vmatrix} 4/5 \\ 1/5 \end{vmatrix}$.

We verify this solution by calculating the average payoff, using the initial game matrix

| | $\frac{4}{5}$ | $\frac{1}{5}$ | 0 | |
|---|---|---|---|---|
| $\frac{3}{5}$ | 30 | 10 | 30 | 26 |
| $\frac{2}{5}$ | 20 | 50 | 50 | 26 |
| 0 | 25 | 25 | 50 | 25 |
| | 26 | 26 | 30 | |

(At the left and top the strategy probabilities are shown, at the right and bottom – the average payoff on selecting a given pure strategy as compared to the found mixed strategy of the adversary). Verification shows that the strategies $B_3$ and $A_3$ have justifiably been excluded. Thus the doctors should apply X-ray therapy with a probability of 3/5 surgery with a probability of 2/5, and they should not use the chemical preparation at all. The payoff of the game equals 26. For the most unfavourable statistics of the disease the average chance of recovery of the patient in the optimal strategy is 26%. It is pointed out that the method $A_1$ shows a greater probability of recovery, although at first sight it would seem that the results of using it would be less favourable.

# 13 Learning

The process of learning usually consists in the transmission of either knowledge, or methods of solving problems, or both, from teacher to learner. Two radically different methods can be applied for teaching people to solve problems. The first consists in communicating to the learner the algorithm for solving problems. The second is based on learning by example.

Thus, teaching people or machines how to perform arithmetic or logical operations is achieved by explaining the appropriate algorithm or by compiling a programme which realises this algorithm. However, the solution of problems such as reading, involving the recognition of letters or numbers, printed in various type faces or handwritten, is not taught by explaining the structure of these symbols or describing the mechanism of recognition. Such mechanisms may even be unknown to the teacher who uses learning by example.

Learning by use of examples is of enormous importance to the survival of many species in the animal world. By teaching their offspring to acquire food and avoid danger, animals do not communicate to their offspring methods of solving these problems, but they do always use the method of teaching by example. The major part of habits and methods which enable people to solve a variety of everyday problems are also acquired by observation or by analogy, and without necessarily obtaining a description.

Until quite recently, the solution of a variety of problems by automatic machines, including such well-developed automatic machines as digital computers, was in accordance with a programme which explicitly contained the algorithm for solving each problem. It can be assumed that man, after determining the structure of the automaton or working out a programme for the digital computer, *teaches it* to solve problems of a certain type using the first method of teaching. This method has proved unsuitable for such problems where the algorithm is unknown to us, even though we can solve these problems intuitively. For instance, man can easily learn to distinguish a cat from a dog or to recognise his friend or to catch a ball in flight, but he cannot write a programme which would enable a machine to do the same thing.

The desire to expand the potentialities of machines has brought about attempts by scientists and engineers to reproduce in machines the ability to learn as a result of having been given examples of solving various problems. This work has already yielded positive results.

## 13.1. Pattern recognition

The first successful attempt to teach machine by way of examples was the development of an automaton which is capable of learning to recognise visual images, such as geometric patterns, letters, numbers and other symbols. The problem of pattern recognition is as follows. There exists a set $X$ containing a very large number $n$ of various objects

$$X = \{x_1, x_2, \ldots, x_n\},$$

which belong to a relatively small number $m$ of given classes $X_1, X_2, \ldots, X_m$. We will consider that the automaton A solves the problem of recognition if it always, or almost always, relates each object $x$ to the given class $X_i$.

If, for instance, the set $X$ represents an agglomeration of all the possible configurations of numbers, then the problem consists in relating each configuration to one of the ten classes: 0, 1, 2, . . ., 9, each of which represents the *pattern* of the appropriate number. The set $X$ may contain different geometrical figures amongst which it is necessary to distinguish patterns belonging to the class of triangles, squares, circles, etc.

An automaton intended for the recognition of visual patterns should obviously contain input equipment which receives information about the object to be recognised. The input equipment of such an automaton is called the *receptor field.* It can be formed from a mosaic of photocells onto which the image to be recognised is projected. If the receptor field consists of $r$ elements, each of which can be in one of two states (excited or unexcited), than the number of possible configurations at the input will be $n = 2^r$. The number of possible configurations $n$, even for a small number $r$ (of the order of 10), is so large (of the order of millions) that it is practically impossible to place information on each individual configuration into the computer memory.

The output equipment of a pattern recognition automaton should contain $m$ outputs. From a knowledge of which of the inputs is excited, it is possible to judge the class in which the automaton has classified the object presented to it.

For the automaton to be capable of learning it must process a sufficiently large number $N$ of possible internal states $z = \{z_1, z_2, \ldots, z_N\}$, which should include such states in which the automaton A classifies objects in a required manner.

The problem of teaching such an automaton consists in presenting to the automaton a relatively small number $l \leqslant n$ examples of relating objects $x$ to certain classes $X_i$, bringing the automaton A into one of the states $z$ in which it will realise the required classification of the objects, including also those objects which have not been presented to it during the process of learning. It has been found that if such a state of the automaton is chosen in which it will correctly classify a sufficiently large number $s$ of objects selected at random from the set $X$, then it will also classify sufficiently well, with a certain probability, all other objects of the given set $X$. Thus the process of teaching an automaton A to recognise the general outlines of patterns consists of the following.

By acting on its receptor field, $l$ objects chosen at random from the set $X$ for the purpose of learning are presented to the automaton A, Fig. 13.1. The teacher, a reference automaton $A_r$ (man may also play the role of the reference automaton), shows the automaton A to which class of the learning sequence $x_{j1}$, $x_{j2}, \ldots, x_{jl}$ each of the objects belongs.

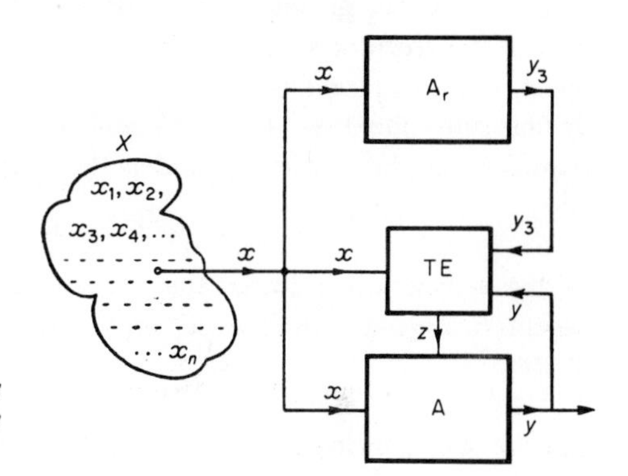


Fig. 13.1. *Sketch showing the process of teaching an automaton to recognise patterns.*

The teaching equipment TE compares the reaction $y_r$ of the reference automaton $A_r$ with the reactions $y$ of the learning automaton A during the process of teaching and changes the state $z$ of the automaton A so that its reactions should coincide with the reactions $A_r$ as often as possible. Then a randomly selected test sequence of objects $x_{k1}, x_{k2}, \ldots, x_{ks}$ is presented to the automaton. If the automaton did not make a larger number of errors on the second test sequence than is permissible, it is considered that the learning process is terminated; otherwise further teaching of the automaton A is carried out, followed by further examination by test sequences. This procedure is continued until the required reliability of recognition is achieved or until it is found that the given automaton cannot be taught to recognise the given objects with the required reliability.

The effectiveness of the process of learning depends largely upon the following three factors: (a) the method of mapping the object into a configuration on the receptor field of the automaton, i.e. what the outlines of the real objects presented at the input of the automaton A are and how they are coded; (b) the multitude of transformations of the input configuration into the output reaction of the automaton, i.e. the number of possible differences in its states; (c) the algorithm of the operation of the learning equipment.

The mapping, or *representation*, should be selected so that the input of the classifying equipment contains information which is adequate for classification. The variety of transformations should be large enough to enable the automaton to learn solutions of a sufficiently wide class of problems, but not too large so as not to produce excessive difficulties in finding the necessary transformation. The operating algorithm of the teaching equipment should ensure the highest possible reliability of recognition with the shortest possible learning sequence.

The problem of getting teaching machines to recognise patterns can be interpreted geometrically. Each configuration on the receptor field, consisting of $r$ receptors, can be represented by a point in the $r$-dimensional space of the receptors, along the coordinate axes of which the degree of excitation of each receptor, $\alpha_1, \alpha_2, \ldots, \alpha_r$, is recorded.

If the number of objects $X$ is divided into given classes, then it is possible to plot in the receptor space a surface which divides this space into areas, each of which contains points representing only objects of a single class. If the set $X$ is divided into the two classes $X_1$ and $X_2$, the problem of recognition can be considered as the problem of finding the equation

$$F(\alpha_1, \alpha_2, \ldots, \alpha_r) = 0$$

of the surface which divides the receptor space into two areas such that all the objects belonging to the class $X_1$ are in the area

$$F(\alpha_1, \alpha_2, \ldots, \alpha_r) > 0,$$

and the objects belonging to the class $X_2$ are in the area

$$F(\alpha_1, \alpha_2, \ldots, \alpha_r) < 0,$$

as is shown diagrammatically in Fig. 13.2. Each state $z$ of the automaton A can be considered as one of the possible variants of the separating surface, and the learning consists in choosing a surface so that it ensures the division of the objects into classes $X_1$ and $X_2$ which are sufficiently near to the division realised by the teacher or *reference* automaton $A_r$.

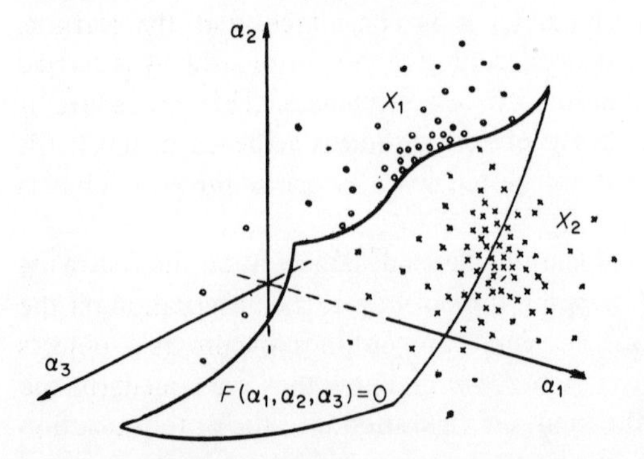


Fig. 13.2. *Surface, separating patterns in the receptor space.*

A large number of algorithms for machines teaching pattern recognition have been worked out. All these are based on comparing the outputs of the pupil and the reference automata and switching over of the teaching automaton (by changing its parameters) into states in which the number of divergences between the outputs of the automata A and $A_r$ decreases.

An example of a simple algorithm for dividing objects into two classes $X_1$ and $X_2$ is the following procedure. The representatives of both classes are coded in the binary code. Then the code combinations obtained are written in a row, one after the other in the form of a table. It is assumed that a certain number of representatives of the classes $X_1$ and $X_2$ exist which were chosen for the process of learning and make-up the learning sequence. For each column of the table of the learning sequence the empirical (experimental) frequency of appearance of 1 and 0 is calculated. For instance, if in the first column of the table there is a single 1 and 9 zeros, then the frequency of 1 appearing is $p_1 = 1/10$, and of zero $p_0 = 9/10$. Then the logarithms of these values are taken and a horizontal table is made with a number of cells equalling the length of the code combination. Into the top row of the table the differences $R = \log p_{01} - \log p_{02}$, and into the bottom of the table the differences $R_1 = \log p_{11} = \log p_{12}$, calculated for each column of the table with the learning sequences classes $X_1$ and $X_2$ are recorded. The procedure of constructing such a table imitates the process of learning in the machine.

Then follows the test. The finished table is added to the table with the code combinations in such a way that the number of their cells along the horizontal should coincide, and the following operation is performed: from the row $R_0$ of the table, first the numbers against which the code contains zeros, then from the sequence $R_1$ the numbers against which there is the figure 1 are written out. All these numbers are added. If their sum is larger than zero it is considered that the given object belongs to class $X_1$, and if it is less than zero, to the class $X_2$.

The theory and technique of designing systems which learn to recognise patterns are still in their infancy. However, even now it is possible to build devices and programmes for digital computers which permit the successful solution of such difficult problems of learning machines as the recognition of handwritten letters and figures, interpretation of geological data, recognition of some speech sounds, and diagnosis of some diseases. In a number of cases the machine proved capable of learning to solve the problem of recognition much better than Man can.

## 13.2. Learning behaviour

As a model of a system which learns, it is convenient to consider the probabilistic automaton A. The behaviour of the automaton is characterised by the probability of a certain reaction observed at the output of the automaton to a stimulus applied to its input. We shall limit ourselves to simple situations, when a finite number of possible stimuli $x_1, x_2, \ldots, x_n$ are known and all the possible reactions $y_1, y_2, \ldots, y_m$ expressing the behaviour of the automaton can be described by the probability matrix $M$ (Table 13.1).

Table 13.1

| | $y_1$ | $y_2$ | ... | $y_m$ |
|---|---|---|---|---|
| $x_1$ | $p_{11}$ | $p_{12}$ | ... | $p_{1m}$ |
| $x_2$ | $p_{21}$ | $p_{22}$ | ... | $p_{2m}$ |
| ... | ... | . . | ... | ... |
| $x_n$ | $p_{n1}$ | $p_{n2}$ | | $p_{nm}$ |

The $p_{ij}$ elements of this matrix characterise the probability that this automaton A will respond in a state $z$ to the stimulus $x_i$ by the reaction $y_j$. If the set $y_1, y_2, \ldots, y_m$ contains a complete selection of all the possible reactions and the automaton always responds to the input stimulus with some output reaction, then the sum of probabilities of all the reactions to each stimulus equals one:

$$\sum_{j=1}^{j=m} p_{ij} = 1 \quad (i = 1, 2, \ldots, n)$$

To each of $n$ possible states $z_1, z_2, \ldots, z_n$ of the automaton A there corresponds a certain matrix $M(z)$ which determines its behaviour.

Each individual act of behaviour of the automaton A is determined by the pair $(x_i, y_j)$, i.e. the $j^{\text{th}}$ reaction to the $i^{\text{th}}$ stimulus, and to each such pair an evaluation $\alpha_{ij}$ can be made to correspond, which characterises the effectiveness of the behaviour of the automaton. Therefore we can consider the process of learning by the automaton to be its transition in such states where the probability of obtaining high values increases.

The described model of a learning automaton is useful when studying processes of learning in certain forms of behaviour of man and animals, and also for building artificial learning control devices.

It has been shown that probabilistic automata can learn purposeful behaviour if they are 'rewarded' or 'punished' by the teacher or the medium, and if every time when after a certain transition the automaton was 'punished' the probability of transition into this state decreases, but in all other cases it increases. In this way it became possible in particular to teach a probabilistic automaton to find quickly the minimum of a function of two variables. This problem reduces to a search for the optimum conditions of operating a system whose effectiveness depends on two control variables. The automaton selected the direction and magnitude of the rate of displacement of the representative point on the plane $X_1$, $X_2$ (Fig. 13.3) in accordance with the changes in the function $y = R(x_1, x_2)$.

The lines of levels $y$ are shown in Fig. 13.3. The 'punishment' of the

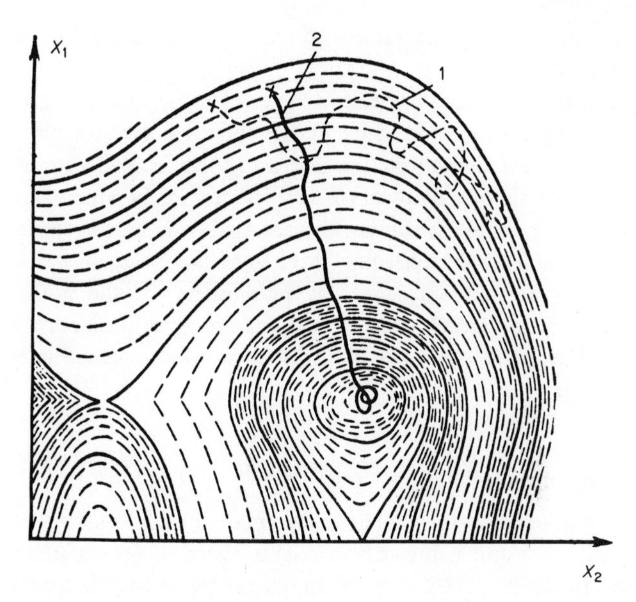


Fig. 13.3. *Search by the automaton for the minimum of a function of two variables* y = R $(x_1, x_2)$.

automaton was smaller the faster the representative point approached the minimum $y$. Experiments have shown that the effectiveness of such a system of learning, which can be judged by comparing the trajectory 1 of movement of the representative point before learning with the trajectory 2 after learning. It can be seen from Fig. 13.3 that the automaton learned to search purposefully for the necessary values of the variables $x_1$ and $x_2$.

Similar methods can be used for teaching automata to solve many complex problems and to develop a purposeful behaviour in various complex situations.

The effectiveness of the learning behaviour of an automaton can be considerably increased if between the stage of learning by examples and the test, a *training* period is inserted. The training may consist, for instance, in verifying the performance of the automaton many times on one and the same learning sequence, the reactions of the medium or the reward and punishment by the teacher being used for additional teaching of the automaton to develop the required behaviour.

It would be impossible to teach either automata or even people who develop complex forms of behaviour, if the pupil had to master the methods of complex behaviour only in terms of rewards or punishments.

In the final analysis, for an automaton to be capable of learning complex behaviour the automaton itself must be very complicated and the number of its possible states must be very large. In such a case, however, it would be an extremely difficult task to find amongst the enormous number of states, one of the few states in which this complex automaton has the required properties, since this would involve the extremely laborious process of sorting out the states of the automaton.

This task can be made considerably easier if, during the process of learning intermediate sub-goals are used which permit the division of learning into stages. During the first stages the automaton learns to solve relatively simple problems, and then on the basis of structures already formed in the automaton which are adapted for solving simple problems, the automaton is taught to sove problems of increasing complexity at each new stage until the final aim is achieved. This learning strategy for an automaton is reminiscent of the normal order of teaching people who at each stage of learning are given problems of increasing complexity which they can solve due to the knowledge gained in the previous stages of learning.

### 13.3. Teaching machine

Teaching people consists in organising the process, as a result of which people master certain facts and habits which enable them to use their knowledge to solve various problems. These may be simple problems such as typing, punching cards, or more complex ones such as the translation of a text from one language to another or finding faults in a radio receiver or solving mathematical problems. Whatever the nature of the activity for which man prepares himself by learning, the aim is to accumulate some knowledge and some 'skill'.

Until recently people were taught only *by people* who occasionally made use of films, sound recordings and visual aids. In the light of modern achievements in cybernetics teaching can be considered as a controlling process, obeying laws which are characteristic of control processes; it is a process which can be optimised and automated to a certain extent by means of appropriate technical equipment.

*The technical equipment used for certain stages of teaching by cooperation with the pupil, without the direct participation of a teacher, is called a teaching machine.*

The teaching is usually arranged in accordance with the diagram shown in Fig. 13.4. It consists of elements of the type O which represent stages of teaching, elements of the type E which represent examination, and connections which designate the sequence of the teaching stages. The diagram contains two types of connection: C, which is the correct answer to examination questions and I, the incorrect answer. It can be seen from Fig. 13.4 that the sequence of learning depends on the results of examinations and permits the transition to the next stage of learning only if the results of the examination are positive. Otherwise the previous stages of learning have to be repeated until the pupil has mastered the material sufficiently well.

Such a scheme of teaching can be realised not only by a teacher but also by a teaching machine programmed in accordance with Fig. 13.4. The required teaching sequences are presented to the pupil by projecting onto a screen or

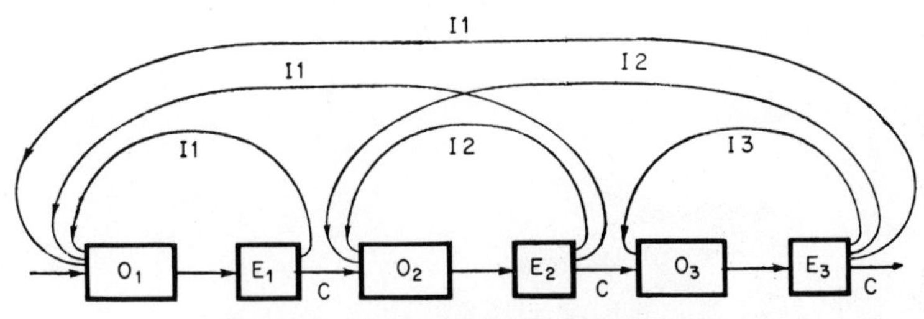


Fig. 13.4. *Diagram of the teaching process.*

onto a cathode ray tube (television receiver) the necessary information (text, drawings, photographs). The test is by one of the following two methods: presentation of alternative replies, or self-testing. In the first method each question is accompanied by a set of possible answers of which one is correct, and the pupil has to indicate, by pressing an appropriate button, which of the answers he considers to be the correct one. In the second method the pupil is required to write his answer and then to press a button which will produce the correct answer on the screen, and after that he can compare the answers. Depending on the results of the test the teaching is continued is accordance with the diagram Fig. 13.4.

For teaching *habits*, various types of *trainers* are used which simulate the conditions of problems which the learner must solve. This is used in the training of pilots of aircraft and ships, for teaching to tune electronic circuits, working with punches (perforators), etc.

The most successful teaching of habits is achieved if the character and tempo of teaching are chosen to fit the individual features of the pupil: his abilities, temperament, accumulated experience in solving similar problems. The individual approach to the learner can be realised by means of *self-adapting* teaching machines. One of the principles used in the design of self-adapting teaching machines consists in the automatic measurement of a set of problems with a sequence and speed depending on the nature and frequency of errors made by the pupil.

For instance, when a pupil is learning to type, the words or symbols given to him during his training can be so varied that they more frequently contain symbols for which the pupil makes errors. In addition, the speed of dictation can be automatically reduced if the number of mistakes made by the pupil over a given period exceeds a certain value. In this way the 'pupil – teaching machine' system establishes a dynamic equilibrium characterised by a certain relationship between the accuracy and speed of solution of the problems at which the teaching process is optimal.

Teaching a large number of pupils, giving individual attention to each of them, can be organised by using a digital computer programmed according to the task and method of teaching. Obviously, in this case special apparatus must be fitted for interchanging information with the pupil and for monitoring the teaching.

*(a) AutoTutor Teaching Machine used in secondary schools for the teaching of mathematics.*

*Teaching machines. (Photographs by courtesy of ESL Bristol.)*

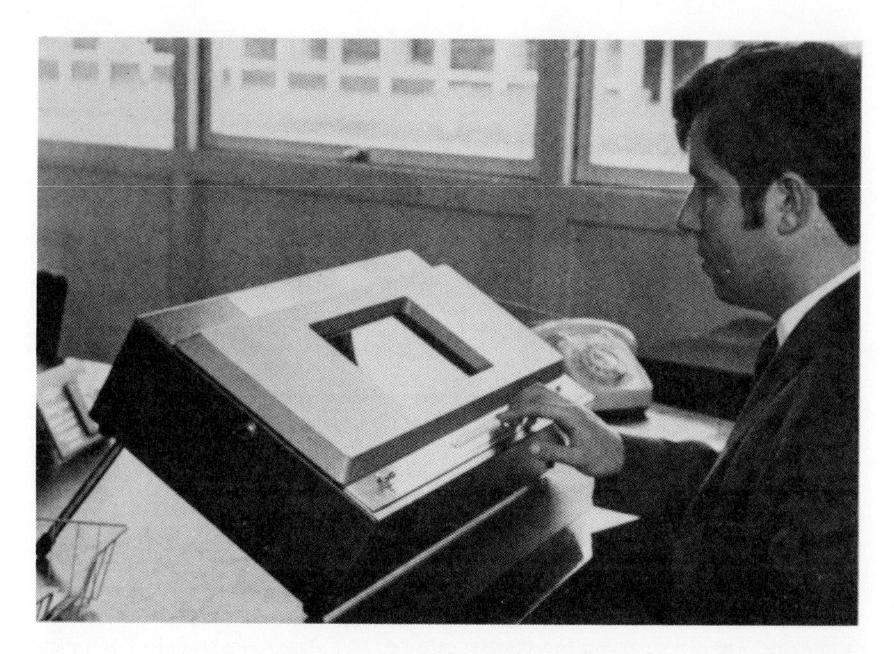

*(b) Bristol Tutor Teaching Machine used in management training.*

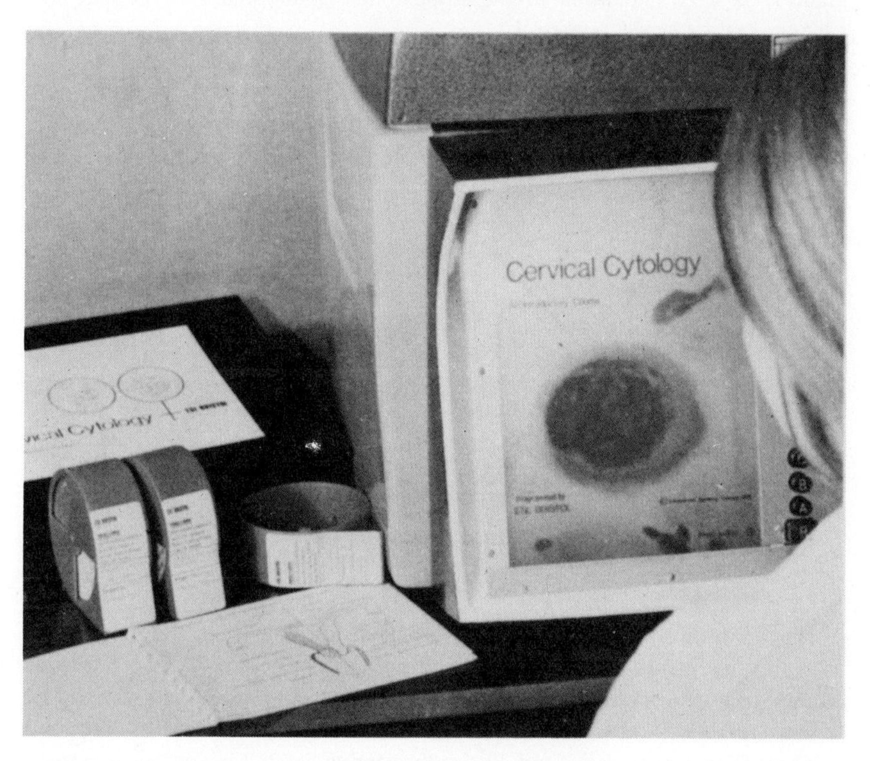


*(c) AutoTutor Teaching Machine used in the training of technicians in cervical cytology.*

## Exercises

1. In a two-dimensional space with the coordinates $x_1$ and $x_2$, two classes of points are given. In Table 13.2 the coordinates (im mm) of the representatives of these classes are given.

Work out graphically and write the equations of the lines separating the two classes.

Table 13.2

| | | | | | | | | | | | | |
|---|---|---|---|---|---|---|---|---|---|---|---|---|
| *1st Class* | | | | | | | | | | | | |
| $x_1$ | 30 | 35 | 40 | 35 | 35 | 45 | 45 | 50 | 55 | 60 | 55 | 60 |
| $x_2$ | 25 | 75 | 60 | 45 | 35 | 40 | 55 | 65 | 70 | 60 | 50 | 65 |
| *2nd Class* | | | | | | | | | | | | |
| $x_1$ | 30 | 40 | 45 | 55 | 65 | 55 | 65 | 70 | 80 | 70 | 85 | |
| $x_2$ | 15 | 20 | 30 | 30 | 35 | 40 | 45 | 25 | 20 | 55 | 45 | |

*Solution.* Fig. 13.5 shows the construction of the separating straight line on the plane $x_1$ $x_2$. The equation of the straight line on the plane can be written as follows: $x_2 = kx_1 + b$ where $k = \tan\ \alpha$. For the given case $k = \tan 45^\circ = 1$, $b = -10$ and the equation of the straight line is: $x_2 = x_1 - 10$.

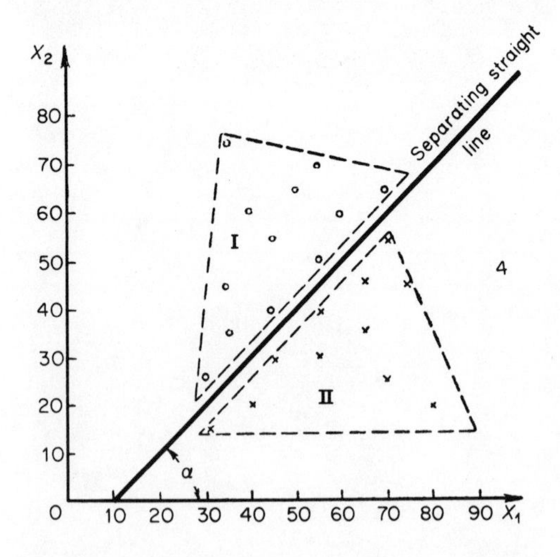


Fig. 13.5. *To Example No. 1.*

2. When drilling an oil well the problem arises of separating oil- and water-bearing strata on the basis of indirect geophysical data. These data are entered in a table and represented in a binary code. As a result, an $n$-dimensional binary vector is obtained for each stratum. Table 13.3 gives the code combinations for the oil- and the water-bearing strata (in the given case the dimension of the vector $N = 18$). Using the algorithm given in Section 13.1,

Table 13.3

*1st class: oil-bearing strata*

| No. of stratum | *Code: binary vector of the stratum* | | | | | | | | | | | | | | | | | |
|---|---|---|---|---|---|---|---|---|---|---|---|---|---|---|---|---|---|---|
| | 1 | 2 | 3 | 4 | 5 | 6 | 7 | 8 | 9 | 10 | 11 | 12 | 13 | 14 | 15 | 16 | 17 | 18 |
| 1 | 0 | 1 | 0 | 0 | 0 | 0 | 0 | 0 | 1 | 0 | 1 | 0 | 0 | 0 | 0 | 1 | 0 | 0 |
| 2 | 0 | 0 | 0 | 0 | 1 | 0 | 0 | 0 | 1 | 0 | 0 | 0 | 0 | 1 | 0 | 0 | 0 | 1 |
| 3 | 0 | 0 | 1 | 0 | 0 | 0 | 1 | 0 | 0 | 1 | 0 | 0 | 0 | 0 | 1 | 0 | 0 | 0 |
| 4 | 0 | 1 | 0 | 0 | 0 | 0 | 0 | 1 | 0 | 0 | 1 | 0 | 0 | 0 | 0 | 1 | 0 | 0 |
| 5 | 0 | 0 | 0 | 1 | 0 | 0 | 0 | 0 | 1 | 0 | 0 | 0 | 0 | 1 | 0 | 0 | 0 | 1 |
| 6 | 0 | 0 | 0 | 1 | 0 | 0 | 0 | 1 | 0 | 0 | 0 | 0 | 0 | 1 | 0 | 0 | 0 | 1 |
| 7 | 0 | 1 | 0 | 0 | 0 | 1 | 0 | 0 | 0 | 1 | 0 | 0 | 0 | 0 | 1 | 0 | 0 | 0 |
| 8 | 0 | 0 | 0 | 0 | 1 | 0 | 0 | 0 | 1 | 0 | 0 | 0 | 0 | 1 | 0 | 0 | 1 | 0 |
| 9 | 1 | 0 | 0 | 0 | 0 | 0 | 0 | 1 | 0 | 0 | 0 | 0 | 0 | 1 | 0 | 0 | 0 | 1 |
| 10 | 0 | 1 | 0 | 0 | 0 | 0 | 0 | 1 | 0 | 0 | 1 | 0 | 0 | 0 | 1 | 0 | 0 | 0 |

Table 13.3 *contd.*

*2nd class: water-bearing strata*

| No. of stratum | *Code: binary vector of the stratum* | | | | | | | | | | | | | | | | | |
|---|---|---|---|---|---|---|---|---|---|---|---|---|---|---|---|---|---|---|
| | 1 | 2 | 3 | 4 | 5 | 6 | 7 | 8 | 9 | 10 | 11 | 12 | 13 | 14 | 15 | 16 | 7 | 18 |
| 1 | 0 | 0 | 1 | 0 | 0 | 0 | 0 | 1 | 0 | 0 | 0 | 1 | 0 | 0 | 0 | 0 | 1 | 0 |
| 2 | 0 | 1 | 0 | 0 | 0 | 0 | 0 | 0 | 1 | 0 | 1 | 0 | 0 | 0 | 0 | 1 | 0 | 0 |
| 3 | 1 | 0 | 0 | 0 | 0 | 0 | 1 | 0 | 0 | 0 | 0 | 1 | 0 | 0 | 0 | 0 | 1 | 0 |
| 4 | 1 | 0 | 0 | 0 | 0 | 1 | 0 | 0 | 0 | 1 | 0 | 0 | 0 | 0 | 1 | 0 | 0 | 0 |
| 5 | 1 | 0 | 0 | 0 | 0 | 1 | 0 | 0 | 0 | 1 | 0 | 0 | 0 | 0 | 0 | 1 | 0 | 0 |
| 6 | 0 | 0 | 0 | 0 | 1 | 0 | 0 | 1 | 0 | 0 | 0 | 1 | 0 | 0 | 0 | 1 | 0 | 0 |
| 7 | 0 | 1 | 0 | 0 | 0 | 1 | 0 | 0 | 0 | 1 | 0 | 0 | 0 | 0 | 1 | 0 | 0 | 0 |
| 8 | 1 | 0 | 0 | 0 | 0 | 0 | 1 | 0 | 0 | 0 | 1 | 0 | 0 | 0 | 0 | 0 | 0 | 0 |
| 9 | 0 | 1 | 0 | 0 | 0 | 0 | 1 | 0 | 0 | 0 | 1 | 0 | 0 | 0 | 0 | 1 | 0 | 0 |
| 10 | 0 | 0 | 0 | 1 | 0 | 0 | 0 | 1 | 0 | 1 | 0 | 0 | 0 | 0 | 1 | 0 | 0 | 0 |

imitate the teaching process on the codes of Table 13.3 and then check whether this algorithm correctly divides the strata presented for teaching. Finally, determine which strata represent the codes given in Table 13.4. (Note: If, in the process of solving the problem, it is necessary to calculate log 0, the value should not be taken as equalling $\infty$ but, say, 10).

Table 13.4

| No. of stratum | *Codes* | | | | | | | | | | | | | | | | | |
|---|---|---|---|---|---|---|---|---|---|---|---|---|---|---|---|---|---|---|
| | 1 | 2 | 3 | 4 | 5 | 6 | 7 | 8 | 9 | 10 | 11 | 12 | 13 | 14 | 15 | 16 | 17 | 18 |
| 1 | 1 | 0 | 0 | 0 | 0 | 1 | 0 | 0 | 0 | 1 | 0 | 0 | 0 | 0 | 0 | 1 | 0 | 0 |
| 2 | 1 | 0 | 0 | 0 | 0 | 0 | 0 | 0 | 1 | 0 | 0 | 0 | 0 | 1 | 0 | 0 | 0 | 1 |
| 3 | 0 | 0 | 0 | 0 | 1 | 0 | 0 | 0 | 1 | 0 | 1 | 0 | 0 | 0 | 0 | 0 | 1 | 0 |
| 4 | 0 | 1 | 0 | 0 | 0 | 0 | 1 | 0 | 0 | 0 | 0 | 0 | 1 | 0 | 0 | 0 | 1 | 0 |
| 5 | 0 | 0 | 0 | 1 | 0 | 0 | 0 | 0 | 1 | 0 | 0 | 0 | 0 | 1 | 0 | 0 | 0 | 1 |
| 6 | 0 | 0 | 1 | 0 | 0 | 1 | 0 | 0 | 0 | 1 | 0 | 0 | 0 | 0 | 0 | 1 | 0 | 0 |

*Solution.* The table of coefficients for calculating whether a stratum belongs to the first or second class is of the type shown in Table 13.5.

Table 13.5

| | 1 | 2 | 3 | 4 | 5 | 6 | 7 | 8 | 9 |
|---|---|---|---|---|---|---|---|---|---|
| $R_0$ | +0.176 | −0.067 | 0 | −0.051 | −0.051 | +0.109 | +0.109 | −0.067 | −0.176 |
| $R_1$ | −0.602 | +0.125 | 0 | +3.301 | +0.301 | −0.477 | −0.477 | +0.125 | +0.602 |
| | **10** | **11** | **12** | **13** | **14** | **15** | **16** | **17** | **18** |
| $R_0$ | +0.125 | 0 | +0.155 | 0 | −0.301 | 0 | +0.125 | +0.051 | −0.222 |
| $R_1$ | −0.301 | 0 | 10 | 0 | +10 | 0 | −0.301 | −0.301 | +10 |

By verifying on the codes of Table 13.3, it is found that the algorithm is erroneous on the strata 3, 5 and 7 (Table 13.3, oil-bearing strata) and on strata 2 and 10 (Table 13.3, water-bearing strata). The remainder will be found correct. In Table 13.4 the codes 2, 3 and 5 belong to oil-bearing strata, and the codes 1, 4 and 6 to water-bearing strata.

3. We are given an automaton which may be in one of the four states: 1, 2, 3 or 4. In the states 1 and 2 the automaton performs the operation a, and in the states 3 and 4 – the operation b. After the operations a and b the automaton will receive punishment from the surrounding medium with the probabilities $P_a = 0.1$ and $P_b = 0.9$ respectively. The graph of transitions of the automaton

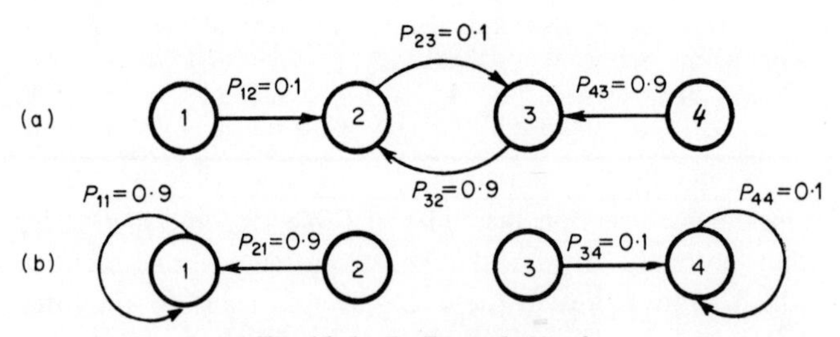


Fig. 13.6. *To Example No. 3.*

from one state into another after punishment by the medium is plotted in Fig. 13.6(a) and the graph for the case when there is no punishment is plotted in Fig. 13.7(b). Considering both graphs simultaneously, it is possible to calculate the probabilities $P_{ij}$ of transition of the automaton from the $i^{\text{th}}$ state into the $j^{\text{th}}$ state (these probabilities are given on top of the arrows in Fig. 13.7).

$$p_{11} = 0.9 \qquad p_{32} = 0.9$$

$$p_{12} = 0.1 \qquad p_{34} = 0.1$$

$$p_{21} = 0.9 \qquad p_{43} = 0.9$$

$$p_{23} = 0.1 \qquad p_{44} = 0.1$$

The probabilities of the other transitions equal zero. The matrix of transitions of the automaton (averaged for the two input signals – punishment and absence of punishment) is written in the form of a table (Table 13.6).

Table 13.6

| | $i$ | | | |
|---|---|---|---|---|
| $j$ | 1 | 2 | 3 | 4 |
| 1 | 0.9 | 0.9 | 0 | 0 |
| 2 | 0.1 | 0 | 0.9 | 0 |
| 3 | 0 | 0.1 | 0 | 0.9 |
| 4 | 0 | 0 | 0.1 | 0.1 |

The probability of the automaton being in any arbitrary state at the time $(t + 1)$ is

$$p_i(t+1) = \sum_j p_j(t)\, p_{ji},$$

where $P_{ji}$ is the probability of transition of the automaton from the $j^{\text{th}}$ state to the $i^{\text{th}}$ state, $P_j(t)$ is the probability of the state $j$ at the time $t$. After a certain number of cycles

$$p_i(t) = p_i(t+1).$$

Moreover, the sum of the final probabilities equals unity.

Write the system of equations and calculate the final probability of each state of the automaton. Compare the automaton under consideration with the automaton described in Section 11.2. Calculate the mean value of the punishments of the automaton and compare it with the mean value of the punishment of an automaton which is incapable of teaching (with equal probability of transitions, regardless of the reactions of the medium).

*Solution.* The system of equations can be written as follows:

$$p_1 = p_1 p_{11} + p_2 p_{21} + p_3 p_{31} + p_4 p_{41},$$

$$p_2 = p_1 p_{12} + p_2 p_{22} + p_3 p_{32} + p_4 p_{42},$$

$$p_3 = p_1 p_{13} + p_2 p_{23} + p_3 p_{33} + p_4 p_{43},$$

$$p_4 = p_1 p_{14} + p_2 p_{24} + p_3 p_{34} + p_4 p_{44},$$

$$p_1 = p_2 + p_3 + p_4 = 1.$$

Inserting the values $p_{ji}$ from the transition matrices and solving the system, we obtain: $p_1 = 0.81$, $p_2 = 0.09$, $p_3 = 0.09$, $p_4 = 0.01$.

The automaton considered is of the same type as the automaton in Section 11.2, except that it has four states instead of two. During adaptation to the environment, the automaton will receive the average number of punishments

$$p^*_{\text{pun}} = (p_1 + p_2)\, p_a + (p_3 + p_4)\, p_b = 0.18.$$

instead of $p_{\text{pun}} = (p_a + p_b)/2 = 0.05$ for an automaton where all the states may occur with the same probability. If the automaton has a larger number of states the average number of punishments will decrease, tending to a value $p_a$, i.e. the automaton will almost always be in a state in which it is less likely to be punished.

# 14 Large Systems

The concept of a 'large system' arose as an expression of a systems approach to the formulation and solution of control problems, which is characteristic of cybernetics. This concept was not introduced for the purpose of classification of systems (dividing them into 'large' and 'small'), but in order to work out a *method for studying* the behaviour of control systems, a consideration based on taking into account all the complexities of such a system.

A distinguishing feature of the systems approach is also that to a greater or lesser extent uncertainty in the behaviour of both the system as a whole and in its component parts is taken into consideration. This uncertainty is the result of random disturbances and participation of people in the system, because random factors of this type cannot be ideally compensated for by control equipment.

The 'large system' concept, in contrast to ordinary local concepts, is based on the fact that any system is considered with its inter-relationships with other systems as a part of some large system.

Therefore, *the object of studying the theory of large systems is a control system considered as a set of inter-connected sub-systems unified by a common purpose*. The presence of separate parts, participation of people, machines and natural surroundings, the presence of material, energy and information inter-connections between parts of the system and external connections between the system considered and other systems, are characteristic of such a system.

The systems approach is necessary for scientific study of many problems of economics, biology and engineering. For instance, it would not be possible to draw justified conclusions about a system, such as an ant hill, without considering it as a large system. To develop satisfactory control of a power system it is necessary to consider as a large system, for example, the natural sources of energy (rivers, location of chemical and nuclear fuels, solar energy, wind energy), power stations, sub-stations, transmission lines and power consumers.

The theory of large systems is in its infancy, and the basic concepts and terminology have so far not been firmly established. Therefore this chapter will deal only with a few problems which, however, are important ones and were chosen more to explain the tasks of this theory than to describe the results achieved by it.

## 14.1. Problems of controlling a large system

The control problems arising in large systems are very varied; they can be arbitrarily divided into two classes: *operational* and *functional.* Operational problems are those associated with the selection of a structure of connections between individual parts of the system, planning the tactics and strategy of the behaviour of a system as a whole and of its parts, the sub-systems, and problems of analysis of the behaviour of the system, evaluation of the results of its operation. Functional problems are in the first instance those relating to the realisation of plans and strategies worked out in the solution of operational problems, when inevitably situations arise which were not foreseen in the planning.

If, for instance, we speak of controlling a production plant, then the *operational tasks* consist of the following:

(1) *Controlling the resources.* Accumulation and maintenance of reserves (manpower, material and financial).

(2) *Realisation of complicated sets of operations.* Construction or re-construction of enterprises, creation and utilisation of the production of new parts.

(3) *Maintaining the working ability of the system.* Maintenance of the equipment and structures, repairs, provision of rest and medical services for the personnel.

(4) *Utilisation of resources.* Manoeuvering with manpower, material and financial resources which participate in the operation of the system.

(5) *Selection of routes.* Control of the movement of flows of materials, semi-finished products, components, distribution of power, control of the movement of information.

(6) *Development of the plant.* Changeover to increasingly more efficient methods, spheres and scales of activity.

(7) *Selection of competition.* Competition between organisations, the fight with nature.

(8) *Marketing.* Studying demand, working out the nomenclature of the manufactured products, organisation of publicity, price control.

(9) *Organisation of labour.* Training of personnel, improving the system of remuneration of labour, improving working conditions.

*Functional problems* in a system for production control consist of ensuring that the technological processes proceed satisfactorily, that the technological operations are carried out in accordance with a predetermined sequence, which involves the coordination of the operating conditions of the individual production sections.

Control systems developed by nature ensure a high efficiency in the control of large natural systems, to which living organisms and their societies undoubtedly belong (shoals of fish, herds of cattle, flocks of birds, swarms of bees and ant hills).

Living organisms solve such operational problems as the construction of a shelter (nest, lair), rearing offspring, laying in seasonal food reserves, fighting the enemy. Functional problems consist in ensuring the survival of the body by feeding, breathing, rest and looking after the young.

The whole arsenal of control methods can serve for controlling large systems, including regulation, adaptation, formulating and realising plans and programmes, teaching games. However, in large systems these problems are more complicated than in the simpler systems considered in the preceding chapters, due to the extreme complexity of the controlled object itself. By definition, a large system is a conglomeration of many inter-connected sub-systems and therefore the number of values characterising this state is very large. Moreover, there are inevitable elements of uncertainty in the behaviour of the system itself and there is the effect of random stimuli on the system from other systems connected to it. Therefore we must conclude that a model of a large system is a dynamic system of a very high dimension with randomly varying structure and parameters.

### 14.2. Effectiveness criterion

To effectively control any system it is necessary to have available a criterion which expresses the degree of correspondence between the behaviour of the system and the control task, as was shown in Chapter 6.

In contrast to the traditional formulation of the control task, in the theory of large systems criteria which define the effectiveness for the solving of such control problems cannot be considered as given. For each control task and each concrete system it is necessary to define the criterion whose numerical value characterises the effectiveness of control. The problem of formulating a criterion applicable to a large system is made more complicated by the fact that any system is a sub-system of some other system of a higher rank, of which it forms a component part. However, the problems of control of part of a system are determined by the control of the system to which it belongs. Therefore in theory an accurate and strict formulation of the control problem and a definition of the effectiveness of the control cannot be achieved for a concrete system without introducing the social-economic factors on a global scale.

Obviously such an approach in the pure form would mean that no practical solution of such problems can be achieved due to their extreme complexity.

However, taking into consideration that only the first steps of increasing the level of investigation exert a strong influence on the nature of the control problem and the effectiveness criterion, it is possible to find sensible levels of investigation for concrete problems, replacing the effect of increasingly higher ranks by corrections based on rough evaluations.

Control of systems would be very primitive and short-sighted if the effectiveness criteria were only to consider the instantaneous state of the system. It is obviously also inadvisable to apply a solution which would bear fruit only after several centuries, but until then would require continuous expenditure. On the other hand a very short-term strategy which does not allow losses at present in favour of considerable advantages in the near future is also unreasonable. It is obvious that selection of an effective criterion of control should be based on a sensible compromise between the advantages (or disadvantages) and the periods required to achieve them. Obviously the decisions arrived at will depend not only on which of the factors enter into the expression for the effectiveness criterion and to what extent, but also on the weighting assigned to the individual factors depending on the period required for obtaining corresponding gains and losses.

One possible way of considering the spread of gains and losses with the progress of time is the introduction of some function $\theta(t)$ chosen in accordance with the nature of the problem. This should be a monotonically decreasing function which should enter as a factor in the expression for the effectiveness criterion. For instance, it is necessary to decide on the selection of one of two variants $A$ and $B$ of building a new plant with various anticipation graphs $R_A(t)$ and $R_B(t)$ of the changes in income with the progress of time (Fig. 14.1), given over a period $(0 - T)$. Here the negative values $R$ denote expenditure. The

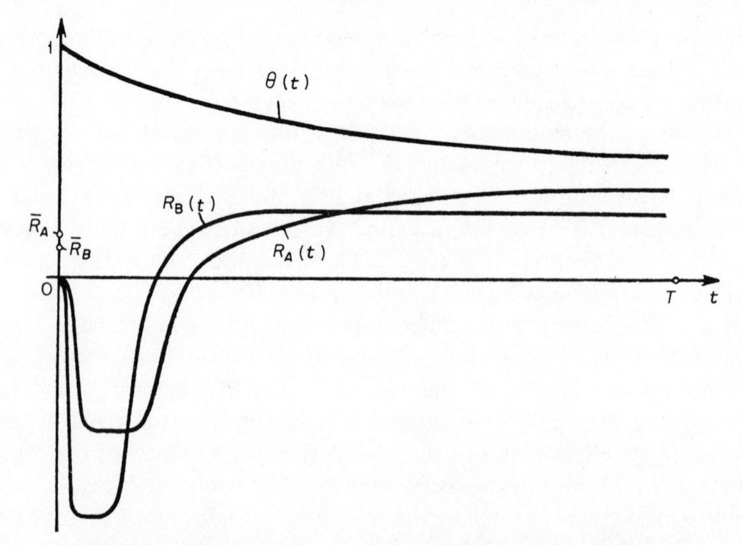


Fig. 14.1. *Graphs of the variations of income and loss functions.*

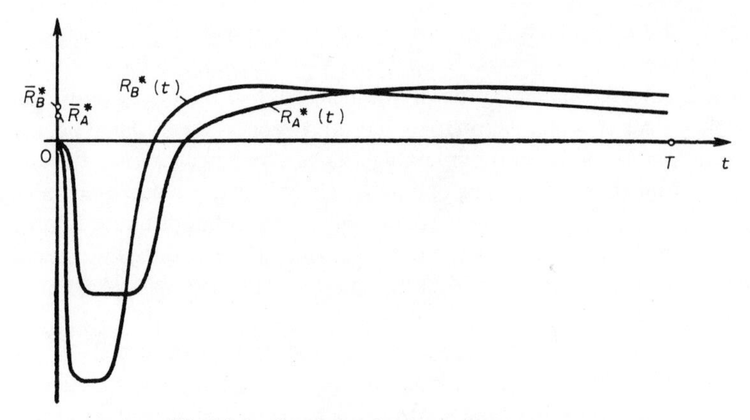


Fig. 14.2. *Graphs of the derived income values.*

selection criterion is the average income over the period $T$. Let us assume that we do not know the shape of the function $\theta(t)$, plotted in Fig.14.1. Multiplying each ordinate $R(t)$ by the appropriate ordinate $\theta(t)$, we obtain graphs of *derived* income values $R^*_A(t)$ and $R^*_B(t)$ as shown in Fig. 14.2. From these graphs the average values $R^*_A$ and $R^*_B$ of the income over a period of time $T$ can easily be determined, and these can serve as a basis for the decision. In the given example the variant $B$ appears to be more favourable because $\bar{R}^*_A < \bar{R}^*_B$ despite the fact that the average income $\bar{R}_A > \bar{R}_B$.

The result is obtained by taking into consideration the fact that in the alternative $A$ a large part of the income materialises much later.

It can be seen from this example how important it is when making a decision to take into consideration the time that is likely to elapse before gains are achieved.

## 14.3. The structure of control systems

One of the important and complicated problems of controlling large systems is the determination of a rational structure for the control system. Since the control of a complex set of inter-connected objects is involved, at first glance it may appear advisable to design the control system in accordance with the principle of *centralised control*.

In a centralised control system all the information on the state of each controlled plant and on the external influences (stimuli) on the system and its individual components is fed into a central control point. On the basis of information on the state of the system and on the control tasks, this central point generates control signals for each of the objects forming part of the system. An inexperienced person may even consider such a control system ideal, because in this case all of the information on the system is concentrated in one control point and therefore the possibility exists in principle of calculating accurately the values of the effectiveness criteria and thus of ensuring optimal control. However, in reality such a point of view is wrong. In the first instance a centralised control system can hardly be realised. For effective control of even one such object it is necessary to obtain and process a very large quantity of information, and for systems containing a large number of objects the quantity of information increases correspondingly. Therefore it would be necessary to collect in the central control point enormous quantities of different information and ensure that this is effectively processed. One of the main tasks of processing the obtained information would be to determine the optimal operation of the system.

As was shown in Chapter 11, search for the optimal operation of a system reduces to the problem of determining the extremum of a function which characterises the criterion of the control effectiveness. The difficulty of finding an extremum increases very rapidly on increasing the number of arguments of the function, i.e. from the dimension $n$ of the state space of optimising systems. In a centralised control system the task of searching for the optimum must be solved in a space of a very high dimension ($n$ may be of the order of hundreds or thousands), which would lead to insurmountable computational difficulties, as was shown in Section 10.5. Moreover, even if it were possible to find the optimum regime for one or another type of complex system, this would require an excessive amount of time and the control actions would arrive with a very great time lag. Therefore the capacity of the data processing equipment limits the possibility of effective centralised control of a complex set of objects.

It is also pointed out that a distinguishing feature of a system with centralised control is a high degree of *rigidity* of the structure, because adaptation to both random changes and changes caused by the evolution of the system and the environment, does not take place in the individual parts of the system but only in the central control point. Centralised control permits stabilisation of a system over a long period, suppressing both fluctuations and evolutional changes in the individual parts of the system without reconstructing them. However, in the final analysis this may be damaging to the system because contradictions between the unchanged structure of a system and changes associated with evolution increase to global dimensions and may require such a radical and sharp reconstruction as would be impossible within the framework of the given structure and would lead to its disintegration.

Centralised control reduces the operational *reliability* of a system. Errors in the operation of the central control point cannot be corrected and react sharply on the state of the entire system. Thus, a system with centralised control is in an unfavourable position compared to other systems. In biological systems for instance, there is no such centralisation. Imagine what would happen if the brain had to control every act of metabolism in each cell of the body. The impossibility of doing this is obvious. Only because the life activity of each cell is almost completely subjected to autonomous control by the nucleic acids, is the body capable of functioning so successfully and of adapting itself to such a fine degree and with such great flexibility to external changes as well as changes inside the body itself.

In the history of human society centralised control has led to social and economic backwardness and sooner or later such a social structure was doomed to failure. A classical example of such centralisation was the economy of the Incas in South America. There the entire social, economic and private life was strictly regulated. Any activity such as the time allotted and the methods used in agriculture, was carried out only on the instructions of the chief controller, the Sapa-inca, and this had to be executed without question. As a result, this great empire with an army of 200,000 men, was conquered without any great difficulty by a detachment of 168 men of the Spanish Conquistador Pizarro.

The above mentioned drawbacks of a centralised structure can to a considerable extent be overcome by using a *hierarchical structure* in the control system. A distinguishing feature of hierarchical structures is the successive division of the system into sub-systems between which a relationship of subordination is established. The control equipment of higher order controls larger sub-divisions of the system, each of which has its own control equipment. Each such sub-division is in turn broken up into smaller ones which also have appropriate control devices, and so on, right down to the elementary sub-divisions of the systems where further sub-division would be inadvisable. An example of a hierarchical structure of control is shown in Fig. 14.3. Breaking down of the system into subordinated parts is carried out in such a way that each part contains objects which are most closely associated with each other. In other words, breaking down is carried out by using 'weak' bonds.

In systems with a hierarchical control structure the lower order equipment should decide on relatively simple local control problems which are within the capacity of control devices with a limited data processing capacity. Then the

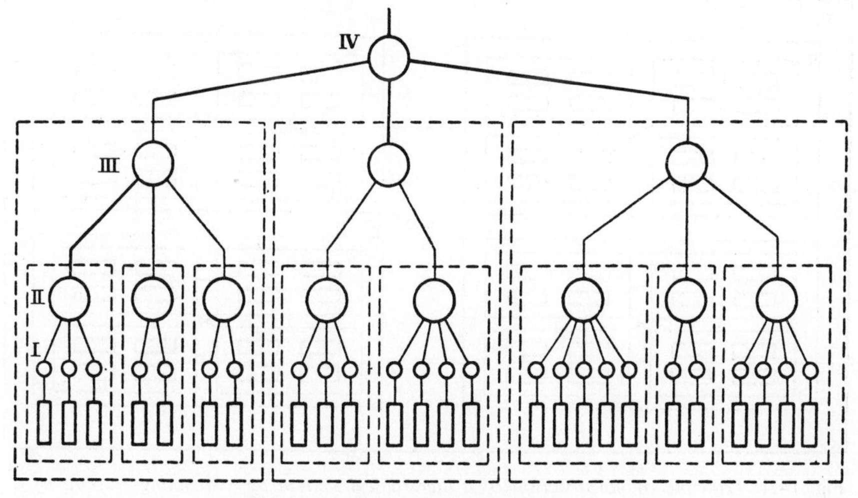


Fig. 14.3. *Hierarchical control structure.*

control devices of the next order will be left to deal only with those control problems which have to be solved in order to coordinate the work of the elementary objects and which can be solved on the basis of less detailed information on the states of the individual objects. The same applies to the control equipment of higher orders, and therefore the volume of information which they have to process is greatly reduced and can be made to correspond to their information handling capacity.

Information on the state of the control system and of its individual parts is transmitted from the system of the lower order to the systems of the higher order in increasingly generalised and systematised form, so that the control equipment of the first order gets the most detailed information on the state of the objects, and as we go over to higher and higher levels this information is generalised in accordance with the nature of the problems.

The control commands in systems with a hierarchical structure are generated by the control equipment of the higher order in the most generalised form and are made more concrete and detailed as they move to the control equipment of the lower ranks. Systems with a hierarchical structure were formed during the natural development of biological, engineering and economic systems. They can be observed at present in all complicated systems, such as the national economy.

One should not think that a hierarchical structure will necessarily pre-suppose the singling out of control devices of a higher order in the form of particular parts of the system, which are 'in a higher position' than the control devices of lower order. On the contrary, in many cases a structure would be best where some of the equipment of the first order together form the equipment of the second order, some of the equipment of the second order in turn form the equipment of the third rank, and so on, as shown in Fig. 14.4.

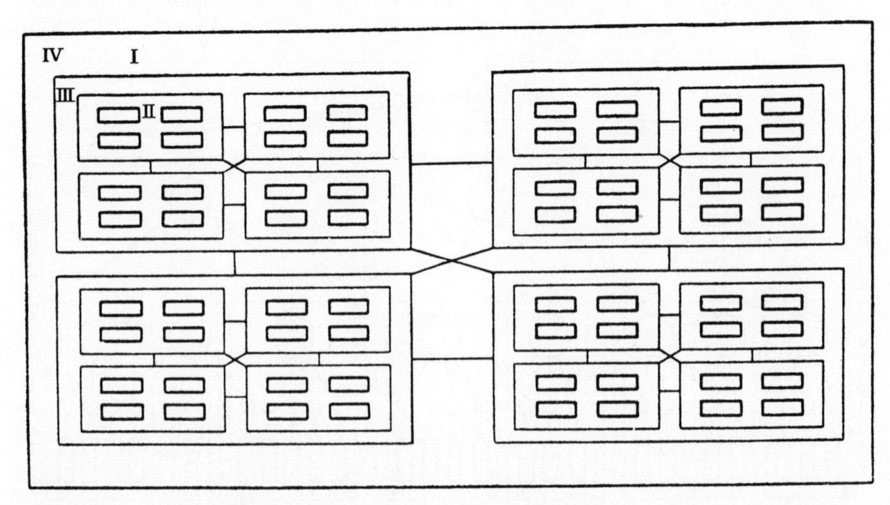


Fig. 14.4. *Principle of construction of a hierarchical system by associating systems of a lower order with systems of a higher order.*

The possibility arises of making decisions on the basis of voting, as shown for the first time by J. von Neumann. This greatly increases the reliability of operation of the system.

The work of the Soviet scientists Kolmogorov and Ofman show that a system in which all the elements have absolutely 'equal rights' and each element is connected to only a few others, can in the presence of a sufficiently large number of elements solve any problem, however complex, which is solved by any other types of equipment (i.e. such a system is a so-called universal automaton).

There is reason to assume that a hierarchical structure based on successive association of sub-systems is used in the human brain.

## 14.4. Design of a hierarchical structure

Frequently the operation of such large systems as, for instance, a factory, is very difficult or impossible to describe mathematically. Nevertheless the system has to be controlled so that the results of its activity satisfy certain criteria. Control based on using strict mathematical methods is occasionally so complicated that it cannot be used even with the available modern large computers. In such situations it is possible to use the idea of simulating the action of the human controller who takes decisions which are partially intuitive. It has been found that most people subconsciously use the same methods which help them to solve all the complex problems that arise; such methods are called *heuristic*. Some heuristic methods are suitable only for solving a certain limited problem, but other heuristic methods exist which are suitable for solving large classes of problems.

One general heuristic method which can be used for controlling large systems is the following. Let us imagine a control device of a large system in the form of a block A (Fig. 14.5) on whose input information is fed on the aim of the

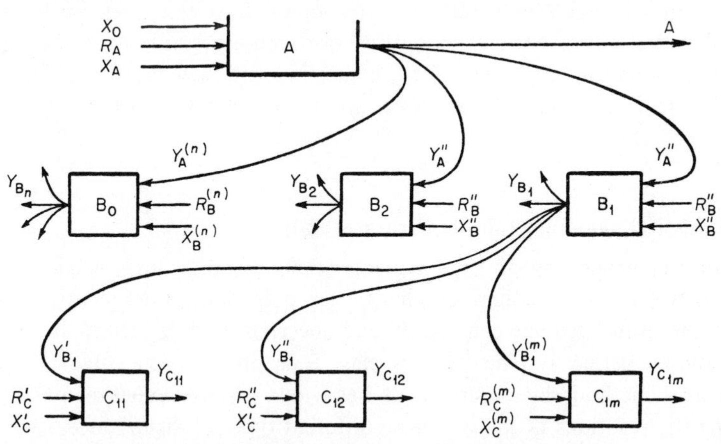

Fig. 14.5. *Diagram of the control of a large system.*

control $X_0$, on the current state of the system $X_A$ and on the constraints imposed on the control $R_A$. The result of the operation of the block A is an output or an output operator $Y_A$, which may be in the form of commands, instructions, routines, etc. If this type of operator is inadequate for effective control of the system, then the control equipment A can break it down in some manner into a set of simpler operators $Y'_A, Y''_A, \ldots, Y^{(n)}_A$ and transmit these for decoding and providing greater detail into the blocks $B_1, B_2, \ldots, B_n$ of a lower control level. Now the set of operators transmitted to the next level will become a set of targets for the blocks $B_1, B_2, \ldots, B_n$. At the same time to accomplish the targets these blocks require more detailed information $X'_B, X''_B, \ldots, X^{(n)}_B$ on the state of the system and on these the constraints $R'_B, R''_B \ldots, R^{(n)}_B$ can be imposed.

The results of the operation of the blocks of the second level are the operators $Y_{B_1}, Y_{B_2}, \ldots, Y_{B_n}$, which again, if necessary, can be separated into components and addressed as sub-targets (sub-routines) into the blocks of the next level. This hierarchical structure can be continued until the control action which is fed into the control system is adequate for the required operation of the system, to make the existing state $X_A$ approach the target state $X_0$. It is pointed out that for a control system organised in this way it is not necessary to send all the input information that it receives into the blocks of the lowest level or to send the information from each block of a given level to all blocks of a higher level. For instance the data on the current state of the system $X_B$ should contain information from the lower level blocks $C_{11}, C_{12} \ldots$ which are directly connected to the block $B_1$ and the information which is still required for forming the compound (statement) operator $Y_{B_1}$.

In this way the information on the state of the system, received by each of the higher levels, has already been filtered by the lower levels. This appreciably reduces the volume and the processing time of the information fed into the blocks and permits finding a solution or a control action within an acceptable time and an acceptable expenditure of resources. It is true that the solution obtained in this way is not always optimal nor even necessarily near to optimal, but in most situations the constraints on the time and resources available are so great that it is necessary to sacrifice optimality in favour of obtaining an acceptable solution.

### 14.5. Statistical modelling (Monte Carlo method)

Study of the properties of large systems, their reactions to changes in the characteristics of the medium or changes of individual components, is very difficult due to the extreme complexity and uncertainty of its behaviour.

To study existing systems of this class it is obviously possible to apply experimental methods by observing its behaviour (passive experiments) or by observing the reactions of the system to different types of disturbances specially generated for the purpose of study (active experiments). Usually experimental investigations on real systems are extremely expensive and limited by the permissible tolerances of the operating conditions which exclude the possibility of detecting the behaviour of the system under unusual conditions, for instance during failures or accidents. Moreover, such methods are not applicable for redeveloped systems, the properties of which have to be evaluated accordingly to how the system is built.

Due to the above difficulties, the *method of statistical modelling* is often used for studying the behaviour of large systems.*

The method of statistical modelling is based on simulating random events in a model of the system under investigation. Statistical modelling is usually realised by a digital computer. Data on known properties of the system to be simulated and of the external medium are fed into the computer. Random factors are simulated by a source of random values with given statistical properties. Such a source is a table of random or pseudo-random numbers generated by the computer on the basis of a special algorithm.

An important feature of the method of statistical modelling is that each *individual* solution of the problem does not in itself characterise the system but is a random quantity. Only as a result of obtaining a *large number* of individual solutions and *averaging* them, can the sought relationships be established.

The method of statistical modelling permits successful solution of such problems as, for instance, determination of the optimal periods between re-setting of equipment to ensure that the quality of the manufactured product remains within permissible limits, forecasting the size of queues in queueing systems in such things as transportation systems, optimal distribution of parameter tolerances of intermediate products in flow production.

For the purpose of illustration the method of statistical modelling will be used to solve a relatively simple problem.

Let us assume that of a large number of components which are available to an assembly worker, half the quantity has positive deviations from the nominal dimensions and half negative deviations from the nominal dimensions. During assembly, each component is made up of three parts. The correct functioning will be impaired only in parts which are made up of three components all having positive deviations. It is necessary to determine the probability of obtaining a correctly functioning part if the assembly worker takes the parts from the appropriate container at random. The problem can be solved analytically. Denoting by $p$ the probability that the component taken at random will have a positive deviation from the normal, then the probability of assembly of a satisfactory part $p_{sat}$ can be calculated by the formula:

$$p_{sat} = 1 - p^3 = 1 - \left(\frac{1}{2}\right)^3 = 0.875.$$

* This method is also called the method of statistical tests or the Monte Carlo method (originating from the town of Monte Carlo which has the world's largest gambling casinos, in particular roulette).

This problem can be solved by statistical modelling by tossing three coins, considering for instance heads as 'plus' and tails as 'minus'. If simultaneous tossing of three coins produces three heads, then it will be assumed that the assembled mechanism will not function correctly, whilst for all other combinations (two heads and one tail, or one head and two tails, or three tails) the assembled part will function correctly. If the coin-tossing test is repeated many times and the probability $p_{sat}$ is estimated, it can be seen that this probability will be near to 0.875.

If statistical modelling is effected by means of computers the solution of very complex problems can be achieved within an acceptable time, including problems of the type considered in the earlier example but with a larger number of components which are assembled into complicated units. The Monte Carlo method is also very effective for obtaining numerical solutions of many problems which are not probability problems, for instance calculation of multi-dimensional integrals, solution of complex systems of differential equations, finding an optimal variant during planning, and many other problems.

### 14.6. Queueing

In solving problems relating to the control of large systems, cases are frequently encountered where it is necessary to take into consideration the throughput capacity of engineering equipment, whose optimal control must be ensured. Usually for solving such problems one of the sections of the theory of probability – the theory of queueing, is used.

For instance, in complicated production units, each of the types of equipment (reactors, machine tools) should carry out a predetermined volume of work within a given time, i.e. to service, supply, machine, fabricate a certain quantity of the starting material. Similarly, each of the sub-divisions of the plant (sections, bays, shops, etc.) can be considered as the appropriate servicing division, which over a period of time has to service (carry out, machine, manufacture) a series of operations (flow of orders, components, raw material, assemblies, etc.). Obviously, economic units at higher levels – plants, firms, branches of the industry, economic councils – can also be considered as queueing systems.

In planning the operation of such systems it is necessary to take into consideration the throughput capacity, the number of orders which can be satisfied per unit of time, the queueing time of the orders, etc. Clearly, the technical-economic indices of the operation of individual units and sub-units will depend on these characteristics. The accuracy of forecasting the number of orders which can be served will determine whether the plan has been correctly worked out and whether it can be accomplished; the waiting time determines the required number of bunkers and other equipment for storing semi-finished products waiting to be handled, and on this also depends the number of circulating equipment and the speed of turn-round.

Let us consider a simple example of a queueing problem. A country telegraph office receives 'rush' telegrams. The frequency of such telegrams $\lambda$ is 2 telegrams per hour ($\lambda = 2$). Each telegram is immediately given to a postman for delivery to the addressee. Let us assume that the average time of delivery to the addressee and return of the postman to the telegraph office is $\mu = 2.5$ hours. An important characteristic of this sytem of queueing is the average number of postmen en route at the same time ($a$), and the probability that $k$ postman ($p_k$) will be occupied at the same time. The average number of simultaneously occupied postmen increases with increasing density of the telegrams being delivered and with increasing distance of the addressees from the telegraph office.

Since the number of telegrams per hour and the delivery time are independent random variables, the average number of postmen $a$ required will be the product of the density (frequency) of telegrams and the average delivery time:

$$a = \lambda\mu = 5.$$

Cases when a very large number of postmen are simultaneously engaged in delivery work or when none of them are so engaged, are not very likely for the given example. The accurate value of the probability that $k$ postmen will be engaged simultaneously in delivery work can be determined according to the formula:

$$p_k = \frac{a^k\, e^{-a}}{k\,!}$$

This is referred to as the Poisson distribution. The shape of this function for $a = 5$ is shown in Fig. 14.6. The most likely case is that simultaneously $a$ and $(a - 1)$ postmen will be engaged in delivery, i.e. 5 and 4 postmen respectively. This probability $p_5 = p_4 = 0.75$. The probability that all the postmen will be free is $p_0 = 0.0067$.

If the working time totals 1000 hours, then only for 6.7 hours will all the postmen be free at the same time.

Most practical problems of the theory of queueing are so complicated that either their solution cannot be obtained analytically or the formulae obtained

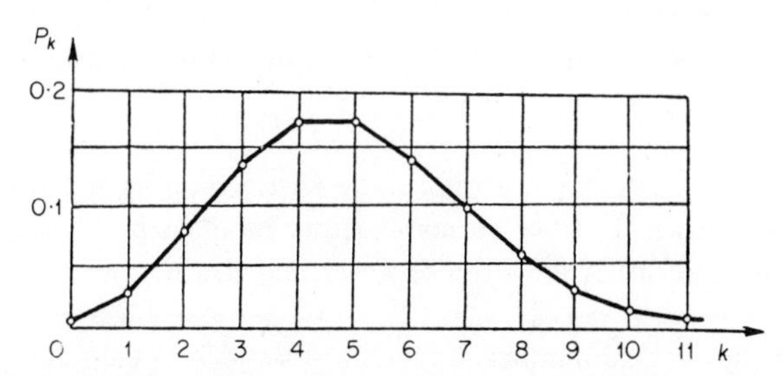


Fig. 14.6. *Distribution of the probability of* k *postmen being busy at the same time.*

are excessively cumbersome and unsuitable for calculations. In such cases the Monte Carlo method, which can be realised on fast computers, is extremely effective.

**Exercises**

1. Enumerate the operational and functional problems arising in the control of the power system of a large undertaking. (The power system consists of a power station, sub-stations, steam boilers, compressors, distribution lines for electricity, heat and air, pumping stations, and a despatcher table).

*Solution. Operational problems*: manoeuvering manpower, material and financial resources; reconstruction of the management of the system; maintenance of equipment, repair and installation work, re-setting; despatcher control (distribution of various types of power in the individual subdivisions of the undertaking, transportation of fuel into the undertaking and of processed materials out of it, exchange of information within the system and outside); provision of rest and medical services for the personnel; training of personnel; improvement of working conditions; participation in competitions inside the undertaking; emergency work.

*Functional problems*: ensuring that the individual units operate at the required conditions; adherence to a predetermined sequence of switching on and off, sets, lines, substations, pumps, etc; compliance with the load graphs.

2. A system for supplying electricity to several counties which is being designed, contains many power stations and power transmission and distribution systems. It is assumed that control machines (computers) are extensively used in the system. What control structure should be chosen for the given system, and why, is a control (computer) machine used? (See Fig. 10.7).

*Solution.* It is advisable to choose a structure as shown in Fig. 10.7(b) which in addition to a universal control machine also contains local control equipment. Such a control system will have several levels of hierarchy and the global control equipment will be able to control elastically the system as a whole without interfering in the details of the control of each individual power station and sub-station.

3. One of the departments of a publishing house receives the manuscripts of books and the average number of manuscripts received per month is two. On receipt, each manuscript is sent to a reviewer and the average reviewing time is two months.

Determine the average number of reviewers working at the same time and the probability that this number is not in excess of nine people.

*Solution.* The average number of reviewers $a = \lambda\mu = 5$. The time during which the number of simultaneously occupied reviewers does not exceed 9 is composed of the time when all the reviewers are free at the same time, and when 1,2 . . . . 9 inclusive are busy simultaneously. Therefore the required probability (in accordance with Fig. 14.6) is

$$p_{\leqslant g} = p_0 + p_1 + p_2 + \ldots + p_g \approx 0 + 0.03 + 0.08 + 0.14$$
$$+ 0.175 + 0.175 + 0.15 + 0.10 + 0.07 + 0.03 = 0.95$$

Therefore 9 reviewers will be adequate with a probability of 0.95.

# 15 Operational Control

Operational problems which arise in large systems have specific features and therefore a specific approach is required for their solution. One of the important features of operational problems is that the control is usually a team of executives, with the appropriate technical means and equipment required to achieve a certain aim. The object to be controlled may be a contracting organisation which builds industrial undertakings; a design office which develops a project for a new motor car, or a team of scientists who solve a specific problem.

*We will define as an operation, the activity of a team of executives aimed at reaching a set target.* An operation can always be divided into *stages*, the execution of which is necessary and sufficient for the entire operation to materialise. The individual stages of the work, the components of the operation, are inter-linked by various conditions which limit the selection of the sequence of completion and by overall manpower, material and financial resources.

To control an operation effectively it is necessary to solve problems of two types: (1) to work out the optimal plan for carrying out the operation, and (2) to ensure that the realisation is near to the optimal plan under changing conditions.

A scientific approach to the solution of these problems is possible only if a mathematical model is available which reflects sufficiently clearly the properties and characteristics of the object and is suitable for investigation by formalised methods. One of the most convenient and most extensively used models for solving problems of operational control is the *network model*.

In this chapter we will consider some methods of operational control based on using network models (PERT).

## 15.1. Network models

The network model of operations (or 'network' for brevity), is a particular type of graph, namely the *finite directed graph without contours*. To explain the terms used in this definition we must deal briefly with information relating to the theory of graphs.

By graph, we understand a set of points, or nodes, on a plane which are connected by continuous lines. If the connecting lines are provided with arrows indicating differences of passage in the direction of the arrows and against the direction of the arrows, such a graph is called a directed graph. The points which are connected with the lines are called the *vertices of the graph*. The graph is called *finite* if it has a finite number of nodes. An example of a finite directed graph is shown in Fig. 15.1. To each node we assign a number. The connecting line will be denoted by the respective number of the nodes $(i, j)$ which it connects. The sequence of connecting lines where the end of each preceding connecting line (edge) coincides with the beginning of the next one is called a *path*. The *contour* is the path where the initial node coincides with the end one. No contours (or loops) are permitted. The graph Fig. 15.1 contains the contour 1 - 3 - 4 - 1 and therefore by definition it is not a network.

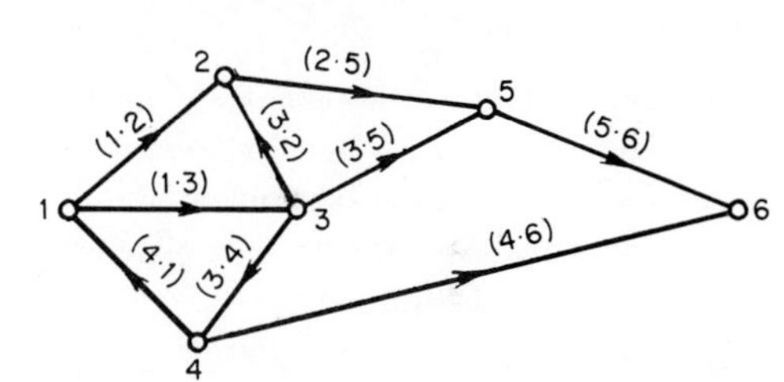


Fig. 15.1. *Finite directed graph.*

In network operation models the edges (connecting lines) usually represent individual stages of the operation – work; the nodes represent *events* denoting termination of a certain stage of the operation. In the network considered, two events can be distinguished: the *beginning* and *final* events. The concept *work* in network models can be understood in a number of senses: it may designate a particular work which requires labour, material resources and time, or an *anticipation* which does not require expenditure in labour or resources but takes a certain amount of time, or, finally, *fictitious work* requiring no expenditure in labour or resources and no time, but indicates the impossibility of starting some work before a certain event has occurred.

As an example of a network model we will consider a relatively simple operation consisting in installing a motor onto a base plate. The operation comprises the following work:

1. Ordering of the base plate.
2. Manufacture of the base plate.
3. Transportation of the base plate to the site.

4. Preparation of the foundation under the base.
5. Construction of the casing for the base.
6. Concreting the base.
7. Hardening the concrete.
8. Fitting the base plate.
9. Ordering the motor and its delivery from stores.
10. Transportation of the motor to the site.
11. Installation of the motor.

Based on the sequence of operations determined by technological factors, a network can be plotted as shown in Fig. 15.2. The configuration of this network reflects the conditions of admissibility of the beginning of each operation in accordance with the rule: the operation starting from the $i^{th}$ node of the network can only begin after all the work which enters into the $i^{th}$ node has been completed. In the example under consideration the node (event) 6 'contains' two operations: 'transportation of the base plate' and 'hardening of the concrete' and from it follows 'fitting of the base plate'. This means that fitting of the base plate can start only after it has been delivered and after the concrete has set.

The events represented by the nodes of the network have the following meaning in the given case:

0. Decision on starting the work. *Beginning of the operation.*
1. The base plate has been ordered.
2. The base plate has been manufactured.
3. The foundation for the base plate is finished.
4. The casing is finished.
5. The concrete has been poured.
6. All is ready for fitting the base plate.
7. The motor has been ordered.
8. All is ready for installing the motor.
9. Motor fitted. *End of the operation.*

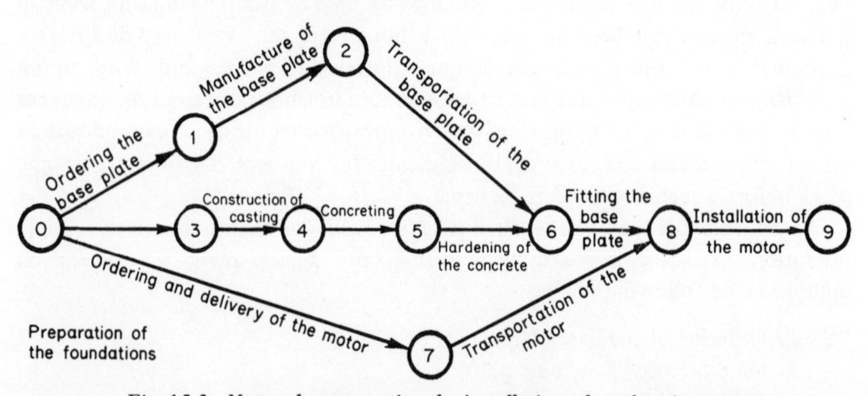


Fig. 15.2. *Network representing the installation of an electric motor.*

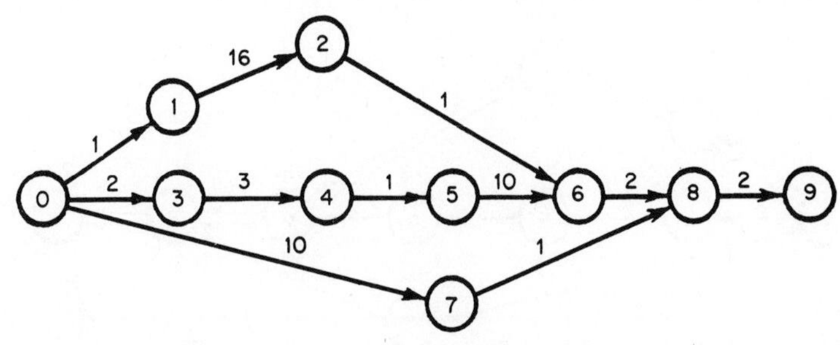


Fig. 15.3. *Network graph with time characteristics of individual operations for installing an electric motor.*

The work items which compose the operation are characterised by certain values of time and resources which are necessary for completing them. We shall consider only the time evaluation of networks and therefore we can only take into consideration the time $t(i, j)$ expended for each item of work $(i, j)$, situated between the event $i$ and the event $j$. The time of performing an item of work can hardly ever be forecast with a high degree of accuracy; it is only possible to estimate the probability of completion within certain time limits. However, for rough calculations some average anticipated duration of the work can be applied which will be used to characterise the work in network models. A network model with time characteristics of the work items for the example considered is given in Fig. 15.3. The anticipated duration of the work is entered in terms of days above each connecting line of the network.

Network models so constructed are being used for planning operations and re-planning them during the process of realisation.

## 15.2. Network analysis

One of the important indices determined from a network model is the value $T_{ei}$ – the *earliest time* of the $i^{\text{th}}$ event occurring, i.e. the minimum time separating the given event from the beginning.

The $i^{\text{th}}$ event is considered as occurring when *all* the work entering into the $i^{\text{th}}$ node of the network has been completed. Therefore the event does not occur until the *last* of the work items entering into the given node has been completed. If several paths lead from the initial event to the event under consideration, then the time up to the occurrence of the event will be determined by the duration of the *longest path*. Thus in Fig. 15.3 two paths: 0 - 1 - 2 - 6 and 0 - 3 - 4 - 5 - 6 lead to event 6. The duration of these paths equals:

$$T(0, 1, 2, 6) = t_{0.1} + t_{1.2} + t_{2.6} = 1 + 16 + 1 = 18 \text{ days},$$

$$T(0, 3, 4, 5, 6) = t_{0.3} + t_{3.4} + t_{4.5} + t_{5.6} = 2 + 3 + 1 + 10 = 16 \text{ days}.$$

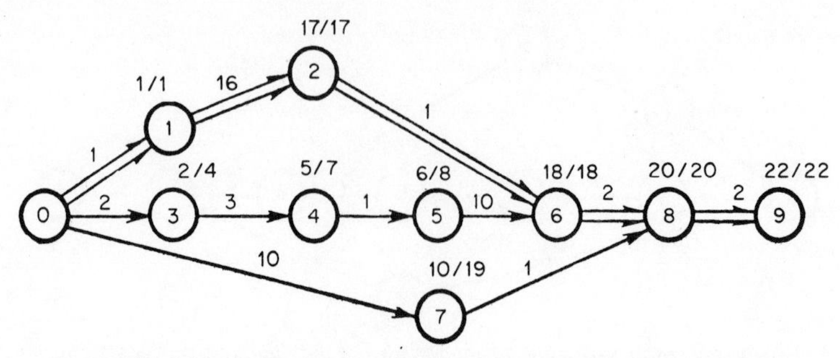


Fig. 15.4. *Critical path for the example considered (Figs. 15.2 and 15.3).*

Obviously event 6 cannot occur before 18 days and $T_{e6}$ = 18 days. Finding in this way the value of $T_e$ for each event, including the final $k^{th}$ event, it is possible to determine $T_K$, the time to completion of the operation. The time $T_K$ is the duration of the longest path which leads from the initial event to the final event. This path is called the *critical path.* In each network there exists at least one critical path. In the network under consideration the critical path passes through the events 0 - 1 - 2 - 6 - 8 - 9. In Fig. 15.4 the critical path is highlighted by double lines; its duration $T_K$ = 22 days.

The critical path occurs only for a part of the events and work items contained in the network, and therefore it can be seen that the remaining events and work items which are not in the critical path can be accomplished with some delay without affecting the time required for completion of the whole operation.

The availability of spare time for accomplishing a part of the work and the time of occurrence of some events, are important factors which are taken into consideration in the planning of operations and controlling their realisation. To reveal available reserves in time, we introduce the concept of the *very latest permissible time of occurrence of the i $^{th}$ event - $T_{si}$.*

The very latest permissible time for starting the $i^{th}$ event is the *maximal* time which separates the $i^{th}$ event from the beginning, *without upsetting the time* $T_K$ of completion of the entire operation.

The value of $T_{si}$ for any event can be determined by deducting from $T_K$, the duration of the longest path, from the $i^{th}$ event to the $k^{th}$ event. For event 7 of the network shown in Fig. 15.4, $T_{s7}$ can be determined from the following simple calculation:

$$T_{s7} - T_k - T(7, 8, 9) = T_k - (t_{7,8} + t_{8,9}) = 22 - (1 + 2) = 19 \text{ days.}$$

$T_{e7}$, the earliest time of occurrence of event 7, equals 10 days. Consequently, although event 7 can occur $T_{e7}$ = 10 days after the beginning of the operation, it could be extended to $T_{s7}$ = 19 days without endangering the period of completion of the entire operation. If event 7 occurs 19 days from the beginning of the operation, and the work (7, 8) requires 1 day, then event 8 will occur 20 days from the beginning of the operation, i.e. at the same time as if event 7 had occurred earlier, and the entire operation can be completed in the same 22 days.

The order of calculation of the values of $T_{si}$ for each event of the network is the same as for determining the $T_{ei}$ values, except for the difference that for the longest path to the event under consideration the counting is not done from the initial but from the terminal event and the duration of this path is deducted from $T_K$. In Fig. 15.4 the earliest possible (numerator) and the latest (denominator) permissible periods of occurrence of individual events are recorded.

The difference between the latest permissible and the earliest possible time of occurrence of the $i^{\text{-th}}$ event determines the value $R_i$ – *spare time available for the occurrence of the given event.*

$$R_i = T_{si} = T_{ei}.$$

For events on the critical path the $T_s$ and $T_e$ values are the same, and for these the time reserve is zero, which is obvious since any extension of the longest path will make it even longer.

Determination of the critical path and calculation of spare time are the main tasks of network analysis, which must be solved in every case where network models are used for planning and re-planning of operations. Usually the complexity of a network is much greater than in the example described here. Usually hundreds and even thousands of events and operations must be taken into consideration. Therefore the critical path, reserves and other network indices are often calculated by computers using programmes specially written for the purpose.

### 15.3. Operational planning

In most cases the moment of completion of an operation is the date for reaching the target of a given operation. This is because the individual operations are not isolated, but form a link in a chain of other operations whose purpose is the achievement of a target of a higher order, or a system which realises the operation under consideration. Therefore one of the tasks of operational planning is the determination of the *order* of carrying it out, as well as detecting those conditions under which the task will be accomplished with the required probability no later than the target date.

Since the date for completing the operation, $\theta_k$, is determined by the date of beginning of the operation, $\theta_0$, and the period of its execution, $T_k$, and $T_k$ depends on the resources used, $\theta_k$ can be influenced by the date of $\theta_0$ and the amount of resources expended to realise the operation.

If for the given resources the period of completing the operation is $T_k$, then the data for beginning the operation, which ensures completion, is determined by the date

$$\theta_0 = \theta_k - T_k.$$

The duration of the operation, $T^0$, can be reduced not only by increasing the resources but also by redistribution of the individual operations. Since the period of carrying out the operations equals the duration of the critical path, which is determined by the duration of accomplishment of work on the critical path, shortening the time $T_k$ can be achieved by shortening the duration of these critical operations. For this purpose additional resources can be shifted from the tasks where spare time is available to tasks on the critical path. Such redistribution of resources will shorten $T_k$ until all time reserves are exhausted, i.e. until all the paths leading from the initial to the terminal event have become critical.

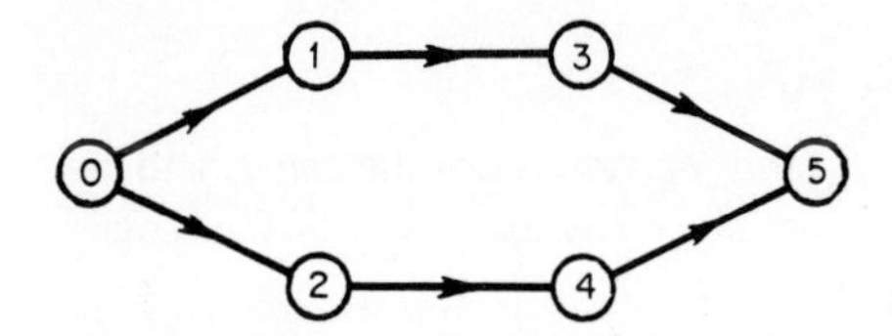


Fig. 15.5. *Network model of operations.*

The condition that all the tasks are critical is not sufficient for achieving an optimal plan. Let us consider, for instance, the network Fig. 15.5. Data on the labour content and maximum number of people who can be used for each operation are given in Table 15.1.

Table 15.1

| *Number of operations* | *Labour content, man days* | *Maximum permissible number of people.* |
|---|---|---|
| 0 – 1 | 12 | 4 |
| 0 – 2 | 12 | 4 |
| 1 – 3 | 4 | 1 |
| 2 – 4 | 4 | 1 |
| 3 – 5 | 12 | 4 |
| 4 – 5 | 12 | 4 |

Let us assume that 4 people are available for carrying out the operation. Two people will be assigned to the operations 0 - 1 and 0 - 2. After 6 hours we change over to performing the tasks 1 - 3 and 2 - 4 (it is pointed out that only two people are working). Finally, 4 hours later, we assign two people each to the tasks 3 - 5 and 4 - 5. The operation is accomplished in 16 hours, and both paths are critical. The obtained solution is not optimal. An optimal plan would be, for instance, the following: 4 people are assigned to carry out 0 - 1 (3 hours), 3 people to carry out 0 - 2, and one person to carry out 1 - 3 (4 hours), then 3 people are assigned to the operation 3 - 5 and one to the operation 2 - 4 (4 hours); finally 4 people are assigned to the operation 4 - 5 (3 hours). In this way the operation is completed in 14 hours.

Fig. 15.6 shows the graphs of utilisation of resources; for the first plan see Fig. 15.6(a), and for the second (optimal) plan see Fig. 15.6(b). The reason for

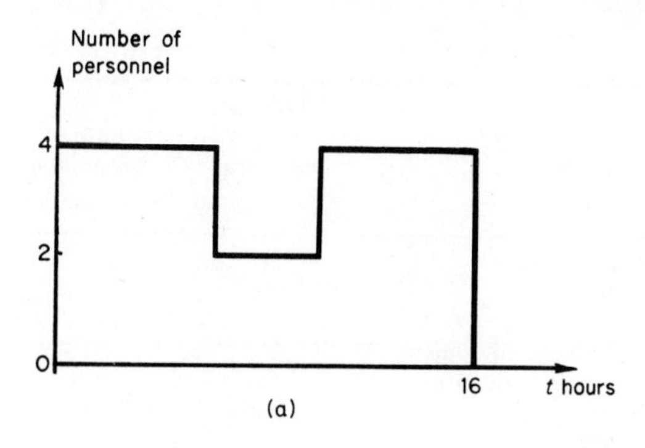


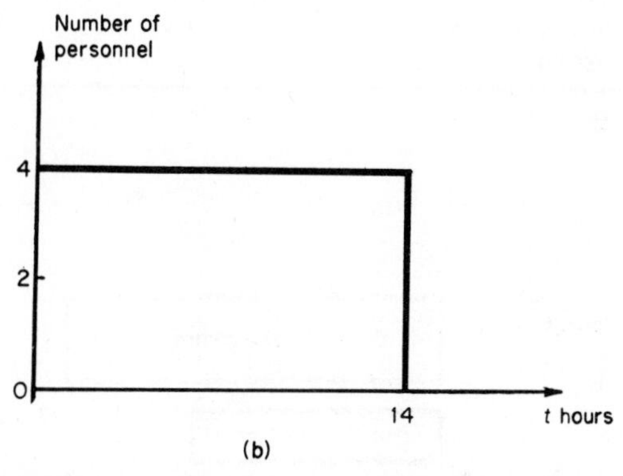


Fig. 15.6. *Graphs showing the utilisation of resources: (a) for the first plan; (b) for the second (optimal) plan.*

the first plan not being optimal is that it is not possible to use more than one person each for the operations 1 - 3 and 2 - 4.

The time taken in performing the operation does not always determine the effectiveness of a plan. Optimality is often judged by a different criterion, for instance the cost of the operation or the maximum number of people required to carry it out.

Let us consider the problem of optimal distribution of resources on the following simple example. Let us assume that it is necessary to carry out the electrical installation work in a sub-station under construction, and the labour required for this and the completion times specified for the network are entered in Table 15.2.

The sequence of operations and the graph of the required number of installation personnel $L(t)$ are shown in Fig. 15.7. It can be seen that the maximum number $L_{max}$ of personnel is 45, if their number remains constant throughout the operation. However, the number $L_{max}$ can be sufficiently reduced if graphs $l_i(t)$ are worked out for the number of operatives engaged in each operation for which the total requirement of personnel $L(t)$ would be more uniform. For the case under consideration, such a graph is shown in Fig. 15.8. It

Table 15.2

| *No. of the operation* | *Designation* | *Volume of work, man days* | *Date of beginning* | *Date of termination* |
|---|---|---|---|---|
| 1 | 2 | 3 | 4 | 5 |
| 0 – 1 | Installation of control panels | 200 | 1.4 | 20·4 |
| 0 – 2 | Installation of high voltage equipment | 600 | 1.4 | 30.4 |
| 1 – 3 | Fitting up the control panels | 200 | 20.4 | 10.5 |
| 1 – 4 | Installation of the control circuits | 450 | 20.4 | 20.5 |
| 2 – 5 | Setting of the high voltage equipment | 300 | 30.4 | 30.5 |

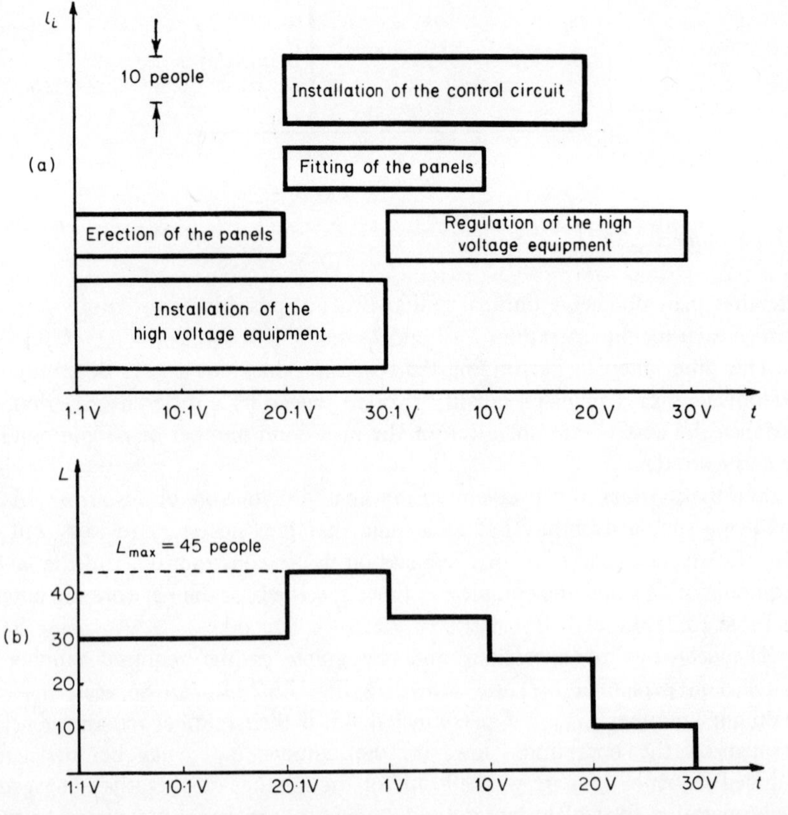


Fig. 15.7. *Example of optimal distribution of resources: (a) sequence of electrical installation work; (b) graph of the number of fitters required.*

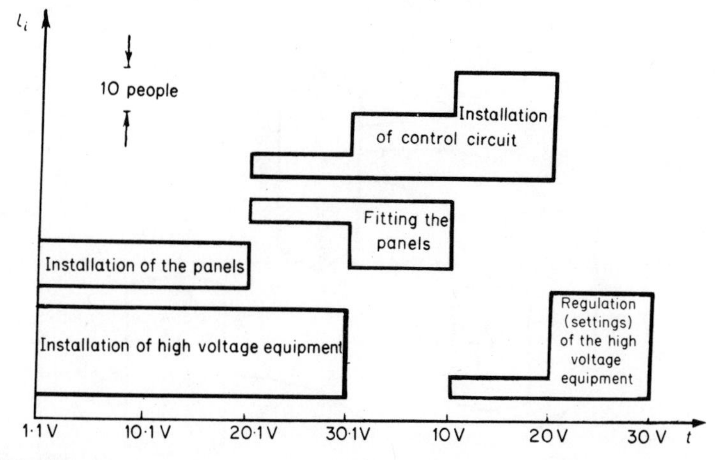


Fig. 15.8. *Sequence of electrical installation work for optimal distribution of the installation personnel.*

can easily be seen that in this case the number of installation personnel $L_{max}$ is reduced to $L^*_{max} = 30$ people. This shows that optimising the distribution of resources, in the given case the manpower, ensures that the work is carried out with fewer people than would be required for the initial, less optimal, plan.

## 15.4. Operative control

When operations are put into practice, deviations from the plan usually occur because many factors have not been taken into consideration in the planning stage, for example the duration of execution of individual operations may deviate from the calculated values, delays may occur owing to the illness of operatives, delays in deliveries of materials. It is often necessary to carry out unforeseen work and some operations which were planned may prove unnecessary. Due to all these factors, it is not sufficient to draw up an optimal plan; it is also necessary to monitor its execution and apply operative control.

Operative control is realised by changing predetermined tasks and redistribution of the resources. For this purpose it is necessary to periodically re-plan the operation as the work progresses, taking into consideration the tasks already accomplished, the real availability of resources and the possible changes in the target dates.

Periodic re-planning of operations is carried out by determining the critical path, the reserves in time and the optimal distribution of resources, using the same methods as were used during the initial planning with the only exception that for the work already accomplished the anticipated dates of completion are replaced by the real dates and the anticipated availability of resources by those really available and at the disposal of the man who controls the operations.

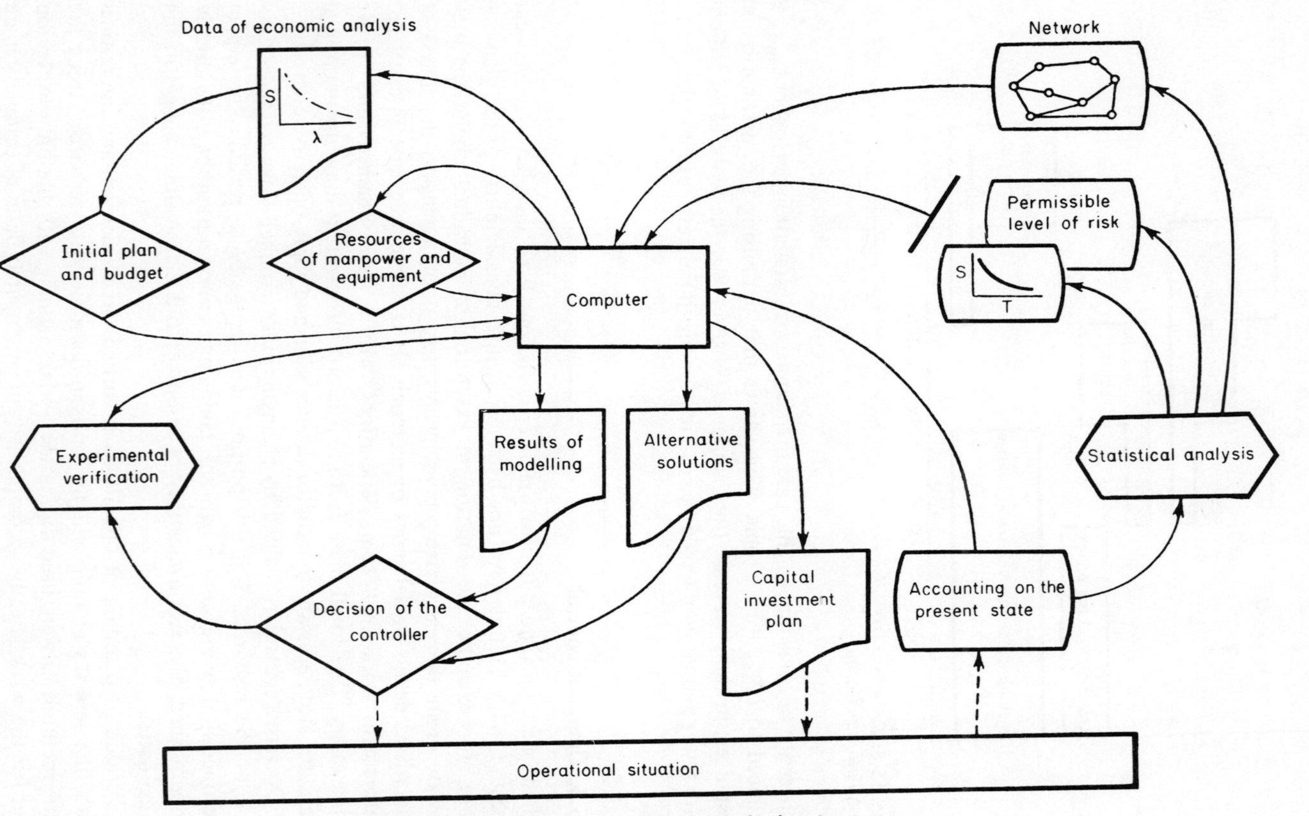


Fig. 15.9. *Block schematics of a network planning system.*

In this case the network model will no longer be static but a *dynamic* model, which changes according to the data obtained on the state of the control system. If the initial plan represented a certain *programme* to be realised, operative control is based on feedback which carries information on the real state of the system and transforms it into a *closed-loop system*.

The structure of a system of operational control, using a network model, is shown in Fig. 15.9. It can be seen that the basis for decision-making is data obtained by calculation of the characteristic of the network model of the operation. These calculations are made by the operations centre (staff) either manually or by computer. The data on the real state of the operation, on the available resources and on the directives, are also regularly fed into an operations centre and used as a basis for working out alternative solutions which are submitted to the controller of the operations. Since the network is a model of the operations, it facilitates tests to detect the consequences of one or another type of solution. Playing on the network with various alternative solutions permits forecasting the further progress of operations for various alternative actions on the part of the controllers and in this way to select the most suitable solution.

Network planning and control systems are being extensively used and enable a considerable improvement in the organisational control of industry, construction, transportation, and scientific and engineering development work.

## Exercises

1. Plot a network containing the operators: 1 - 2, 1 - 3, 2 - 3, 3 - 4, 2 - 4, 3 - 5, 4 - 5. Verify whether this network contains contours.

*Solution.* See Fig. 15.10. The network does not contain contours.

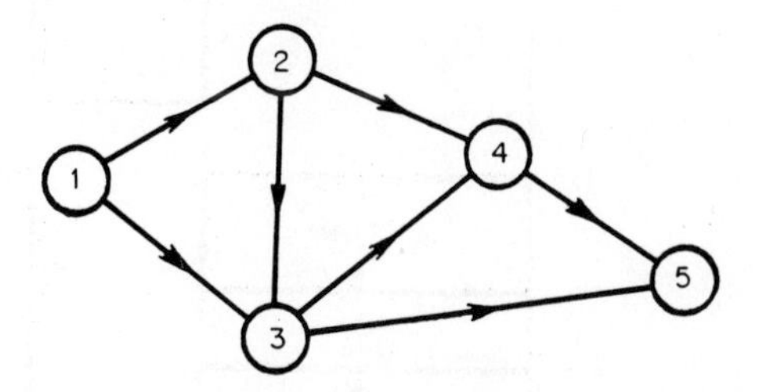


Fig. 15.10. *To Example No. 1.*

2. In the network plotted in accordance with Example No. 1, find the earliest time of occurrence of each event, if the duration of the work is as given in Table 15.3.

Table 15.3

| *Operation* | 1 – 2 | 1 – 3 | 2 – 3 | 3 – 4 | 2 – 4 | 3 – 5 | 4 – 5 |
|---|---|---|---|---|---|---|---|
| *Duration, days* | 5 | 3 | 1 | 3 | 2 | 5 | 2 |

*Solution.* See Table 15.4.

Table 15.4

| *Number of event* | 1 | 2 | 3 | 4 | 5 |
|---|---|---|---|---|---|
| $T_{e}i$ | 0 | 5 | 6 | 9 | 11 |

**3.** Find the critical path in the network shown in Fig. 15.11. Calculate the spare time available for events 4 and 5.

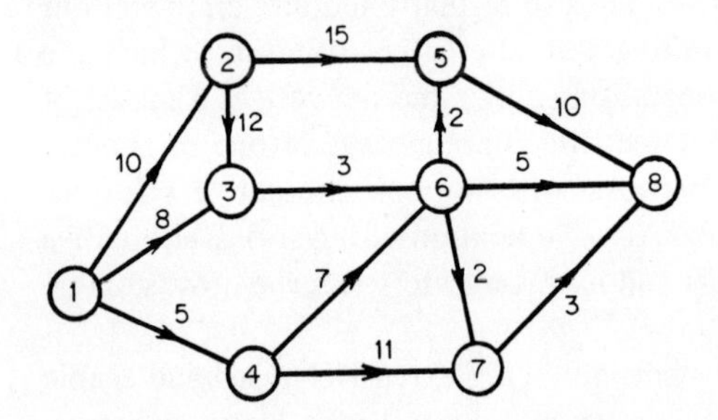


Fig. 15.11. *To Example No. 3.*

*Solution.* (a) The critical path passes through the events: 1 - 2 - 3 - 6 - 5 - 8, its duration is 37 days.

(b) The total spare time for the event 4 is 8 days, for the event 5 – zero (because this event is on the critical path).

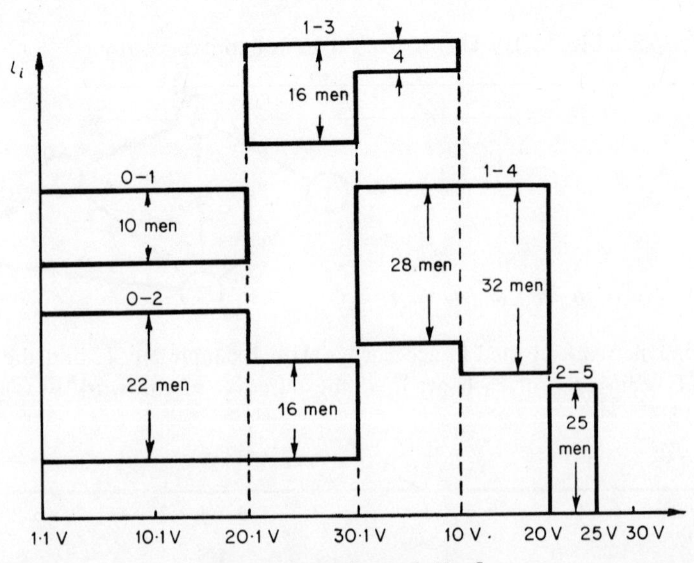


Fig. 15.12. *To Example No. 5.*

**4.** Determine whether the duration of execution of the operation will change if in the network Fig. 15.11:

(a) the duration of the work (operation) 2 - 5 increases to 18 days;

(b) the duration of the work (operation) 6 - 8 decreases to 3 days.

*Solution.* (a) The duration of the operation increases by 1 day.

(b) The duration of the operation does not change.

**5.** Find the optimal distribution of resources for the complex of operations shown in Fig. 15.8, if the labour available on the installation work increases to 600 man days and the work on regulating the high voltage equipment decreases to 125 man days.

Determine the difference between the requirement in personnel in the case of uniform and optimal graphs of using personnel for each operation.

*Solution.* In the case of uniform graphs the number of operatives for each operation is 60 people; in the case of using optimal graphs it is 32 people. The difference is 28 people. One of the alternative solutions is shown in Fig. 15.12.

# 16 The Brain

The high performance of control systems which ensures the continuance of life in the animal world has attracted the attention of scientists for many centuries. The control in living organisms is one of the most jealously guarded secrets of nature, into which science has so far penetrated very little. The problem of control in living organisms is being studied from various aspects by representatives of various scientific disciplines. Physiologists and psychologists study the structure of the control systems on the basis of investigations into the behaviour of organisms, histologists by elucidating the cell structure; biophysicists and biochemists investigate the physical and chemical processes forming the basis of control of biological systems. Cybernetics also contributes to the study of control systems in living organisms. The task of cybernetics is to understand and reproduce the features of living control mechanisms in terms of the general relationships in control theory.

The birth, development and activity of each living organism are controlled by a great variety of control systems which differ in the way they solve problems and the methods applied in their solution. The life of each individual cell cannot be visualised without a control system which reconstructs the cell's structure under conditions of continuous breaking up and ensuring the fulfilment of the cell functions. The complex and highly efficient system required for ensuring homeostasis is necessary for in order to control the processes in the body in such a way that certain vital quantities in the organism should not exceed permissible limits. Adaptation of the body to changing conditions of the environment also requires the continuous participation of very fine control mechanisms.

The methods used in various living organisms for exchanging information between the individual organs and the transmission of control signals from the control centres to the motor organs may differ greatly. It is known that these functions are accomplished by the *hormone systems*, i.e. systems based on the use of specific chemical substances produced by certain organs (e.g. the thyroid) and act on other organs, to regulate their functioning. Transmission of information and control commands is also by signals circulating in the *nervous system*, which in addition is responsible for storage and complex processing of information which ensures the living ability and purposeful behaviour of the body. In this chapter elementary information is given on the control processes realised by the nervous system.

## 16.1. Neurons

The nervous system consists of individual cells, called nerve cells. Various types of nerve cells are classed under the general term *neuron*. The neuron is a specialised cell with the properties of being excitable and able to conduct signals in a certain direction. A schematic diagram of a neuron is given in Fig. 16.1. It consists of the cell body, containing the *nucleus* and the *cytoplasm*, which are enclosed in a shell, a *membrane*, from which emanate branching extensions,

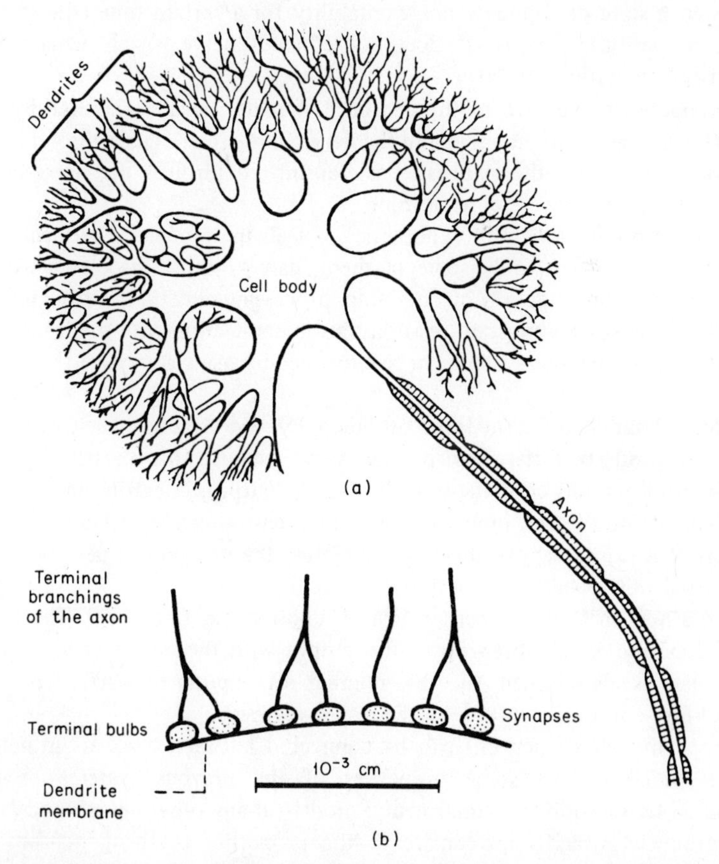


Fig. 16.1. *Schematic diagram of a neuron: (a) cell body; (b) ends of the axon.*

*dendrites* and an axial extension, the *axon*. The terminal branches of the axon are in contact with other cells through the *synaptic contacts (synapses)*. From the axon there are other branchings, *collaterals*, which are also in contact with other cells.

Three types of neuron are distinguished: sensory, motor neurons and association neurons.

Excitation of a neuron depends on the nerve signals which are applied to the synapses, which may be either exciting or inhibiting ones.

The number of synapses on one neuron varies between a hundred and a thousand. The state of the neuron membrane (its electric potential) depends on the magnitude and duration of the signals fed to the synapses. When the potential of the membrane reaches a certain *threshold* value, about 40 mV, a neural impulse occurs – an activity wave, which propagates over the nerve fibre of the axon. The excitation which is transmitted over the fibre represents an electrochemical process; the speed of its propagation depends on the diameter of the fibre and varies between 1 and 150 m/sec. After carrying the pulse, the nerve fibre is in a state of complete non-excitability for a certain time (the so-called refractory period), that is, it does not conduct nerve signals whatever the intensity of excitation.

A characteristic feature of the neuron is that the signals generated by it do not differ in magnitude; the signal in the nerve fibre either equals zero or has the maximum value. In other words, it is transmitted through the nerve cell by means of a binary code – 'all or nothing'.

From the point of view of cybernetics, not all the properties and features of neurons are important, only some of them, namely those associated with the performance of the function of obtaining, processing and transmitting information. Therefore some important relationships in connection with the information processes in the nervous system can be explained by using a very rough model of a neuron.

A logical model of a neuron, proposed by the American scientists W. S. McCulloch and W. Pitts, has proved very useful. This model represents multi-input threshold equipment with a single output, operating in the binary code. The action of each input on the formal neuron can be either +1 or –1. If the sum of actions is above the threshold, then the neuron will become excited and a signal will appear on its output.

Such a neuron realises a certain logical function, see Chapter 9. The type of logical function which inter-relates the output with the signals at the input of the model is determined by the number of inputs of each type, their connections, and the magnitude of the response threshold.

Formal models of neurons can be connected between brackets in network circuits which possess some properties of the nervous systems of living organisms. In addition to formal neural models, many other neural models have been produced which approximate to the properties of living neurons to a certain extent. In these models it is possible also to represent such features of neurons as the space-time summation of signals, refractory period (time inexcitability), accommodation (adaptation to the nature of the stimuli), and spontaneous excitability. These models permit the working out of hypotheses as to the functioning of neurons. However near the behaviour of a model may be to that of a neuron, it does not allow conclusions to be drawn on the structure of the neuron from the structure of its model. This is because of the impossibility in principle of judging the internal structure of a 'black box' from the reactions of the output to the input value, as was shown in Chapter 3.

## 16.2. The nervous system

The *central nervous system* of man (Fig. 16.2) is composed of the brain and the spinal cord, in the same way as for all other vertebrates. The brain has the following main parts: (1) the medulla oblongata; (2) the cerebellum; (3) the midbrain; (4) the corpus striatum; (5) the cerebrum. Signals are received by the central nervous system from sensory elements, *receptors*, and are transmitted to the motor elements, *effectors*, by means of a network of nerve fibres forming the *peripheral nervous system*. Each part of the nervous system, as well as the nervous system as a whole, is made up of neurons, connected to each other by synaptic contacts, which connect the axon of one neuron with the cell body of another. The estimated number of neurons which make up the nervous system of man is of the order of $10^{11}$, of which about half are in the corpus striatum. The bodies of the nerve cells of the frontal brain are mainly concentrated in its surface layer, the *cerebral cortex*, where apparently the main centres responsible for the conscious activity of the brain are located.

A lucid, although very rough and primitive idea on the structure of the information links formed by nervous systems, which also ensure the control of the body, is conveyed by Fig. 16.3. Data on the external medium collected by the sensory organs, transmitted to the receptors and then over the nerve fibres to the peripheral nervous system, are channelled into the appropriate sections of the central nervous system. Here the information obtained is processed, evaluated, compared with the stored information and, if necessary, commands are formed which are transmitted over the nerve fibres to the effectors which act on the motor organs (muscles, glands). The activity of the motor organs is monitored by feedbacks: internal, which verify the intensity of the action of the motor organs, and external, which verify the final effect of the realisation of commands.

It can be seen that the structure of the system described here has many characteristics in common with the structures of the automatic control systems considered in Chapter 7.

Information is collected by means of the appropriate receptors and effectors, and commands are sent not only regarding the interaction of the body with the medium but also for ensuring the normal functioning of its internal organs and systems.

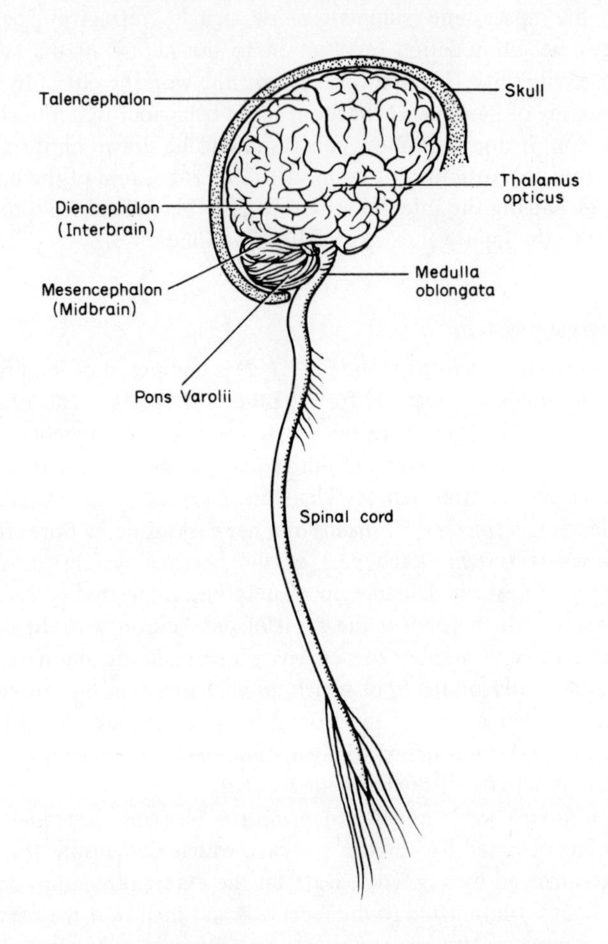


Fig. 16.2. *Sketch of the central nervous system of man.*

A simplified model in the form of a nerve network proposed by McCulloch and Pitts, consisting of an accumulation of inter-connected neurons, has made it possible to use formal methods for studying the nervous system. Each nerve network is composed of input neurons, internal neurons and output neurons. The processes in such networks are considered as processes in discrete instants of time $t = 1, 2, \ldots$, in the same way as processes in finite automatons (see Chapter 9). The state of each input neuron, whether excited or not, is determined by the reactions of its adjoining perceptual organ to the state of the medium. The state of each internal and output neuron of the nerve system at time $t$ is determined by the state of its inputs at the previous instant of time $(t - 1)$. The neuron will be excited if at least $h$ of its excited inputs (synapses) were excited at the previous instant and if not one of the inhibitory inputs has been excited. The value $h$ is called the *threshold* of this neuron.

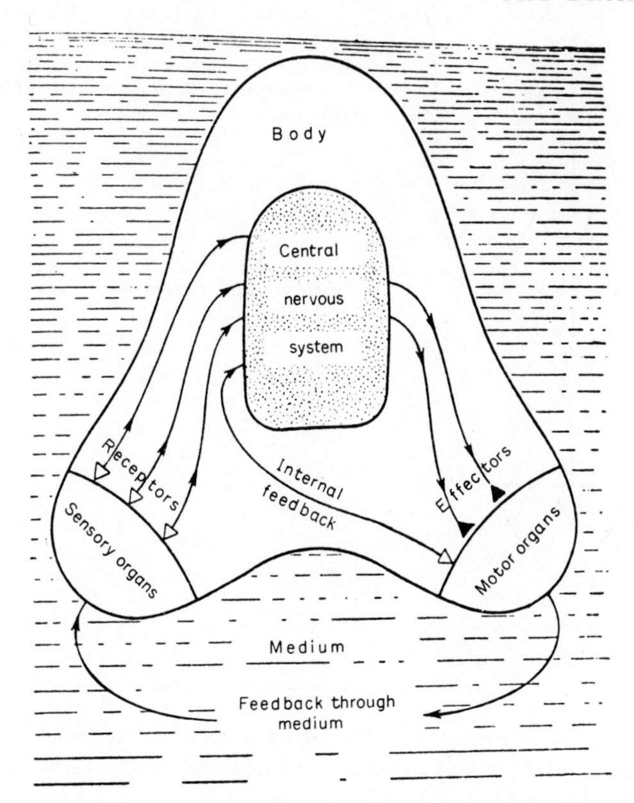


Fig. 16.3. *Structure of the information links effected by the nervous system.*

Simple examples of nerve networks are shown in Fig. 16.4. Here $X_i$ denotes the state of the input, $Y$ of the output, and $Z$ of the internal neurons; the exciting synapses are designated by black dots and the inhibitory ones by circles; the numbers in the triangles depicting the neurons give their threshold values.

The network shown in Fig. 16.4(a) realises the conjunction operation, since the logical function which determines the condition of excitation of the output neuron for a threshold value $h = 3$ can be written as follows:

$$Y(t) = X_1(t-1) \& X_{-2}(t-1) \& X_3(t-1) \& \overline{X}_4(t-1) \& \overline{X}_5(t-1). \quad (16.1)$$

It can easily be seen that if the threshold changes, for instance to $h = 2$, the function which realises the network 16.4(a) can be expressed as:

$$Y(t) = \{[X_1\ (t-1) \& X_2(t-1)] \vee [X_1(t-1) \& X_3(t-1)] \vee [X_2(t-1) \& X_3(t-1)]\} \ \& \overline{X}_4(t-1) \& \overline{X}_5(t-1). \quad (16.2)$$

The operation of logical addition (disjunction) is realised by the network shown in Fig. 16.4(b) when the threshold $h = 1$

$$Y(t) = X_1(t-1) \vee X_2(t-1) \vee X_3(t-1). \quad (16.3)$$

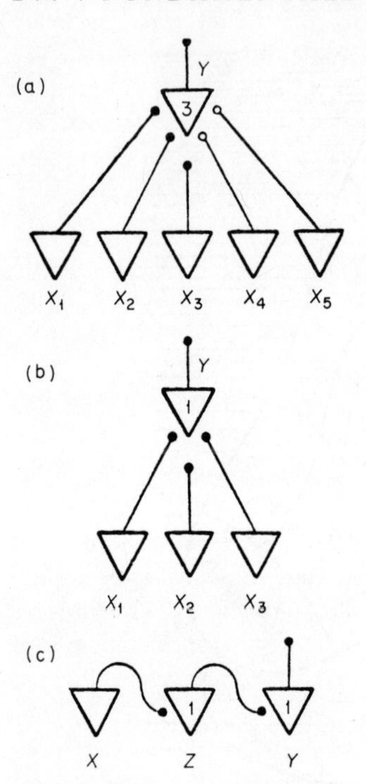


Fig. 16.4. *Simple principles of nerve networks: (a) network for the conjunction operation; (b) network for the disjunction operation; (c) network for delaying by 2 instants.*

The delay by $k$ instants is achieved by a network consisting of an input, output and $(k - 1)$ internal neurons with a threshold $h = 1$. A delay network for two instants is shown in Fig. 16.4c.

$$Y(t) = Z(t - 1),$$

$$Z(t) = X(t - 1),$$

and therefore

$$Y(t) = X(t - 2). \tag{16.4}$$

If the nerve network contains loops (feedback), then its state at any given time depends on its initial state, i.e. on the state of the network at the instant when it formed. Thus, for instance, the network shown in Fig. 16.5(a) has

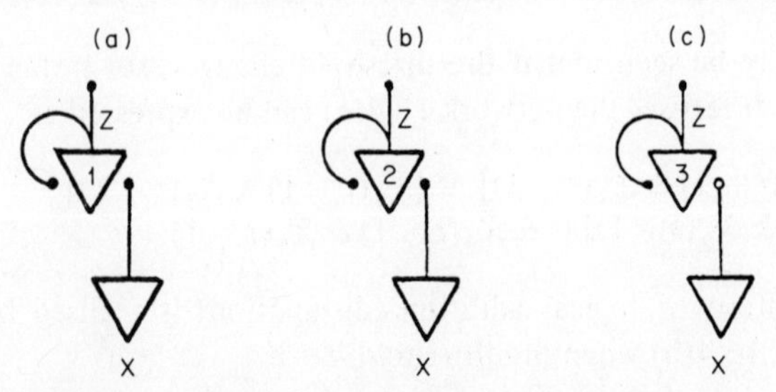


Fig. 16.5. *Examples of nerve networks with loops.*

the following properties. If at the initial instant of time $(t - 1)$ the neuron X or Z were excited, then the neuron Z will be in an excited state during all the subsequent moments due to the loop excitation of Z in the $i^{th}$ instant which will cause the excitation of Z in the $(i + 1)^{th}$ instant. The excited state Z will also be observed if the input neuron X was excited at least once at any time between the initial instant and the time considered.

The neuron Z of the network shown in Fig. 16.5(b) will be excited in a moment following the moment considered, only if it was excited in the initial moment, and the neuron X will remain excited from the initial moment up to the moment considered.

The network shown in Fig. 16.5(c) differs from that of Fig. 16.5(a) in that the synapse which connects X and Z is an inhibitory one. The excited state Z in such a network can occur only if it was excited in the initial instant, but the neuron X has not been excited a single time.

Networks with loops differ in principle from networks without loops as the former are capable of remembering some events (the state of neurons at a certain instant of time) over a very long period, whilst networks without loops conserve the traces of events only for a number of instants not larger than the number of synapses in the longest chain of the series-connected neurons contained in the given nerve network.

Compared with the initial state, a change of the state Z in the networks shown in Fig. 16.5(a) and (c) is due to the fact that some time in the 'history' of this network an event occurred – the input neuron was excited, and this state Z is conserved 'forever' so long as this network is in existence. The network shown in Fig. 16.5(b) is capable of 'memorising'. From the moment it was created the input neuron was always, without exception, in the excited state.

Up to now it has been assumed that the neurons and their inter-connections operate faultlessly. This assumption is obviously not realistic. As with artificial equipment such as finite automata, living nerve networks are subjected to various factors which may cause malfunctioning, i.e. actions which are not in accordance with the initial structure of the connections and the threshold values of the individual elements. It is more realistic to assume that for each element of the network there is some probability $\epsilon > 0$ of malfunctioning. However small the value of $\epsilon$, it is necessary to accept the fact that errors may considerably change the properties of nerve networks. As an example of the fundamental changes which are introduced by the assumption of malfunctioning in the properties of the systems, we will consider the long-range memory organ, the neuron with a loop, shown in Fig. 16.5(a). In the case of faultless functioning, such an organ, once excited, will always remain in the excited state.

Let us assume now that this organ may make an error with the probability $\epsilon$. If we consider the state of this organ starting from $S$ instants onwards after it has been excited and we select $S \gg 1/\mu$, then it is clear that the probability that during this time the organ may make a mistake as a result of which the excitation could be lost or that it may become excited spontaneously, is near to 1.

Therefore from the state of this organ at the moment of time under consideration, no conclusion can be drawn on its state in the sufficiently distant past. If the network contains a large number of inter-connected elements, and if the error in one or several of them leads to an error in the system as a whole, then the probability of the malfunctioning of the circuit increases and becomes appreciable even for very low values of $\epsilon$.

Under these conditions the reliable and stable operation over many years of the nervous system, which contains thousands of millions of individual neurons, each of which will obviously not have an infinitely high reliability, becomes incomprehensible. However, it has been found that structures of circuits exist which have a higher reliability than the reliability of their components. Some methods of increasing the reliability were proposed and substantiated by the American scientist J. von Neumann. One such method is the use of *multiple lines*. Each signal in the network is transmitted to a bundle of $N$ lines. Taking some confidence level $\delta$ $(0 < \delta - 0.5)$, we consider that a signal is present if the number of lines in the excited state is larger than $(1 - \delta)N$ and that a signal is absent if the number of lines in the excited state is below $\delta N$. The state is indeterminate if the number of excited lines is within the range $\delta N$ to $(1 - \delta N)$.

A necessary element of systems with bundles of lines is a *mixer*, Fig. 16.6, which realises the function

$$Y(t+1) = (X_1(t) \,\&\, X_2(t)) \vee (X_2(t) \,\&\, X_3(t) \vee (X_1(t) \,\&\, X_3(t)). \qquad (16.5)$$

that is, an organ which reacts to a 'majority vote'; it is excited if in the previous cycle at least two of its three inputs were in the excited state.

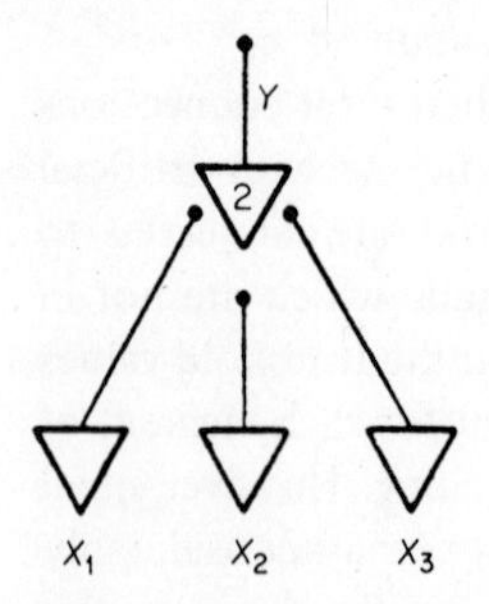


Fig. 16.6. *Mixer.*

The principle of increasing the reliability of a network by means of multiple lines can clearly be seen from the circuit shown in Fig. 16.7, intended for realising the logical function $Y = R\,(X_1, X_2, X_3)$. Here the number of lines in the bundle $N = 4$, and a mixer C is used for recovery of the signal. As the confidence level we assume $\delta = 0.25$, which means that if no more than a quarter of the lines in the bundle (i.e. not more than one line) is excited, we consider that there is no signal, and if no less than $(1 - \delta)N = 3$ lines are in the excited state, we assume that a signal is present. It can be seen from Fig. 16.7 that if, as a result of an error, one of the lines in the bundle emanating from the circuit $R$ (repeated 4

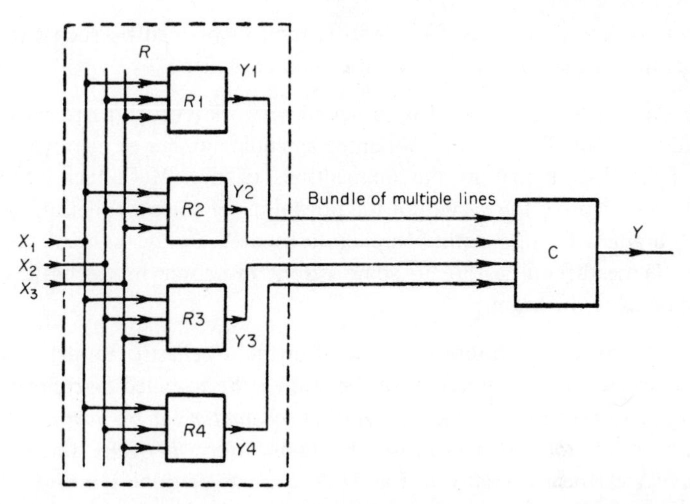


Fig. 16.7. *Improvement in the reliability of networks by means of multiple lines.*

times) is excited, this will not produce an excited state at the output of the mixer. If, however, for the signal '1', one of the lines in the bundle emanating from $R$ proved to be in the non-excited state, this will not stop the output from the mixer being in the excited state. This circuit, however, does not increase (improve) the determinateness when two lines in the bundle are excited.

There is a possibility that a certain clustering of neurons in the nervous systems of animals and man can function as regenerative organs and ensure reliable operation of complex nervous structures.

The described properties of neural networks point to an equivalence between neural networks and finite automata in the sense that for any neural network an automaton can be chosen whose behaviour (reaction to a sequence of input signals) will be similar to that of the neural network under consideration. Neural networks made up of formal neurons have a number of characteristics similar to those of the nervous systems of animals and human beings. They enable modelling of the regulator functions, reflex and conditional-reflex mechanisms, and some anomalies caused by nervous diseases. It does not follow, however, that finite automata are equivalent to a living nervous system and that any function of the brain can be simulated by finite automata; the McCulloch-Pitts nerve networks are only simplified models of a very complex system, the brain, although obviously it cannot be stated that the opposite is true i.e. that the brain can simulate McCulloch-Pitts nerve networks.

## 16.3. The perceptron

The McCulloch-Pitts nerve network represents a deterministic model of a nervous system, i.e. with a fixed structure of connections and threshold values. Such a model corresponds to the concept of the brain as a type of specialised computer. In addition to useful scientific results obtained by such a model, it also leads to considerable difficulties, the main ones being:

(a) The class of problem solved by a neural network is much narrower than the potentialities of the living brain containing an equal number of neurons.
(b) A slight disturbance in the connections of the (McCulloch-Pitts) nerve network may deprive the system of the possibility of correct functioning, whilst the real brain will cope with complex tasks even in the case of very great damage. These difficulties are to some extent overcome in stochastic (probability) models of the brain.

One of the most interesting and fruitful stochastic models used for reproducing some of the functions of the brain is the so-called *Perceptron.*

*Perceptrons are models of a certain class, characterised by having a memory and with a random structure of the connections between the individual elements.* A classical Perceptron, Fig. 16.8, consists of three types of threshold elements:

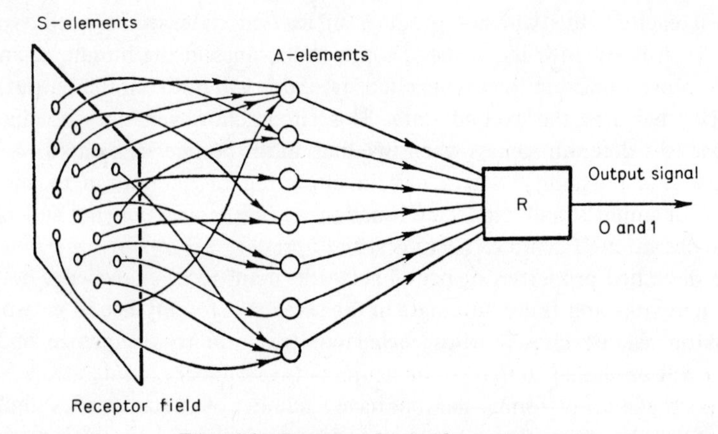


Fig. 16.8. *Diagram of a simple perceptron.*

(1) S-elements – receptors which receive information on the external conditions in a manner similar to the sense organs of a living body;
(2) A-elements – associative elements which receive every signal of some of the S-elements.
(3) R-elements – form the output signal under the effect of an A-element acting on them. The connections between the A-elements and the R-elements are variable, so that the degree of their action on R-elements may increase under the effect of stimulation signals.

Let us assume the receptor field (of S-elements) of the perceptron receives an input. It is assumed that by some method the R-element reacts to what occurs, positively or negatively in a certain sense, and in the case of a positive reaction a stimulation signal is generated. Then the 'useful' connections in the Perceptron will gradually intensify and increase the frequency of the positive reactions of the Perceptron to situations perceived by the receptor field. In this way the Perceptron can be taught to exhibit a certain behaviour imposed on it by the system of stimuli, however intensive the connections at the beginning of the process of learning, and regardless of the fact that the connection circuits of its S-elements with the A-elements and of the A-elements with the R-elements were of a random nature.

It is pointed out that the character of the behaviour of a Perceptron, as that of any other learning machine, is established under the influence of a system of stimuli and is determined by the reaction considered as being a good one (a positive one). If the stimulation signal is received from man, the 'teacher', then the behaviour of the Perceptron will gradually approach what the 'teacher' desires of it. If the stimuli are obtained by comparing the reactions of the Perceptron with those of any other system to the same situation, then the Perceptron will imitate the behaviour of that system. Without systematic stimulation there will be no purposeful change in the behaviour of the Perceptron.

The Perceptron is the first automaton capable of learning pattern recognition (see Chapter 13). However, automata of the Perceptron type are also capable of solving many other problems, including problems solved by the living brain, if their structure is made complex enough. They will not only learn to recognise and classify a situation in the external medium as perceived by the receptors, but also to evaluate a situation and generalise concepts. Consequently Perceptrons possess the property of perception, although only in a very primitive form, and therefore they are very attractive and fruitful for simulating some intellectual functions of the brain. Obviously the similarity between the Perceptron and the brain is purely *functional* and does not apply to the structure and the mechanisms of operation.

## 16.4. Thinking

Despite the fact that many processes in the brain have been studied, the essence of cognitive phenomena such as thought, fantasy, memory, remain unclear, although everyday experience evidences the fact that these phenomena exist, and many people have formed intuitive notions on them. Numerous investigations have established a certain role of the nervous system and in particular of the brain in realising this intellectual activity of organisms.

From the information given in Section 16.2 on neural networks, consisting of formal neurons, it can be seen that such networks possess a variety of properties and complex behaviour, despite the fact that the properties of neurons of which they are composed are very simple. It can be stated with justification that the enormous intellectual potentialities of the brain are not determined by the properties of the individual neurons but by the complicated structure formed by the thousands of millions of its nerve cells.

Study of the evolution of the brain in the process of development of living organisms on the earth, the elucidation of its functions and role in ensuring their survival, lead to the conclusion that the brain has developed as a specialised organ for survival by adaptation of the body to a medium which is varied and changes with the progress of time.

However, adaptation would be impossible without information about the environment. Therefore the function of the brain as an organ ensuring adaptive behaviour consists primarily in processing the information obtained on the conditions surrounding the organism and the state of the organism under these conditions. Thus the brain can be considered as a very complicated cybernetic system which brings about the control of the organism on the basis of processing the information it receives.

The extremely high complexity of the internal structure of the brain makes its detailed study very difficult, and therefore it is attractive to use the 'black box' approach. All the physiological techniques of study of the higher nervous activity of animals and man are based essentially on the 'black box' principle.

I. P. Pavlov and his school havc laid thc foundations of this branch of science, based on objective evaluation of the reactions of the body on various complex stimuli. As a result of their investigations some important relationships governing the operation of the brain have been discovered. One of the important results of the work of Pavlov is the development of an experimental technique of generation and extinction of a *conditioned reflex*. In contrast to the *unconditioned reflex*, which represents the innate ability of the body to react in a certain way to external stimuli, the conditioned reflex is a reaction of the body to a conditioned stimulus which indirectly, in association with an unconditional stimulus, produces a certain response reaction.

In the classical investigations of Pavlov on dogs, the conditioned reflex was produced by a bell. The response was that of salivation; the bell was switched on only before food was given to the dogs. The dogs then developed an association between the conditioned stimulus, the bell, and the unconditioned stimulus, the food. Further development of this technique by complicating the set of actions enabled revelation of more complex relationships of the operation of the brain.

The fundamental importance of studying the mechanisms of generation of conditioned reflexes to the theory of higher nervous activity is due to the fact that the brain is considered in cybernetics as a learning system, and the conditioned reflex as an elementary act of learning. This point of view is based not only on the observed learning ability of living organisms, but also on the obvious inadequacy of the memory in the genetic carrier (about 50,000 genes in man) to transmit to offspring information on the structure of the brain, which consists of millions of neurons.

The living brain is not the only one capable of learning. Artificial systems, in particular automata as described in Chapters 9 and 12, are also capable of learning to fulfil such biologically important functions as the classification of complex situations and adaptation. It is very easy to realise in artificial systems a

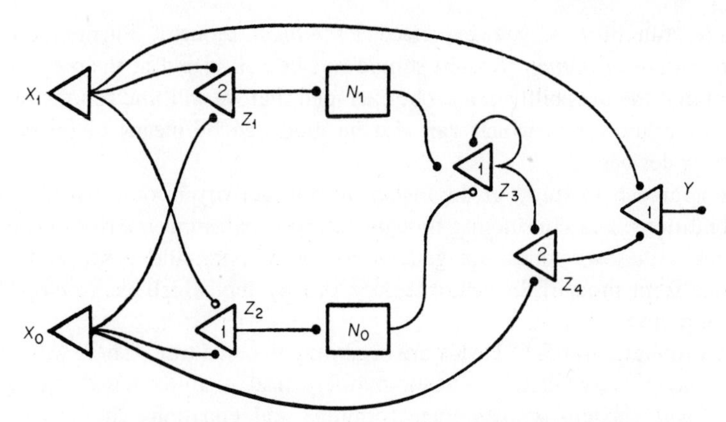


Fig. 16.9. *Model of a conditioned reflex with a storage device.*

model of a conditioned reflex. One of such models, Fig. 16.9, contains *accumulators*, in addition to ordinary formal neurons. Accumulators have the property of transmitting pulses to the output after receiving from the input a number of pulses equalling the threshold of the accumulator. This can be realised by means of a pulse counter described in Chapter 9.

The threshold of the accumulator changes in a random manner; when the counter is full up it reverts to zero and begins to store the pulses again. The unconditioned stimulus is perceived by the receptor $X_1$, and the conditioned one by the receptor $X_0$; the output effector neuron $Y$ controls the reactions on the unconditioned stimulus. It can be seen from Fig. 16.9 that the internal neuron $Z_1$ will transmit pulses into the accumulator $N_1$ every time the conditioned stimulus $X_0$ is intensified by the unconditioned stimulator $X_1$. After a sufficiently large number of coincidences (determined by the threshold $N_1$) the memory neuron $Z_3$ is excited and the stimulus of the receptor $X_0$ will produce (via the neuron $Z_4$) the reaction $Y$.

If the conditioned stimulus $X_0$ feeds signals without intensifying stimuli sufficiently frequently (presumably from $X_1$), then the neuron $Z_2$ will transmit into the accumulator (storage device) $N_0$ a sufficient number of pulses for exciting its output, which will remove the excited state of the neuron $Z_3$ and by so doing, destroy the earlier formed conditioned reflex.

The model of the conditioned reflex can be developed into a matrix scheme which permits establishing a conditioned reflex relationship between some of the many stimuli and reactions.

It cannot be assumed that the multitude of complex phenomena of thought can all be reduced to conditioned reflexes. The finer mechanisms of human thought are undoubtedly based on a great variety of information processing methods, most of which still remain to be discovered and understood.

Of great importance from the point of view of strict materialistic understanding of thought is the work of A. N. Kolmogorov, who demonstrated convincingly that it is possible in principle to make artificial systems which can perform the same functions as are expressed by *human* thought. Furthermore, the specific nature of *human* thought should not be considered as the only form of thought and the possibility cannot be excluded that we will find (on the earth or in space) other forms which can also be modelled by means of information processing devices.

The approach to study of the higher nervous activity in objective physiological conditions, and considering thought phenomena as information processing, complies with laws which are general for both living and artificial systems, removing from thought the veil of secrecy and mystery which has developed as a result of prejudice.

The prophetic words of Pavlov are beginning to come true. 'There will come a time – maybe very distant – when mathematical analysis based on natural science, will encompass with huge formulae and equations all the balancing (between the body and the medium) achieved by the brain, including ultimately also itself'.*

**Exercises**

**1.** Fig. 16.10 shows a neuron with six synapses, all of which are stimulating. What logical function $Y(t + 1)$ can this neutron realise with respect to the inputs $X_1, X_2$ and $X_3$ at the time $t$, if the threshold value $h = 1$?

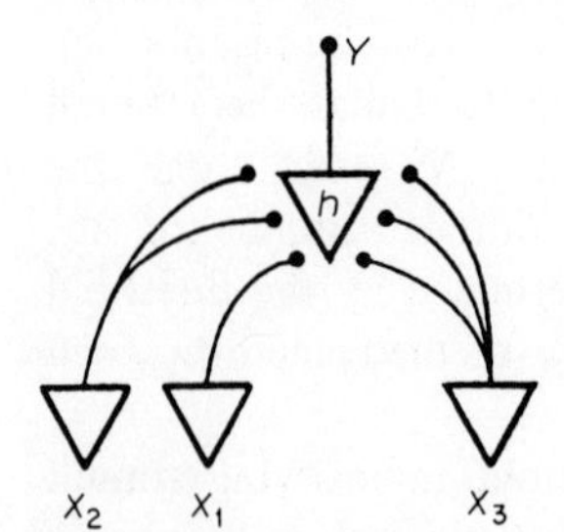


Fig. 16.10. *To Example No. 1.*

*Solution.* $Y(t + 1) = X_1(t) \vee X_2(t) \vee X_3(t)$.

**2.** (Continuation). How does the function $Y(t + 1)$ change if the threshold value $h$ changes from 2 to 6?

*Solution.*

(a) $h = 2$: $Y(t + 1) = X_2(t) \vee X_3(t)$;

(b) $h = 3$: $Y(t + 1) = X_1(t) \& X_2(t) \vee X_3(t)$;

(c) $h = 4$: $Y(t + 1) = X_1(t) \& X_3(t) \vee X_2(t) \& X_3(t)$;

(d) $h = 5$: $Y(t + 1) = X_2(t) \& X_3(t)$;

(e) $h = 6$: $Y(t + 1) = X_1(t) \& X_2(t) \& X_3(t)$.

* 'Natural Science and the Brain' – Speech at the 12th General Meeting of Naturalists and Doctors held in Moscow, December 28, 1909.

**3.** What function does the network shown in Fig. 16.11 realise, if the thresholds of both neurons $Z_1$ and $Z_2$ equal $h_1 = h_2 = 1$.

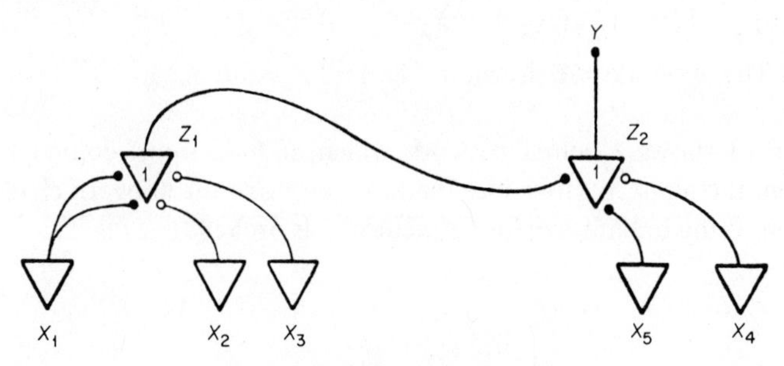


Fig. 16.11. *To Example No. 3.*

*Solution.* The network realises the function

$$Y(t+2) = (X_1(t) \,\&\, \overline{X}_2(t) \,\&\, \overline{X}_3(t) \,\&\, \overline{X}_4(t+1) \vee (X_5(t+1) \,\&\, \overline{X}_4(t+1)).$$

**4.** (Continuation). What value does the function $Y$ assume, if the thresholds of both neurons increase to $h = 2$?

*Solution.*

$$Y(t+2) = X_1(t) \,\&\, \overline{X}_2(t) \,\&\, \overline{X}_3(t) \,\&\, X_5(t+1) \,\&\, \overline{X}_4(t+1).$$

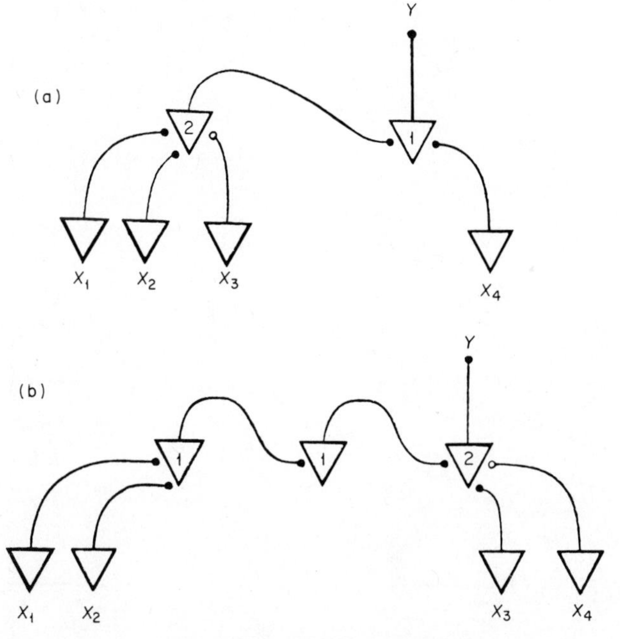


Fig. 16.12. *For Example No. 5.*

**5.** Construct the neural network which realises the following functions:

$$\text{(a)}\quad Y(t+2) = \overline{(X_1(t)\ \&\ X_2(t)\ \&\ X_3(t))} \vee X_4(t+1);$$
$$\text{(b)}\quad Y(t) = (X_1(t-3) \vee X_2(t-3))\ \&\ X_3(t-1)\ \&\ \overline{X_4(t-1)}.$$

*Solution.* The networks are shown in Fig. 16.12, a and b.

**6.** Fig. 16.13 shows a neural network which at the output counts $Y$ pairs of pulses from the input $X$. How does the functioning of the network change, if the connection to the braking synapse of neuron 2 is broken?

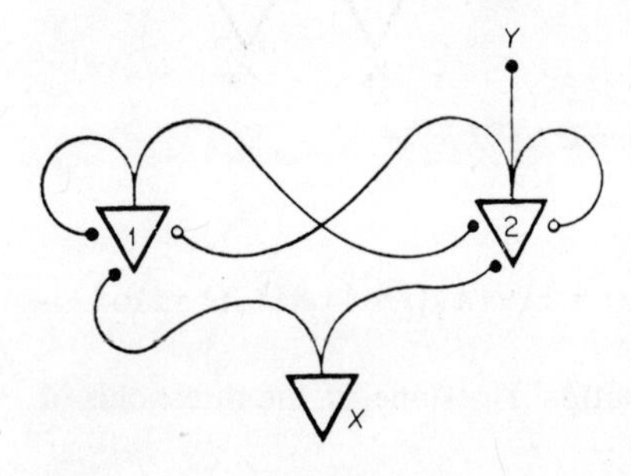


Fig. 16.13. *To Example No. 6.*

*Solution.* In this case, if a series of pulses arrives at the input $X$, then all except the first will arrive with a delay of one instant at the output $Y$.

# 17 Organised Systems

When studying the problems relating to the control of complex systems, attention must be paid to differences in the systems as regards their degree of ordering and organisation. An essential property of any organised system is that it is ordered to a greater or lesser extent. It is impossible to visualise a system more chaotic than moving molecules in thermodynamic equilibrium. Therefore, as the beginning of ordering it is useful to assume a state of such a chaotic system whose entropy $H$ will be maximal and will equal $H_m$.

As a yardstick of the degree of order $R$ of the system, we will take the degree of deviation of its state from that of thermodynamic equilibrium, using for this purpose the 'redundancy' notion introduced by Shannon:

$$R = 1 - \frac{H}{H_m} \tag{17.1}$$

Then the degree of ordering of any system will be evaluated by the magnitude $R$, which varies between 0 and 1 and assumes the value 0 for a system in complete disorder and 1 for an ideally ordered system, in which the state of all its elements is unequivocally determined, and, consequently, the entropy of the system is equal to zero. However, an order differing from zero is not enough to class the system as an organised one. For instance, the planetary system has a high degree of order but cannot be considered as an organised system. If irreversible processes, such as heat transfer, occur in the system and external random disturbances act on it, then its entropy will increase and, consequently, the degree of its order will decrease and may ultimately drop to zero. To compensate for this natural fall of order, to retain the order of the system, it is necessary to receive in some way from the environment a negative entropy – *negentropy,* for instance in the form of information which can be used in re-establishing the order.

Consequently, the second important distinguishing feature of an organised system is its ability to absorb negentropy and use it to maintain or even increase its degree of order. In the latter case the system will not only be organised but it will be *self-organising.*

Finally, a third feature of an organised system is the presence in it of functionally differing inter-connected parts, which allow us to distinguish the structure and purpose of some elements of the system from the structure and purpose of other of its elements, and to establish the nature of interaction between the individual elements and the medium surrounding the system.

In this chapter we will consider some problems of control in organised systems: artificial organised systems – organisation and living organised systems – organisms.

## 17.1. Maxwell's demon

The problem of increasing the degree of ordering and consequently the degree of organisation of the energy aspcct of a physical system was very clearly formulated by Clerk Maxwell in 1871, in his book 'The Theory of Heat', in the form of a paradox which is incompatible with the second law of thermodynamics.*

Imagine, said Maxwell, some being, a 'demon', so perfect that he can follow the movement of each gas molecule which fills a vessel. Then this vessel is divided by a barrier into the two parts A and B, and in this barrier a valve is fitted through which an individual molecule can pass from one part of the vessel into the other when the valve is open (Fig. 17.1).

Let us assume that initially the vessel is completely filled with gas at a certain temperature, which according to the kinetic theory of heat corresponds to a certain average speed of the molecules. Since the movement of the gas molecules is of a random nature, there will be molecules whose speed is above the average and molecules whose speed will be below the average. Then, by opening the valve in the barrier at the appropriate instants of time, the demon can enable the faster molecules to pass from A to B and the slower molecules to pass from B to A, and as a result increase the temperature in the part B and lower it in the part A without expenditure of energy.

The paradox is illustrated if it is considered that in order to control the valve it is necessary to have *information* available on the movement of the molecules, which cannot be obtained without a certain expenditure of energy, which is larger than the gain in energy obtained as a result of sorting the molecules into 'fast' and 'slow' ones. If we consider the system: vessel, gas, barrier and demon as being in a state of thermodynamic equilibrium, i.e. there are no processes of energy conversion or passage from one part to the other in the system, then it is impossible in principle to follow the movement of the molecules because such a

* According to the second law of thermodynamics, any change in an isolated system can only lead to an increase in its entropy.

Fig. 17.1. *Maxwell's Demon.*

system does not contain signals which could serve as sources of information on trajectories and velocities. In order that the 'demon' should be able to estimate the movement of the molecules he must at least see them, and for this purpose it is necessary to provide light. However, the source of light is a system which is not in equilibirum and cannot function without expenditure of energy.

It has been found that the quantity of energy necessary to obtain the required information will more than absorb the gain from utilising it, and therefore there is no infringement of the second law of thermodynamics.

It can be seen from the example considered that even such a primitive ordered system as the separation of 'fast' gas molecules from 'slow' ones is impossible without information, to obtain which it is necessary to introduce negentropy into the system. Obviously, this is also true for more complex systems in which an increase in the order of organisation requires a flow of negentropy from the surrounding medium.

Another interesting feature is that in disproving the possibility of a Maxwell demon, we are led to establishing a direct physical relationship between information and energy. It proves possible to calculate the minimum quantity of negentropy necessary for obtaining each unit of information. If the entropy $S$ is measured in erg/deg and the information in bits, then the increment in entropy $\Delta S$ of the system due to the increment in information $\Delta I$ on its state is approximately

$$\Delta S \approx -10^{-16}\ \Delta I \text{ erg/deg.} \tag{17.2}$$

It can be seen from the relationship (17.2) that in order to exert an appreciable influence on the energy balance of a system in which individual components have the order erg/deg, the quantity of information must reach enormous proportions, of the order of $10^{16}$ bits. In artificial systems we operate with very much smaller flows of information. Thus, for instance, the quantity of information contained in the circuit of a complicated system consisting of 1,000 elements, each of which may contain up to 10 connections with other elements, is altogether $1.33 \times 10^5$ bits, which is less than 1 thousand millionth of one unit of entropy. Such apparently negligibly small influences have nevertheless a great fundamental importance and under certain conditions can prove considerable, in the same way as the influence of velocity on the mass of a body, established by the theory of relativity, also proved important. In particular, for living organisms with a complex structure, the flux of negentropy during transmission of information may prove commensurable with the changes in the entropy of the system.

### 17.2. Control structure in an organised system

Any organised system: a production plant, a medical centre, a telephone network or an insurance company, must not only conserve its state of organisation but also accomplish the appropriate functions. Therefore, in organised systems, two types of control problem are solved: control of the internal organisation of the system, and control of its functioning. For solution of these problems the system must have available appropriate organs, responsible for controlling its functioning and maintaining the system in a state in which it is capable of working. Fig. 17.2 shows a typical control circuit of a production plant, showing the sub-divisions which fulfil the above types of function. The group B contains the services concerned with controlling production, group A the sub-divisions ensuring internal order of the system, and the management C who coordinate the work of groups A and B and also coordinate the activity of the system with processes in the ambient medium. The same flow sheet shows in double lines the material flows, and in single lines the flows of information (information and command signals).

The described structure for controlling the operation of a plant is applicable where the control functions in all the sub-divisions are performed almost exclusively by people and automata play a subordinate role, carrying out only primitive operations of regulating technological processes, accounting, etc. Naturally, the question arises as to whether it is not possible to build a system for controlling a plant directly by cybernetic machines. This would ensure its viability with the minimum participation of personnel in the control process. It has been found that this problem can be solved in principle. Although the time has not yet arrived for machine systems to control complex plant this will become technically feasible and economically advisable in the course of time. But even the theoretical solution of this problem is of great fundamental importance. It can be assumed that an analysis of such systems from the point of view of thermodynamics may be advisable.

In fact many, if not all, production processes are 'anti-entropical', i.e. are associated with lowering the entropy of the initial raw material during its

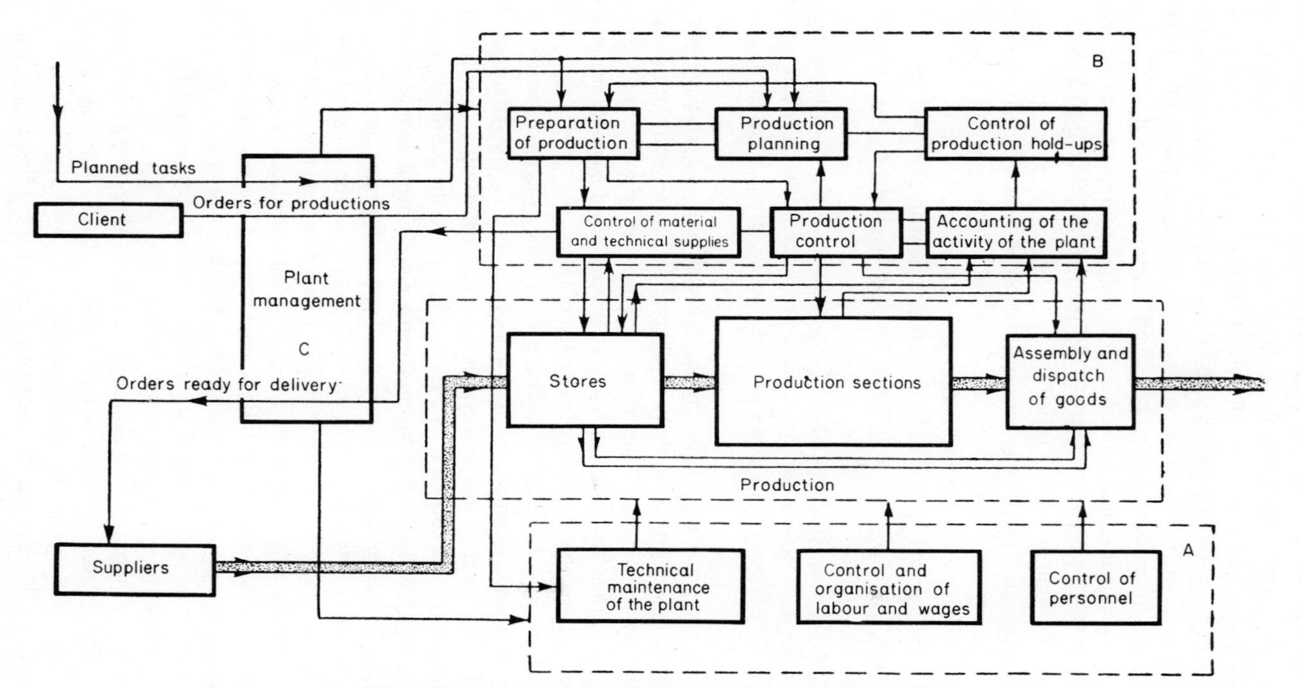


Fig. 17.2. *Typical circuit for controlling the production of a plant.*

conversion into manufactured goods. This is particularly clear in the extraction of useful minerals, separation and purification of substances, manufacture of components of a given shape from blanks, assembly and control of the quality of production. The maintenance of any plant in working condition (repair and replacement of worn-out equipment, reconstruction of production spaces and communication lines, organisation of logistics and transportation of the raw material and the finished products) is also 'anti-entropy' in nature.

Development of a machine system for controlling complex organised systems, such as a production plant, a system which functions under changing conditions, involves the following main difficulties:

(1) The state of the complex system is determined by a very large number of factors, the number of control actions is large and it is not possible to sift all the combinations for selecting the best action within a reasonable time.

(2) Interaction between parts of the system and with the medium proceeds along so many channels that it is not possible to take all these into consideration in the control system.

(3) The properties and characteristics of the system and the environment change with the progress of time according to laws which are not usually known, but in spite of this they have to be known in some measure for performing the control action.

(4) The time required for calculating the optimal control becomes longer, the more complex the control system. Furthermore more accurate results are required the more long-range the forecasting of the behaviour of the system should be. Solutions arrived at as a result of prolonged calculations lose their value if they cannot be obtained and supplied quickly.

The enumerated difficulties of realising a machine system of control can be completely overcome, provided we do not try to build an *ideal* system capable in any situation of finding instantaneously and implementing a strictly optimal control. In reality we do not require a system to be optimal but to be *satisfactory*, which will at least perform a control no worse than man would. The potentialities of man as regards receiving and processing information are limited, and even the most talented leader takes decisions on the basis of a relatively small number of factors and very rough forecasts, and spends a relatively long time in making decisions. This means that he cannot obtain a strictly optimal control. Nevertheless, plants and other complex organised systems controlled by man exist, and many of them develop successfully. This means that machine control systems may also prove viable if they perform the control functions, if not ideally, at least no worse than human beings do.

One possible structure of a machine system for controlling a plant is given in Fig. 17.3. This system is built up of learning, homomorphous (simplified) models of plants and *media* of various orders.

The simplest and most quickly changing models of the first rank are intended for processing relatively small flows of information and to provide, quickly, control signals calculated to operate over short time intervals. The models of the second

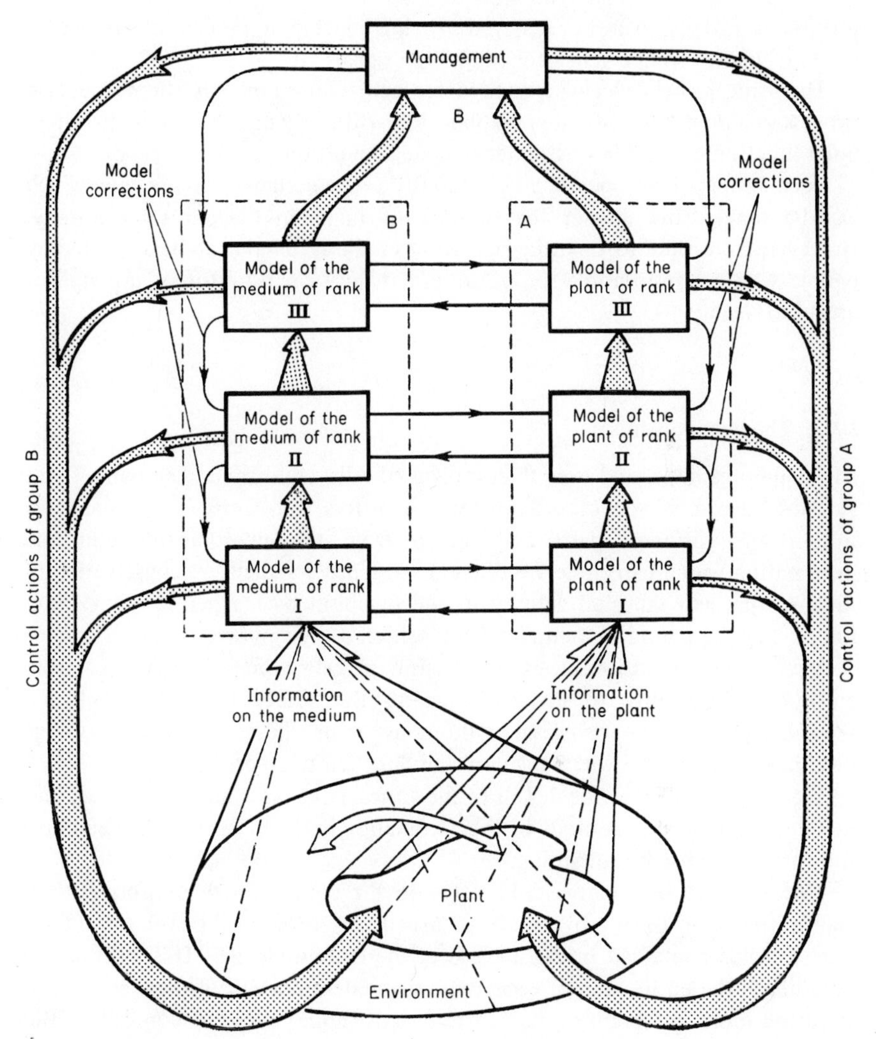


Fig. 17.3. *Example of a possible structure of a machine system for controlling a plant.*

rank take into consideration more data, and calculate the control actions over a longer period of time. The models of the higher ranks work still more slowly but produce control actions for even longer periods. The models of the plant and of the medium of all ranks interact and generate the control signals of the group A, intended for keeping the plant in order, and the group B, for accomplishing its functions with respect to the medium.

In order that models change in accordance with changes in the properties and characteristics of the plant and the medium, they must be made *adaptive*, and for this purpose principles can be used which were described in Chapter 11. The 'initiative' of a machine system, exemplified by a sufficiently large variety of possible solutions, can be achieved by introducing a moderate number of random changes from a generator of random signals.

The same principle can be applied for controlling not only the production process of a plant but also other complicated artificially organised systems which must function in a stable way under changing conditions.

Such systems may operate autonomously over a certain period of time which can be longer, the greater the number of ranks in the control hierarchy. However, from time to time the operation of the system has to be corrected by people who are responsible for ensuring that the system performs in a manner useful to society.

### 17.3. The living organism

All living organisms, and even their individual cells, are organised systems. These systems have a degree of ordering which differs from zero. They consist of functionally different parts and absorb from the environment food and information which carry reserves of energy and order and ensure conservation of the ordering and the vital activity of the organism over a period many times longer than the period of natural decomposition of living tissue.

More complex processes of metabolism in cells, their development from embryos, adaptability to the environment, the ability to learn, think and reproduce, cannot be visualised without assuming the existence of very fine mechanisms of control which regulate and direct all these processes.

We still do not know the details of the structure of these control mechanisms, nor do we know the methods of their formation in the living organism and how they are transmitted to offspring.

Investigations by geneticists have led to the conviction that information is transmitted to offspring by means of a code, 'written' in the structure of the nucleic acids contained in the nucleus of the embryonic cell. This information determines the entire plan of construction and development of the organism, its type and individual features, its ability to exist under changing conditions. But the hereditary information contains only data necessary for forming the structure and ensuring the viability of the organism, which could be accumulated during the process of evolution of the given species. Programmes of the behaviour of the organism in the medium cannot be contained in a ready form in cells, since all the situations which may occur cannot be foreseen. But apparently, the code of hereditary information contains programmes on the construction of mechanisms which can produce reactions in accordance with a given situation, as well as mechanisms which can learn purposeful behaviour.

What can the general structure of these control mechanisms be? Do they represent a centralised or a decentralised control system? How do they ensure coordination of the action of individual cells, tissues, and organs which are vital to the viability of the organism as a whole? We can already try to answer these questions even before science has penetrated deeply into the control processes of living organisms.

The possibility of centralised control of all living processes, including cell metabolism, can be discarded, since it is impossible to concentrate in any organ the huge quantity of information required on the state of the system which consists of millions of such complex objects as cells. It is even more difficult to visualise an organ capable of processing such a quantity of information. Moreover, to transmit information on the state of each cell to a central control would require an information network of a kind which has not been observed in any organism. Consequently the possibility of centralised control in living organisms is unlikely.

It is also difficult to assume the second extreme possibility of building an organism in the form of systems where each cell is individually controlled, i.e. functions performed by cells which are independent of the state of the other cells. This is unlikely because of the remarkable coordination of the functioning of the system as a whole.

These considerations and the observed structure of cells, tissues and organs indicate a hierarchical structure for the control systems of organisms.

There is sufficient justification for assuming that a cell has its own control mechanism, responsible for the metabolism in the cell. Interaction with other cells of the tissue is constructed in such a way that the functioning of the cells is also subordinated to the interests of the entire set of cells.

It can be assumed that the most favourable condition of operation of a cell occurs when it fulfils optimally its functions with respect to the tissue. The same argument applies for the tissue in relation to the particular organ, and for the organ in relation to the body as a whole.

*A harmonic combination of the 'interests' of the parts and the whole* is, apparently, the fundamental principle which determines the structure of the control in living organisms.

Atrophy of tissues and organs when they are not used over a long period, and on the contrary, intensified feeding and development when they are intensively used, are arguments in favour of such a supposition.

## 17.4. Self-organising system

*A self-organising system is a system whose degree of order increases with the progress of time.* At first sight it would appear that an increase in the degree of order contradicts the second law of thermodynamics and consequently either such systems cannot exist or the second law of thermodynamics is incorrect. In reality this is not so, and the possibility of existence of self-organising systems does not contradict the principles of thermodynamics. Not only can self-organised systems not be considered in isolation from their environment, in reality the energy and order cannot be formed from nothing and they can be drawn only from appropriate sources. Assuming that the sources of energy and order are within the system, then in principle the system cannot have the properties of self-organisation since at best (if only reversible processes occur in it) the entropy will not increase, but it can also not decrease. According to (17.1) the degree of order $R_0$ of the system at the time $t_0$ is determined by the expression

$$R_0 = 1 - \frac{H_0}{H_m} \tag{17.3}$$

where $H_0$ is the entropy of the system at the time $t_0$. Let us assume that the entropy of the system at the time $t_1 = t_0 + \Delta t$ equals $H_1$, and its degree of order

$$R_1 = 1 - \frac{H_1}{H_m} \tag{17.4}$$

Let us assume that the maximum entropy $H_m$ of the system does not change. Then the system will be self-organising if $R_1 > R_0$. For this to happen it is necessary to fulfil the condition $H_1 < H_0$, and consequently

$$H_1 - H_0 = \Delta H_1 < 0, \tag{17.5}$$

and since $\Delta t > 0$, then

$$\frac{\Delta H}{\Delta t} < 0 \tag{17.6}$$

The condition (17.6) means a decrease in entropy with the progress of time in a closed system, which contradicts the second law of thermodynamics and consequently is impossible.

This means that if the degree of order of any system is to increase, it must receive energy and order from outside (negentropy). Obviously the entropy of the medium will increase and the increase will be greater than the decrease of the entropy of the self-organising system, and consequently the entropy of some system containing a self-organising system and sources of energy and order will on the whole increase, despite the fact that in one part of it, in the self-organising part, it will decrease.

Consequently self-organising systems can exist only as particular exceptions of the general rule of existence of physical systems, by 'sucking' energy and order from the environment, and as a result causing 'irreparable damage' to it.

We would arrive at the same conclusion by considering the case when the increase in the degree of order of a system is reached by increasing the maximal entropy $H_m$ with the entropy $H$ remaining unchanged, since for this to happen we would have to increase the number of elements of the system (on which $H_m$ depends), which also requires 'food' from the medium.

These conclusions fully agree with our observations on unique known types of self-organising systems: individual living organisms and their societies. The necessity of absorbing energy and order from the environment should not discourage us, since the resources of the universe are virtually unlimited. Even more, we can and should aim at increasing the number of self-organising systems by adding artificial self-organising systems to the natural ones. Even if we do not yet know how to build them, the time will surely come when we will learn to do this and will build remarkable self-organising automata to help mankind.

## Exercises

**1.** The entropy of a single-atom ideal gas is expressed by the formula (in natural units)

$$S = N \ln \frac{V}{N} + CN$$

where $C$ depends on the temperature and mass of the gas atoms, $N$ is the number of atoms, $V$ is the gas volume.

A vessel of volume $V$ is divided by a barrier into two parts: $V_1$ and $V_2$. On the left is gas 1 with $N_1$ number of atoms, on the right is gas 2 with $N_2$ number of atoms. The atomic masses and temperature of the gases are equal (i.e. $C_1 = C_2$). In addition the densities are equal, i.e.

$$\frac{N_1}{V_1} = \frac{N_2}{V_2} = \frac{N}{V}$$

where $N = N_1 + N_2$.

On removal of the barrier the gases intermix due to diffusion. Calculate the increase in entropy involved (the so-called *mixing entropy*). Calculate what quantity of information is required to divide the gases in the reverse direction. What do the obtained results indicate?

*Solution.* The entropy of the entire system prior to mixing equals

$$S_{\text{init}} = \left(N_1 \ln \frac{V_1}{N_1} + CN_1\right) + \left(N_2 \ln \frac{V_2}{N_2} + CN_2\right) = N \ln \frac{V}{N} + CN.$$

Since $\frac{V_1}{N_1} = \frac{V_2}{N_2} = \frac{V}{N}$ and $N_1 + N_2 = N$. After mixing, each gas will occupy the entire volume $V$. The entropy will equal

$$S_{\text{final}} = \left(N_1 \ln \frac{V}{N_1} + CN_1\right) + \left(N_2 \ln \frac{V}{N} + CN_2\right) =$$

$$N \ln V + CN - N_1 \ln N_1 - N_2 \ln N_2$$

The increase in entropy equals

$$\Delta S = S_{\text{final}} - S_{\text{init}} = -N \ln N + N_1 \ln N_1 + N_2 \ln N_2 = -N\left(\frac{N_1}{N} \ln \frac{N_1}{N} + \frac{N_2}{N} \ln \frac{N_2}{N}\right)$$

In order to achieve the reverse separation of the gas it is necessary to determine the 'type' of each gas atom. The probability that a randomly selected atom of the gas mixture belongs to the gas 1 is $P_1 = N_1/N$. Similarly $P_2 = 1 - p = N_2/N$. The information obtained in the test for determining the 'type' of one atom of gas equals $-P_1 \ln p_1 - p_2 \ln p_2$. Since the mixture contains a total of $N$ atoms, the total quantity of information equals

$$I = -N(p_1 \ln p_1 + p_2 \ln p_2) = \Delta S.$$

Therefore the quantity of information necessary for reverse separation of the gases equals the entropy of mixing. This is a particular case of the *negentropy principle of information* according to which obtaining information on the physical system is equivalent to a decrease in entropy.

2. Equipment for separating uranium isotopes separates the initial material with atomic concentrations of the components equalling $c_1$ for uranium-235 (component $X_1$) and $c_2 = 1 - c_1$ for uranium-238 (component $X_2$), into the two fractions $y_1$ and $y_2$. The concentration of $U^{235}$ in the first fraction equals $c_{11}$, in the second fraction $c_{12}$. The part of the first fraction (in numbers of atoms) equals $p$. Obviously, $c_1 = pc_{11} + (1 - p)\, c_{12}$. Calculate the decrease in entropy (the increase in degree of order) as a result of the separation.

*Solution.* Using the result of Example 1, we find that the decrease in mixing entropy per one molecule of the initial material equals (in natural units)

$$\Delta S = -[c_1 \ln c_1 + (1 - c_1) \ln(1 - c_1)] + p[c_{11} \ln c_{11} + (1 - c_{11}) \ln(1 - c_{11})]$$
$$+ (1 - p)[c_{12} \ln c_{12} + (1 - c_{12}) \ln(1 - c_{12})].$$

The obtained result has only a theoretical-informational sense: as can be seen, $\Delta S$ equals the quantity of information $I\,(y,\, x)$ in the fraction relative to the component (the fraction and component are considered as random variables which assume two different values each).

3. What is the source of negentropy for green plants? For animals living on land? For anaerobic bacteria?

*Solution.* The source of negentropy for green plants is the thermodynamic non-equilibrium system consisting of the radiation of the sun, the food substances in the soil and the carbon dioxide in the atmosphere. For animals, a similar system is the food and the oxygen in the air. For anaerobic bacteria – the complex organic compounds which they decompose into simpler ones.

4. What is the source of negentropy for such 'non-living' but stable non-equilibrium systems as the hydrological cycle, geysers, tides?

*Solution.* For the hydrological cycle – the radiation of the sun; for geysers – the internal and external layers of the earth, which have differing temperatures; for tides – the rotation of the earth.

**5.** Into a closed cavity, whose walls and all the objects inside it are maintained at the temperature $T$, a television camera is placed which has the temperature $T_1$. Is it possible to 'see' anything inside the cavity by means of this television camera, if:

(a) $T_1 = T$ (b) $T_1 < T$ (c) $T_1 > T$?

What general principle is illustrated by this example?

*Solution.* (a) No, it is not. The entire cavity will be filled by an isotropic (equal in all directions) equilibrium radiation of the temperature $T$.
(b) Yes, it is. Due to the difference between the temperatures $T$ and $T_1$ the radiation inside the cavity will not be isotropic. The television camera will exchange energy by radiation which is not equal in the various directions. The energy flow will be larger from the bodies with a greater radiation capacity and their image will be brighter.
(c) Yes, it is. In this case the image will be negative with respect to the image obtained in case (b).

*General conclusion*: In order to obtain information it is necessary that there should be no equilibrium between the observed and the observing systems. Systems in a state of equilibrium with each other cannot exchange information.

# 18 Man and Machine

The activity of man over millions of years of his existence on Earth has mainly been to develop the riches and forces of nature for his biological and social interests. Man has learned to hunt, to till the soil, to raise animals; man created industry and transportation, he uses natural sources of energy, in other words he utilises all methods of actively influencing nature. But use of natural resources in the interests of man is unthinkable without a purposeful influence since firstly not a single system can function indefinitely without disturbing its functioning (if we ignore systems whose reliability increases with the progress of time), and secondly, the conditions of the medium change with the progress of time and so do the aims and interests of man. Therefore, however large the achievements of mechanisation and automation, however new and rich the sources of energy available to man, whatever the systems of information processing, man was, is and will always be the decisive link in a complicated chain of interaction between man and nature; he will always be in command of any system created by him and for him.

Technical progress and in particular the development of means of communication, computing and automation, have led to the development of machines which are capable of monitoring the progress of various processes and controlling them. Therefore man acquired the possibility of controlling these processes not directly but indirectly by means of technical equipment which is subordinated to him. The functions of automatic equipment and of the people who participate in the control are so intertwined that it becomes impossible to correctly formulate, let alone solve, many control problems by limiting oneself to the technical aspects of the problem, and ignoring the psychological and physiological factors associated with the participation of man in the control process.

Therefore in cybernetics the scientific man-machine discipline has developed, which investigates the two basic problems: (1) symbiosis between man and machine; (2) distribution of functions between man and the automaton.

The problem of symbiosis between man and machine includes study of the properties of man as a link in the system in which he, together with technical equipment, participates in performing work and the problem of distribution of functions between man and machine requires solution of the problems of harmonic combination of human and mechanical factors.

## 18.1. Symbiosis of man and machine

The simplest form of interaction between man and machine is the accomplishment by man of the functions of an operator who closes a circuit triggering off actions in a control system. The functions of an operator are performed, for instance, by a pilot, by a ship's helmsman, by a driver, a radar station operator, a train driver.

Since the properties of control systems are determined by the properties of all its components, then of all the properties of man-operator for such systems the most important are the dynamic properties of man as a transmitting link in the control chain.

In contrast to machine links of a system, which usually realise a certain law of transformation of signals, man is capable of learning a great variety of actions on obtained signals. He can perform the role of a static or dynamic link; he can correct his activity on the basis of the rate of change of a signal; he can regulate his actions in proportion to the magnitudes of the signals, or realise some nonlinear transformations. Formation of the law or processing signals (control methods used by the human operator) occurs during the process of learning. However, even after completion of the learning, the methods of control used by man do not remain strictly rigid and can be described mathematically only in terms of averages. The class of transformations which man can learn is limited by his physiological properties, i.e. by the range of changes of his dynamic characteristics.

One of the basic indices characterising the dynamic properties of the human operator is the value of the *latent period* – the time from the instant of giving a stimulus, to the instant when the operator develops a reaction. The latent period is not the same for stimuli of various *modalities*, i.e. for stimuli acting on the various sense organs of man (analysers). The range of the values of the latent period $\lambda$ are given in Table 18.1.

It can be seen from this table that the time lag introduced into the control circuit by the operator is much greater than the time lag in electric or pneumatic components, and is therefore a decisive factor in the evaluation of the total time lag in the control circuit.

Numerous investigations have shown that the law of transformation of signals received by the operator into output actions depends greatly on the properties and characteristics of the remaining parts of the system and on the degree of training of the operator.

This 'flexibility' of transformation realised by the human operator enables him to adapt himself very effectively to the character and tempo of processes in a control system, obviously if this is compatible with his physiological limitations which limit his speed of operation. The methods of control realised

Table 18.1

| *Type of stimulus* | $\lambda$, *m/sec* |
|---|---|
| Touch | 90 – 220 |
| Sound | 120 – 180 |
| Light | 150 – 220 |
| Smell | 310 – 390 |
| Taste | 310 –1080 |
| Movement (effect on the vestibular apparatus) | 400 |
| Pain | 130 – 890 |

by a human operator after acquiring the necessary habits are near to the optimal ones. This is shown, in particular, by experiments in teaching people to control a system consisting of two dynamic members. As was shown in Chapter 9, the optimal process of changeover of such a system from any initial state into any given state contains two intervals: during the first the control action is retained on one limit value, in the second on the other limit value.

Recordings of the processes of control of such a system with man acting as a control device are shown in Fig. 18.1. Here $X$ denotes the output of the controlled system and $u$ the control action (position of the lever by means of which man acts on the input of the system). Fig. 18.1(a) shows the process which is characteristic for the first stages of training, when the operator is just training to change the system from the initial state into the given state. The process shown in Fig. 18.1(b) corresponds to the stage when the operator has acquired the habits for effectively controlling the system. A process which is very near to the optimal as regards speed of response is shown in Fig. 18.1(c); it is characteristic for an operator with an adequate period of training. It is interesting that the methods of optimal control were found by the operator during training, without instruction by specialists in the theory of optimal control.

In order that man should be able to control any process, it is necessary that this process be modelled in his brain in the form of sensations and perceptions which carry information about the controlled process. The effectiveness of control by man is determined by the following factors: (a) the quantity of useful information on the controlled process; (b) information from past experience, drawn on for the purpose of processing the obtained information; (c) the method of using the obtained information, i.e. the speed, accuracy, the freedom from error of human reaction to phenomena reflected in his brain as sensations and perceptions.

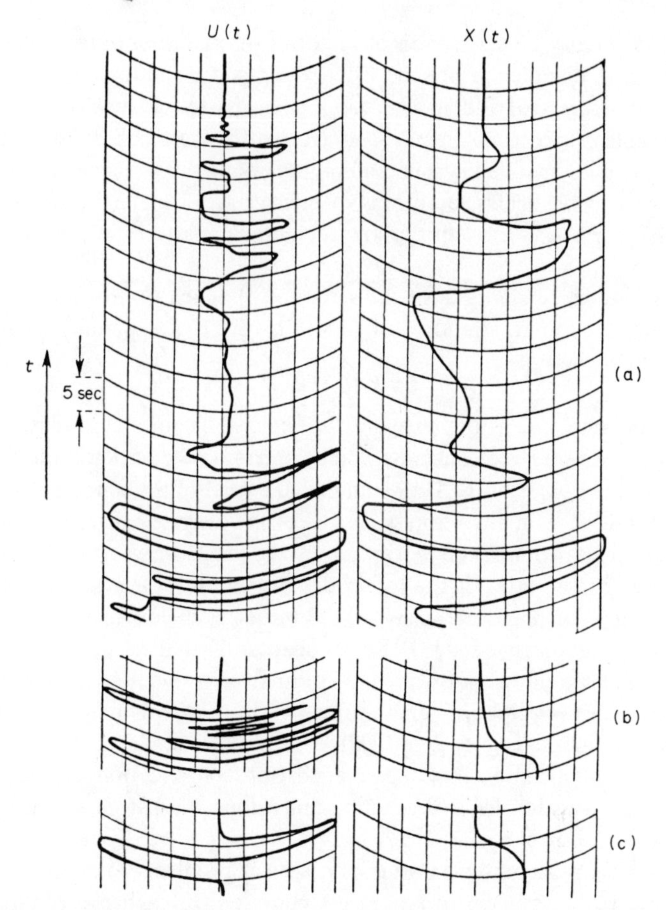


Fig. 18.1. *Learning of a human operator to control a system in an optimal manner.*

The process of perception by the brain of objectively occurring phenomena begins with the effect of the phenomena – stimuli on the receptor analysers, i.e. on the nerve endings of the sense organs. The stimuli which carry the useful information on the controlled process and which require an answering reaction are called *stimulation signals.*

The character and number of signal stimulators greatly influence the control process of the human operator. To carry out primitive forms of work, man must choose from a large number of signal stimulii. The position is different in 'Man and automaton' systems, where selection of the signal stimuli which carry the information on the state of the controlled system is determined in advance by the designer of the equipment, and they are presented to the operator by the technical equipment. In such systems the choice of the controlled coordinates of the system and the form of presentation of the signals are very important.

Selection of a 'language' for interchange of information between man and machine, in other words selection of methods of coding information, is a most important and complex problem. Upon the correct solution of this problem depends the success of the simultaneous participation of man and machine in solving problems posed to them. Most frequently the operator is informed on the state of the controlled system and on the events taking place in it, by means of light or acoustic signals, by means of indications of metering instruments, as well as by formation of appropriate geometrical patterns on a cathode ray screen.

To evaluate the magnitude of the sensation $f$, which occurs under the effect of a light or sound stimulus of intensity $\theta$, the experimentally derived relationship of the type

$$f = K(\theta - \theta_0)^n, \quad (18.1)$$

can be used, where $\theta_0$ is the *absolute threshold* – the minimum sensed value of the stimulus, $K$ and $n$ are constants which depend on the background, the shape, dimensions and colour of the light stimulus, the level of the sound stimulus.

The exponent $n$ in the expression (18.1) is always smaller than one, since numerous experiments on the perception by the sense organs of man have shown that the perception ratio of the intensities of stimuli is always lower than their physical ratio. For the visual analyser, for instance, the exponent $n = 0.33$ after adaptation to darkness and $n = 0.45$ after adaptation to light.

Study of the methods of presenting information to the human operator is the subject of a new discipline – engineering psychology, which basically relies on experimental methods of investigation. Data accumulated by engineering psychology contain important recommendations on selection of the modality and intensity of signals, shapes and dimensions of instrument scales, their positioning on control panels, etc.

One of the most complex but very topical problems in the man-machine relationship is the problem of language to be used for exchange of information between man and computer during 'collective' work on the solution of some problem. It is obvious that the internal language of the machine, the language used for exchanging information between parts of the machine and for its processing, should be selected so as to conform with the properties and features of the components from which the machine is assembled. This language, however, is very inconvenient for direct communication between machine and man. For this purpose it would be much better to communicate in a language used for communication between men, i.e. the written word and speech, the language of mathematical symbols and geometrical patterns. It would be convenient to use such a language not only for introducing the initial data into the machine and for retrieving data from the machine, but also – and this is the main point – for *programming* the operation of computers.

Development of theoretical concepts on the basis of which communication with a machine would approach forms convenient for man, and development of equipment which would realise such forms of communication, have been intensively worked on during recent years.

## 18.2. Division of work between man and machine

One of the central and specific problems in the 'Man and machine' relationship is that of optimal division of the functions between man and automatic equipment within the framework of a unified control system. When designing control systems it would not be advisable to aim at automating all the control functions which can be automated. It would also be inadvisable to apply the opposite concept, namely that 'purely human' thought functions and decision making should under no circumstances be handed over to machines.

An optimal system, including man and machine, should ensure the maximum obtainable effectiveness in operation (compatible with the limitations of the equipment and man) by the best use of the potentialities and advantages of both components. At the present stage of development more sophisticated machines are capable of solving problems which have been formalised inadequately, of taking decisions in unusual situations, and of applying original methods of control.

On the other hand man, as part of the control system, has many inadequacies and limitations such as a limited speed of response to signals, an insufficient reliability when required to carry out monotonous operations, a high dependence of his performance on external conditions, and on his internal psychological and psychic state, fatigue. The most important limitation of the control function of man is his maximum speed of reception and processing of information.

Experimental data indicate that the speed of information processing by man in recognising symbols reaches 50 bits per second. However, this cannot be taken as a basis for calculating the permissible loading of an operator, because his effectiveness of operation depends on a number of factors: on the alphabet of signals used, on the algorithm required for processing the information, etc. In reality, under complicated conditions of operation of the human operator, accompanied by reception of signals and action on control organs, satisfactory processing of the information is realised if the information flow does not exceed 6 bits per second. If the information flow is higher, the loss of input information increases sharply. The operator will no longer be able to perceive the information, which is beyond his maximum throughput capacity, and this will bring about an increase in errors and delays in the reactions to the signals received.

Automatic equipment is inferior as regards flexibility and potential, but superior for performing a limited set of operations, for speed, reliability, absence of fatigue and lack of dependence on noise. The potentialities in using automatic devices for solving control problems widen continuously, and encompass more and more new fields which had hitherto been considered as the exclusive prerogative of man; for this reason it is fundamentally impossible to outline the limits of their development.

The problem of rational distribution of functions between man and automaton can be considered as solved only when we have available a criterion which will permit the selection of the optimal variant of the distribution function in each concrete case, which will satisfy the limitations imposed on the system and at the same time the criterion will reach its optimum value. When dividing the functions between man and machine, it is necessary to take into consideration a number of factors of differing degrees of importance and the selection must be made on the basis of one estimation, therefore the criterion $I$, which takes into account particular evaluations of the system with respect to individual factors, according to their relative importance, is considered the overall criterion. As such an overall criterion the expression

$$I = f(I_1, I_2, \ldots, I_n) \approx \sum_{i=1}^{n} \alpha_i I_i,$$

can be used, where $I_j$ is the estimation with respect to individual indices, $\alpha_i$ are weighting coefficients which take into consideration the importance of the respective indices.

The particular evaluations $I_i$ should comprise, for instance, the effectiveness of control, the reliability of control, cost of the equipment, running costs on the control, fatigue of the operator, the degree of satisfaction of the operator from the work.

One should not think, however, that optimal distribution of functions between man and machine will ever be found and fixed 'once and for all'. In the first instance it is necessary to take into consideration that solution of such problems depends largely on the actual conditions, for instance on the length of time for which the plant is likely to be in operation (once only or multiple action), the purpose of the system (production or military), and many other factors which do not form part of the formal evaluation. Moreover, the functions imposed on the automaton must obviously be selected from a set of functions which the automaton 'can' perform, and this set is continuously increasing. Ultimately the division of work between man and machine depends greatly on economic and social factors, which also change with the progress of time; the cost of manufacturing equipment, reliability requirements, the cost of labour, the views and preferences of people, i.e. the attractiveness to people of carrying out one type of function or another, all of which are subject to change. Therefore division of work between man and machine will always be subject to a variety of factors which change from time to time.

## 18.3. Bio-electric control

Transmission of commands in the nervous system of animals is realised by electric pulses as has been mentioned earlier. For instance, when a man wants to clench his fist, electric pulses will begin to flow into the muscles which perform this movement. The pulse frequency will be higher, the more strongly the man wants to clench his fist. Performance of complex movements, when a man throws a stone at a target or rides a bicycle, is associated with the production and execution of a complex programme of signals which control the work of dozens of muscles.

It has been found that these bio-electric signals, which are used by the body for controlling muscular movement, can also be used for controlling artificial equipment. The necessity for such interaction between the nervous system of the body and man-made equipment arises, for instance, in the design of manipulators or artificial limbs.

The manipulator is a mechanism which reproduces the movements of the human operator. Such mechanisms are used when work must be carried out under conditions dangerous to human life (for instance under intensive radioactive radiation, high temperatures or pressures, under water), also when it is necessary to apply forces in excess of human capabilities. By receiving by means of the respective electrodes the bio-electric signals which are directed towards the muscles of the human operator, the servo systems of the manipulator can be made to work in the way the operator wishes them to work.

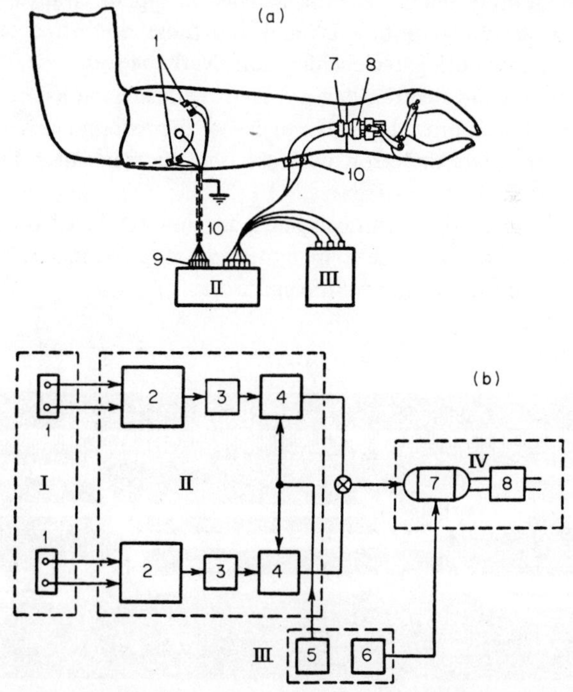


Fig. 18.2. *(a) Semi-industrial (pilot plant) circuit of an artificial hand. (b) Block schematics of the artificial hand with bio-electric control: I. current-collecting unit, II. control unit, 3. power supply unit; IV. drive unit; 1. electrodes, 2. voltage amplifiers, 3. rectifiers, 4. power amplifiers, 5. power supply to the amplifiers; 6. power supply to the motor, 7. motor, 8. reductor gear, 9 and 10. connectors.*

Bio-electric control systems intended for artificial limbs are very important. About ten years ago a group of engineers, physiologists and doctors made an artificial hand which is controlled by bio-currents. The clenching of the fist is performed by a miniature electric motor built into the hand, Fig. 18.2(a). The motor is controlled by the two commands: 'clench the fist' and 'unclench the fist'. These commands are generated in the artificial hand in accordance with the intensity of bio-electric signals taken from the amputated muscles of the forearm of the man who has lost his hand. The spots from which the bio-electric signals are taken (by means of applied electrodes) are so selected that the clenching and unclenching of the fist are controlled by the same groups of muscles which perform this function in a healthy hand. Therefore a person fitted with such an artificial hand will quickly learn to control it by continuing to use habits of coordination of movement which he had acquired prior to the amputation.

Fig. 18.2(b) shows the block schematics of the artificial hand with two channels of amplification and conversion of the bio-electric signals to ensure reversing control of the drive to produce clenching of the fist.

The character and the tempo of the movements of the artificial hand can be judged from the oscillogram recorded for one cycle, Fig. 18.3. The flashes of the pulses representing the bio-electric activity as tapped from the contracted muscles during clenching of the fist and also those of their antagonists, the muscles which release during unclenching, can clearly be seen.

More recently the bio-electrically controlled artificial hand has been perfected to such a degree that it proved possible to ensure the required movements of the fist and also to regulate the effort developed by the mechanism in accordance with the desire of the man wearing it.

Bio-electrically controlled artificial limbs are now produced on an industrial scale and have proved invaluable to many victims of accidents, enabling them to return almost completely to normal life and work.

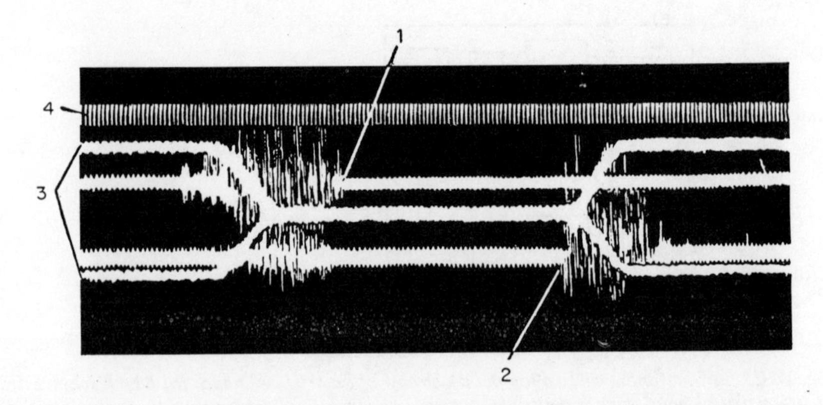


Fig. 18.3. *Cycle of movements of the bio-electrically operated artificial hand: 1. electric signal which contracts the muscles, 2. electric signal which relaxes the muscles, 3. clenching of the fist, 4. time marks (spacing between the cycles 0.02 sec).*

### 18.4. Man and machine

The scientific formulation and solution of the problem of controlling such systems as complicated production plants, the armed forces, economic systems, all involving the participation of many people, inevitably requires elucidation of the laws of the relationship between man and his environment, and this means primarily nature and society. Although not universal, one of the possible and fruitful approaches to the solution of this problem, in addition to the sociological approach, is the approach from the point of view of cybernetics where man is considered as a system whose behaviour is determined not only by his internal motives but is also conditioned to a certain extent by the control actions generated by the environment.

Such control actions are encouragement, discouragement and education. Like any other controlled system, man has certain limitations imposed on the one hand by his own nature and the material world around him, and on the other by society. Limitations of the first type are, for instance, the physical forces of man, the need to breathe, quench his thirst and hunger, and that perception of the conditions surrounding him is limited by the properties of his sense organs.

Acting within the bounds of such natural limitations, in the same way as any other living animal, man tries (consciously or unconsciously) to choose a behaviour which optimises a certain criterion, characterising an overall sensation of satisfaction with life. Although the criterion changes with the progress of time and differs from one individual to another, nevertheless it exists and is fundamentally linked with the degree to which man achieves his goals and satisfactions.

### 18.5. The future of cybernetics

The main purpose of this chapter has been to relate man's development to cybernetic development. It has been made clear that we are going through a man-machine phase of development and that this is crucial to the future both of machines themselves – automata, control systems, etc. – and human society. The most important single feature is for our human societies in the world to understand the significance of this phase of scientific and human development. If there is no understanding, then we are almost bound to make the minimum use of the things that are going on in our environment. This could only be injurious to the whole of civilisation.

Cybernetics is in many ways a key science, if not *the* key science of which much of our future social and organic evolution depends. There are dangers and numerous problems connected with that evolution, and the only way on the short term that we can anticipate these difficulties is a full understanding of what cybernetics entails.

# 19 Outline of Future Prospects

During the very short period of its existence, only two decades, cybernetics has made enormous progress. It had a great effect on the outlook of scientists and engineers, doctors and administrators, generals and statesmen. From the extravagant point of view, which produced enthusiasm in some people and scepticism in others, cybernetics became a serious scientific discipline with a *Weltanschauung* characteristic of its own. It developed its own subject and method of investigation, its own laws and principles. Although it is a theoretical discipline it nevertheless succeeded also in supplying practical results, the importance of which it is difficult to overestimate.

What are the prospects of this science? How does it contribute to our knowledge and our life? What are its tasks in the near and distant future? Do cybernetic machines present a danger to mankind as a potential competitor? These and many other questions associated with the prospects of cybernetics are now disturbing many people throughout the world. Obviously nobody would dare at present to give definite answers to all these questions, but some views on the development of this science can be advanced.

Throughout the entire history of mankind, the technical, cultural and social progress consisted of elemental processes; although these did obey certain rules which follow the laws of development of society, they were not directed into a definite channel and were not controlled consciously and purposefully. Now the position has changed radically. We have entered an era when society can and must formulate the aims of development of engineering, science, art, determine the optimum ways of achieving these aims and consciously direct the effort to realising them.

Many of the problems of development which arise in various spheres of life require the methods and means of cybernetics for their solution.

Modern society must solve complex control problems if it is to maintain its existence and its further development. To improve manufacturing technology, it is necessary to solve intricate problems regarding the automation of production processes. Due to the continuously increasing complexity of transportation systems, new methods of controlling the flow of both freight and passengers is required. To satisfy the increasing requirements of the population in consumer goods and services, it is necessary to introduce operation control of trade, domestic services, medical and sport organisations, etc.

Even more complex are the problems which arise in controlling the economy, controlling the development and execution of large projects, controlling large scientific research projects. There is good reason to believe that the scale of control problems will continue to increase and urgency will continue to be greater and greater. It is obvious that, as the control problem becomes more complex, the role of traditional methods will diminish and cybernetic methods of solving the problems will become predominant. To achieve these aims, new ideas should be developed, which will permit improving control in a number of fields, which so far engineers and scientists are not concerned with, and to develop new means which would enable them to realise these ideas in practice.

## 19.1. Unsolved problems

In cybernetics, as in any other science, new problems are arising continuously as it develops, stimulated by practical requirements as well as the internal requirements of the science itself. Although the few practical tasks of cybernetics which are presented below are the most topical ones in the near future, it may happen, as is often the case, that of greatest importance will be those scientific results which are obtained from unexpected trends.

In the field of control of production, substantial successes have been achieved so far only in automating technological processes. However, such important stages of production as the design of components, structures and systems, preparation and organisation of production as well as organisation of the supply of goods to consumers, both the processes themselves and their control, are still fundamentally in the domain of intuitive methods. Only the first steps are being made to use scientifically justified cybernetic methods of controlling such processes. It is clear even now that problems such as finding optimal solutions during the development of the project can be formalised and carried out by computer more efficiently than manually. In principle the possibility to formalise exists also in the sphere of organisational problems, which arise in all stages of manufacture and at all ranges of the hierarchy of production control starting from the production team, and ending with a very large production unit (combine).

It is possible to visualise in the future a control system for a branch of manufacture designed on the basis of information obtained from a rational and highly automated system of collection and transmission of information on the state of all the elements of the controlled system and on the external (market) situation, consisting of controlling centres with a hierarchical structure which are provided with a system of inter-connected machines for processing the information and giving recommendations to the supervisors who are responsible for specific sub-divisions or for the branch of industry as a whole.

Although systems of this type are possible, and may be advisable, it is obvious that one cannot overestimate the technical difficulties which have to be overcome before such systems can be used effectively and generally. So far no convenient criteria exist for selecting the necessary and adequate information on the state of the system and the medium; methods of optimisation of the operation of very complicated complexes have not been sufficiently developed; scientific methods of formulating optimisation criteria are in the first stages of development; the problem of reliability of operation of complicated complexes has not been adequately investigated; much remains to be done on the problem of lucid languages to be used for exchanging information between individual machines entering the system, and between machines and man.

The realisation of an automatic optimal control system by complicated complexes is virtually unachievable by existing methods because the necessary number of interconnections for a system of the type of an industrial plant runs into astronomical figures. An enumeration of these interconnections would exceed in volume the *Encyclopaedia Britannica*. It is obvious that such a system cannot be designed and even less can it be built.

However, the problem should not be considered as insoluble in principle because such (and even more) complex control organs exist in nature. The brain of man, and even of an animal, fulfils much more complex functions than are required for controlling any production plant, transportation or power system, etc. However, as was shown in the previous chapters, the processes of self-organisation and learning, which produce the high perfection of the living control systems can be realised, and are even being realised, in artificial systems. Therefore, there is reason to hope that further development of work and deeper knowledge of the mechanism of self-organisation and advances in building learning machines open up avenues for 'growing from the embryonic state' towards artificial control systems with unprecedented capabilities and the ability, after appropriate learning and training, to solve the grandiose problems involved in controlling very large systems.

Important and difficult problems arise in the field of *control of resources*. Any economic system requires manoeuvring with resources which are employed in its process of operation: raw materials, power, fuel, materials, equipment, finance and manpower. Since the effectiveness of a system depends strongly on how the resources are distributed, the following problem arises. It is necessary to find and realise for each given state of the system and the medium such a distribution of the limited resources for which the criterion of effectiveness of operation will assume the optimum value. The difficulties involved in solving this problem are associated with the high dimension of the state space of the system, the presence of a large number of local extrema, and the fact that limitations which determine the quantity of the resources of the system change with the progress of time.

Further advances in unravelling the mechanisms of the control processes in living organisms are of enormous importance for *agriculture* and *medicine*. If the secrets of these processes were unravelled we would gain new potentialities in controlling the development of plants and animals, controlling heredity, ensuring rapid breeding of productive species of plants and animals. Even greater is the importance of advances in such problems as healing hereditary diseases, cancer and other diseases associated with disruption of the control mechanism in the living cell and in particular of work associated with combating ageing and solution of problems relating to prolongation of life.

In medical diagnostics, cybernetic methods are likely to become of dominating importance. In many cases dozens of factors have to be taken into consideration for a diagnosis and a man is incapable in principle to do this with the required accuracy. On the other hand, methods used on teaching pattern recognition to machines may become very helpful to doctors and put the diagnosis of diseases on a strong mathematical foundation.

Treatment of many diseases can be considered as control of processes occurring in the body of the patient, and the measures of treatment as controlling actions applied to some complex dynamic system. In this process many important facts established in the theory of control of dynamic systems can be used for working out tactics and strategy for treatment. It is obvious that it will be necessary to take into consideration the extreme complexity, the inadequate knowledge and the specific features of the system which is the living body, but many fundamental relationships of the movement of controlled dynamic systems may prove very useful for evolving such a new approach to the theory and practice of medicine.

The field of *sociological problems* is an extremely important but hardly developed branch of cybernetics. To reveal the laws governing the control of the behaviour of people which form the human communities should be one of the aims of the science which aspires to become a general control theory. It may transpire that the arsenal of the controlling actions which include education, incentive and punishment can be enriched and the utilisation of the results be both more effective and more humane. An accurate analysis will reveal superfluous limitations on the behaviour of people imposed by laws and habits, and dropping these would make the life of people freer and happier.

## 19.2. New means

Even now existing means for information processing do not entirely satisfy practical requirements as regards the fundamental indices: fastness of response, memory capacity and reliability. This discrepancy may increase due to the extreme complexity of the new problems if we do not manage to achieve considerable progress in the design of machines.

An increase in the fastness of performing operations in machines is limited in principle because the speed of movement of the signals cannot exceed the speed of light. Assuming that the minimal length of path of a signal for performing one operation is of the order of one metre, then the maximum speed of response of the machine can be $3.10^8$ operations per second. But even such a high speed is inadequate for solving many problems of a combined nature which require sifting through a large number of alternatives of solution. A substantial increase in the capabilities of the machines should not be expected from increasing the speed of operation of the machine but from changing over from the principle of sequential completion of operations to parallel operation of elements which will ensure simultaneous execution of a large number of operations.

Apparently, the capability of machines can also be increased significantly by using new principles of distribution of information in the memory of the machine and methods of extracting the required data from the memory. Methods of recording and searching data in contemporary machines are very primitive and cannot be compared with the properties of the memory in a living brain. Consequently, improvement of the structure of memory devices will open up considerable, hitherto untapped, potentialities.

A large number of control problems is associated with the study of processes in dynamic systems. For solving problems of this type 'hybrid' machines, combining the principles of operation of analog and digital computers, may prove very effective. In such machines use of analog models permits reducing to a fraction the number of operations which are necessary for obtaining a solution, and the logic and the programmed control give such machines both flexibility and universality.

Even now the reliability of information processing equipment can be made higher than the reliability of the individual elements from which the equipment is built up. This is achieved by including an excess number of elements and, by providing reserves, it is possible to ensure normal functioning of the equipment in spite of failure of individual elements. In future substantial progress can be anticipated in increasing the reliability, on the basis of using another principle – the principle of self-restoration properties of equipment. In such equipment broken communications would automatically become re-established and occurring parasitic connections would be broken. This can be achieved by building elements which have the property of self-restoration, or by using an automatic repair block, which monitors the operation of the individual elements and performs repair or substitution of elements that failed.

The above considerations indicate that many hitherto unutilised reserves exist for improving information processing facilities which will bring about a considerable improvement. However, all this applies to machines with the traditional structure built by traditional methods. Radical changes in the technique of building equipment for information processing equipment may be possible by using particular methods based on principles of self-reproduction, self-perfection and self-organisation. To realise a machine in accordance with these principles, it is necessary to have available a 'medium' which is the source of elementary structural units similar to the manner in which living organisms distribute the food drawn from the surrounding medium and provide the body with molecules that are used for building its organs. Extraction of the required 'molecules' and combination of these can be organised either by means of a definite process – on the basis of a project, a programme or a 'specimen' of a copying machine, or by means of a stochastic process – by encouraging 'useful' and suppressing 'harmful' compounds which occur at random and are evaluated on the basis of appropriate criteria. 'The food' or 'the medium' for such a method of construction (building) of machines can be a store (supply) of finished elements which realise certain logical functions or a network of elements of a single type in which it is only necessary to break the excessive number of connections, or ultimately liquid media in which the necessary structure can form under the effect of physical fields. Although many difficulties have still to be overcome before such methods of designing machines can be realised, in principle the possibility of building such machines is well founded and there is no doubt that it can be done.

A very attractive long-term prospect is the construction of information processing equipment based on hybrid 'mechanical-organic' systems in which an artificial (for instance, a mechanical) 'skeleton' will ensure pre-determined characteristics of the structure whilst living self-organising components will give the system the necessary flexibility for functioning and adapting itself to changing conditions.

### 19.3 The social importance of automation

However great the importance of cybernetics is for individual branches of human activity outlined in the previous chapters, undoubtedly the greatest influence of cybernetics on the development of society so far is in the automation of production. Due to rapid quantitative expansion and a continuously improving quality, automation leads to very radical social shifts, which are unprecedented in history, and are similar to some extent to the first industrial revolution, which was associated in changing over to machine production methods. However, the second industrial revolution is incomparably more grandiose in its consequences. The problems associated with this development are so complex and varied that we have to limit ourselves to the briefest notes on one of the more important of these, namely, a solution of the contradiction between *creative* and *mechanical* work and liquidation of the division of labour generally.

From the time when man evolved from the animal world work was a condition of existence of the human society. Man had to work in the first instance to satisfy his material needs. However, with the development of productive forces the relationships in production changed and accordingly man himself changed. At a certain stage of the social and individual development work becomes a requirement of a healthy man. However, not every type of work is a requirement for an adequately developed person. It is of importance whether the work is creative or mechanical. Creative and mechanical work differ sharply, primarily as regards the process itself. Creative work is distinguished by being non-repetitive and original at each stage. This non-repetitiveness and individuality is not something external, random or compulsory – on the contrary, it is the essence of creative work; only these qualities condition the results of the creative work and form the content of this result. They are the external expression of creativeness, i.e. an activity which produces *essentially* new, original things of value. Really creative work is unthinkable without continuous development of the personality as a whole, without a clear cognition of the entire processes and the ultimate aims of work. Creative work can be (but is by no means always) work on science, arts, teaching or social activity.

In contrast to this 'mechanical' work is distinguished by monotony. The result of mechanical work is achieved by repeating a known sequence of operations. The available elements of originality usually are of an external, random and in the best case auxiliary nature, they do not determine the result of the work and are not a necessary means for achieving the result. Due to its dulling nature it creates in a man a revulsion which is the greater the more the man is developed. Elimination of mechanical work is the task facing modern mankind.

Obviously mechanical work is the necessary consequence of the direct participation of man in productive work, in mass reproduction of material goods, which requires a single type of precisely predetermined sequences of operations. The invention of the first stone axe was a creative act but all the subsequent axes (we do not speak now of further improvements) were manufactured by mechanical labour.

It is therefcre to this end of removing the tedious and dull work from man's shoulders and leaving him free to be human, that cybernetics is dedicated.

## 19.4. Real and imaginary dangers

Estimating the long term prospects of development of cybernetics its founder, Norbert Wiener, expressed the serious fear of a possibility that a group of dictators may get hold of power by means of cybernetic machines which would be capable of controlling individual countries or our entire planet. Fear against the consequences of 'mechanisation' of control was also voiced by the Dominican monk, Dubarel, who wrote the following when Wiener's book *Cybernetics and Society* first appeared:

"Today we are exposed to the risk of creating a mammoth 'world state', where a circumspect and conscious primitive injustice can be the only possible condition of statistical happiness for the masses; the creation of a world which for people in their right senses is worse than hell" (Le Monde, 28 December, 1948).

It would be wrong to simply conceal these opinions. The real situation is sufficiently serious and requires careful analysis, causing us to discard superficial and unjustified optimism. It is easy to understand the anxiety of leading scientists all over the world who fear the social consequences of a rapid development of science and engineering. Never before was the danger so great that the gigantic forces of nature – 'the evil spirit in the bottle' released by man – may turn against him under certain circumstances.

The history of the 20th century contains examples of human solidarity, humanity, fighting for freedom and social justice side by side with the most bestial dictatorship; antibiotics and fighting old age, with gas chambers and concentration camps; triumphs of reason with 'witch hunting', victimising anybody who has a free thought; penetration into the secrets of the universe and the micro-world with hydrogen bombs; fighting for improving the standard of living with a ruinous arms race and military space 'programmes'.

What will mankind benefit from improving cybernetic machines? At present it is hardly possible to give a clear answer to this question. In final analysis, much will depend on mankind itself: on its social maturity, foresight, sense of responsibility, intellect and will.

The question of the danger of mankind being displaced or suppressed by a 'race' of machines, intellectually superior to man can be answered relatively easily. Firstly, the potentialities of people are by no means invariable. In particular, the use of 'amplifiers of understanding' as an extension of the human brain will lift the thinking of man onto a quantitatively new level. The main feature is that those who engage in fantasy on the allegedly inevitable hard fight between people and 'thinking' machines, transfer of the relation between clever beings, much more highly developed than the present-day man, the experience of relationships drawn from the class period of the history of mankind. These are the fight for existence, competition, domination and slavery – concepts which are even now opposed by all progressively thinking people. By pointing out that these attitudes are age old and are characteristic of human nature, the authors of similar fantasies only discover their own limitations. There is no reason to doubt that the future inhabitants of the earth – man and his creations, the thinking machines – will find a reasonable and favourable method of relationships, worthy of such intelligent beings. In any case this problem, even if it will exist, will not be a problem of the near future.

Much more real is the danger of capture of government by an oligarchy using cybernetic machines to achieve their aims. 'The danger to society of machines does not originate from the machine itself, but from its use by man,' said Wiener, *Machine the Master* of Father Dubarel is terrifying not because it may achieve automatic control of mankind . . . . The real danger is that similar machines, although harmless by themselves can be used by man or a group of people for increasing their power on (over) the rest of mankind, or that the political leaders may attempt to control their people not by the machines themselves but by political techniques so very narrow and indifferent to human potentialities, giving the impression that these techniques were indeed produced mechanically' (Norbert Wiener, *Cybernetics and Society*).

Progress in science and technology imposes serious problems on mankind, in particular on scientists themselves. But we cannot hold back science on that account, since it is a product of man's natural curiosity.

If there had been sceptics with a lot of foresight, they would have said that the invention of a stone axe has provided a very dangerous toy which ultimately would lead to the invention of missiles with atomic warheads. It is, however, clear that the possibility of using the fruits of man's genius for evil does not mean that these should not be generated; it only evidences the necessity of fighting for these achievements to be used for good and not for evil. History shows that despite the danger of fire, gunpowder, electricity, atomic energy and many other 'dangerous' inventions, progressive forces of society have managed to control these elements.

The correct conclusion to be drawn from all that has been said above, is that in our time nobody – and in particular not the intellectuals – can shelve responsibility from scientific analysis of the development of society, from influencing actively and purposefully this process; they cannot transfer to other people the responsibility for the social consequences of their activities and for the entire course of modern development of society.

It is incorrect to conclude that society should not use computers for development and realisation of a programme of controlling its own development. The long-term freedom consists of recognising and utilising objective relationships, which is impossible without machines, in view of the enormous quantity of information which has to be processed rapidly and in a complex manner for the purpose of revealing the main tendencies, finding optimal solutions, etc. Control by means of machines will be more perfect than control without machines; ignorance or lack of skill has never provided an advantage in any type of work. It is only necessary to realise that the solutions suggested by the machines may harbour unexpected and fateful consequences similar to the magic talisman from the story of Jacob *The Monkey's Paw*, which fulfils the desires of its master in a ruinous manner. Therefore there is nothing to free man from his responsibility to make choices if he does not want to achieve a society whose stability is maintained by 'circumspect and conscious injustice'.

All this means that we cannot entirely and blindly accept solutions which were adopted on the basis of a criterion which differs from our ideals, our concepts of a desirable organisation of society, on human relations and man itself regardless of who made the decision, whether it is an electronic brain or a bureaucratic machine. The guarantee of progress of human society is in active, free and independent efforts of all thinking and progressive people.

The future depends on our present efforts, on the depth of our understanding of the objective development of society, on our courage, willpower and ability to act. We want to see in the future a really happy society of free and good and noble descendants. Then we will be their worthy ancestors!

# Bibliography

**Chapter 1**

1. Berg, A. I. Cybernetics – The Science of Optimal Control. *Energiya*, 1964.
2. Wiener, N. *Cybernetics or Control and Communication in the Animal and the Machine*. MIT Press and Wiley, 1948 and 1961.
3. Lyapunov, A. A. On some general problems of Cybernetics. Coll. *Problems of Cybernetics*. No. 1. Fizmatgiz, 1959.
4. Poletayev, I. A. The Signal. *Sov. radio*, 1958.
5. Encyclopaedia *Automation of Production and Industrial Electronics. Sovetskaya entsiklopediya*, 1962-1965.
6. Ashby, W. R. *An Introduction to Cybernetics*. Chapman and Hall, 1965.

**Chapter 2**

1. Landau, L. D., Lifshits, E. M. *Mechanics*. Translated from Russian. Pergamon.
2. Peierls, R. E. *The Laws of Nature*, 1955.
3. Engels, F. *Dialectics of Nature*.
4. Yablonskiy, S. V. Fundamental notions of Cybernetics. Coll. *Problems of Cybernetics*. No. 2. Fizmatgiz, 1959.

**Chapter 3**

1. Beer, S. *Cybernetics and Management*. English Universities Press, 1959.
2. Kogan, B. Ya. *Electronic Modelling Devices and their Application to the Study of Automatic Control Systems*. 2nd ed. Fizmatgiz, 1963.
3. Moore, E. F. Speculative experiments with sequential machines. Translated from English. Coll. *Automata*, IL, 1956.
4. Mikheyev, M. A., Kolpakov, P. K. (eds.) *The Theory of Similarity and Modelling*. Izd-vo AN SSSR, 1951.

**Chapter 4**

1. Andronov, A. A., Vitt, A. A., Khaykin, S. E. *The Theory of Oscillations* . 2nd ed. Fizmatgiz, 1959.
2. Landau, L. D., Lifshits, E. M. *Mechanics*. Translated from Russian. Pergamon.
3. Nemytskiy, V. V., Stepanov, V. V. The qualitative theory of differential equations. Ch. 5 of *General Theory of Dynamic Systems*. Gostekhizdat, 1947.

**Chapter 5**

1. Brillouin, L. *Science and Information Theory*. Academic Press, 1967.
2. Fano, R. M. *Transmission of Information: a Statistical Theory of Communication*. MIT Press, 1961.
3. Kharkevich, A. A. *Outline of the General Theory of Communication.* Gostekhizdat, 1955.
4. Shannon, C. E., Weaver, W. *Mathematical Theory of Communication*. Univ. of Illinois Press, 1949.
5. Yaglom, A. M., Yaglom, I. M. *Probability and Information*. Fizmatgiz, 1960.

**Chapter 6**

1. Lerner, A. Ya. *Introduction to the Theory of Automatic Control.* Mashgiz, 1958.
2. Fel'dbaum, A. A., Dudykin, A. D., Manovtsev, A. P., Mirolyubov. *Basic Theory of Communication and Control.* Fizmatgiz, 1963.

**Chapter 7**

1. Lerner, A. Ya. *Introduction to the Theory of Automatic Control.* Mashgiz, 1958.
2. Pugachev, V. S. (ed.) *Fundamentals of Automatic Control.* Fizmatgiz, 1963.
3. Hammond, P. *The Theory of Feedback and its Applications.*

**Chapter 8**

1. Bellman, R. *Dynamic Programming.* Princeton University Press, 1957.
2. Ventsel', E. S. *Elements of Dynamic Programming.* "Nauka", 1964.
3. Lerner, A. Ya. *Principles of the Construction of Fast-Response Servosystems and Regulators.* Energoizdat, 1961.
4. Pontryagin, L. S., Boltyanskiy, V. G., Gamkrelidze, R. V., Mishchenko, E. F. *The Mathematical Theory of Optimal Processes.* Fizmatgiz, 1961.
5. Fel'dbaum, A. A. *Fundamentals of the Theory of Optimal Systems.* Fizmatgiz, 1963.

**Chapter 9**

1. Shannon, C. E., McCarthy, J. (eds.) *Automata Studies.* Princeton University Press, 1956.
2. Gaaze-Rapoport, M. G. *Automata and Living Organisms.* Fizmatgiz, 1961.
3. Kobrinskiy, N. E., Trakhtenbrot, B. A. *Introduction to the Theory of Finite Automata.* Fizmatgiz, 1962.
4. Murray, F. J. *Mechanisms and Automata.* Translated from English. *Kiberneticheskiy sbornik*, No. 1, IL, 1960.
5. Gavrilov, M. A. (ed.) *Structural Theory of Relay Devices.* Collection of articles. Izd. AN SSSR, 1963.
6. Trakhtenbrot, B. A. *Algorithms and Solution of Problems by Machine.* Gostekhizdat., 1957.

**Chapter 10**

1. Kitov, A. I. Electronic Digital Computers. *Sov. radio*, 1956.
2. Kitov, A. I., Krinitskiy, N. A. *Electronic Digital Computers and Programming.* 2nd. ed. Fizmatgiz, 1961.
3. Murphy, G. J. *Electronic Digital Computers.* Van Nostrand. 1959.
4. Turing, A. M. Computing Machinery and Intelligence. *Mind*, 1950, **59**, 433–460.

**Chapter 11**

1. Bellman, R. *Adaptive Control Processes: A Guided Tour*. Princeton University Press, 1961.
2. Tsien, H. S. *Engineering Cybernetics*. McGraw-Hill, 1954.
3. Ashby, W. R. *Design for a Brain*. 2nd ed. Chapman and Hall, 1960.

**Chapter 12**

1. Williams, J. D. *The Complete Strategist*. McGraw-Hill, 1954.
2. Ventsel', E. S. *Elements of the Theory of Games*. Fizmatgiz, 1959.
3. Luce, R. D., Raiffa, H. *Games and Decisions*. Wiley, 1957.
4. McKinsey, J. C. C. *Introduction to the Theory of Games*. McGraw-Hill, 1952.
5. Shannon, C. E. Game-Playing Machines. Transl. from English. *Kiberneticheskiy sbornik*, No. 1, IL, 1960.
6. Shannon, C. E. A Chess-Playing Machine. Transl. from English. Coll. *Papers on Cybernetics and Information Theory*. IL, 1963.

**Chapter 13**

1. Bush, R. R., Mosteller, F. *Stochastic Models for Learning*. Wiley, 1955.
2. Vannik, V. N., Lerner, A. Ya., Chervonenkis, A. Ya. Systems of learning pattern recognition by means of generalised portraits. *Tekhnicheskaya kibernetika,* 1965, No. 1.
3. Tseytlin, M. L. Teaching Stochastic Automata. *Avtomatika i telemekhanika*, 1961, No. 10.
4. Shannon, C. E. Presentation of a Maze-Solving Machine. *Trans. of the Eighth Cybernetics Conference*, Josiah Macy Found, 1952, pp. 173–180.

**Chapter 14**

1. Beer, S. Cybernetics and Management. English Universities Press, 1959.
2. Buslenko, N. P. Mathematical Models of Production Processes. *Nauka*, 1964.
3. Goode, H. H., Machol, R. E. *System Engineering*. McGraw-Hill, 1957.

**Chapter 15**

1. Zukhovitskiy, S. I., Radchik, A. I. Mathematical Methods of Network Planning. *Nauka*, 1965.
2. Burkov, V. N. et al. Network Models and Control Problems. *Sov. radio*, 1967.
3. Rivett, B. H. P., Ackoff, R. L. *A Manager's Guide to Operational Research*. Wiley, 1963.
4. Tsypkin, Ya. Z. Adaptation, Teaching and Self-Teaching in Automatic Systems. *Avtomatika i telemekhanika*, 1966, No. 1.

**Chapter 16**

1. Bernshteyn, N. A. *Ways of Developing Physiology and Problems of Cybernetics Connected with Them*.
2. Braynes, S. N., Napalkov, A. V., Svechinskiy, V. B. *Scientific Reports* (on Neurocybernetics). Izd-vo Mosk. un-ta, 1959.
3. George, F. H. *The Brain as a Computer*. Pergamon Press, 1961.
4. Cybernetics and the Living Organism. Translated from the English. *Naukova dumka*, 1964.
5. Kleene, S. C. Representation of events in neural networks and finite automata. Coll. *Automata*. Translated from English. IL, 1956.
6. Kolmogorov, A. N. Life and thought from the viewpoint of Cybernetics. In the book: Oparin A. I. *Life and Its Relationships with Other Forms of Movement of Matter*. AN SSSR, 1962.

7. Neuman, John von. *The Computer and the Brain*. Yale University Press, New Haven, 1958.
8. Rosenblatt, F. *Principles of Neurodynamics*. Spartan Books, Washington, 1962.
9. Ashby, W. R. *Design for a Brain*. 2nd ed. Chapman and Hall, 1960.

**Chapter 17**

1. Brillouin, L. *Science and Information Theory*. Academic Press, 1967.
2. Foerster, H. von, Zopf, G. W. (eds.) *Principles of Self-organization*. Pergamon Press, New York, 1962.
3. Fel'dbaum, A. A. (ed.) Self-learning Automata. *Nauka*, 1966.
4. Yovits, M. C. (ed.) *Self-organizing Systems*. Proc. of an Interdisciplinary Conference, May 1959. Pergamon Press, New York, 1960.
5. Schroedinger, E. *What is Life? The Physical Aspect of the Living Cell*. 1944.
6. Ashby, W. R. Design for an Intelligence-Amplifier. In *Automata Studies*, Princeton, 1956, 215–234.
7. Ashby, W. R. *Design for a Brain*. 2nd edition. Chapman and Hall, 1960.
8. Gel'fand, I. M. et al. Models of the Structural-Functional Organization of Some Biological Systems. *Nauka*, 1966.

**Chapter 18**

1. Engineering Psychology. Translated from English. *Progress*, 1964.
2. Lilly. *Automation and Social Progress*. Translated from English. IL, 1958.
3. Lomov, B. F. *Man and Engineering*. Izd-vo LGU, 1963.
4. Gaaze-Rapoport, M. G., Lerner, A. Ya, Oshanin, D. A. (eds.) The System 'Man – Automaton'. *Nauka*, 1965.

**Chapter 19**

1. Wiener, N. *Cybernetics or Control and Communication in the Animal and the Machine*. Wiley, 1961.
2. Wiener, N. *Cybernetics and Society*. Houghton Mifflin, Boston, 1954.
3. Wiener, N. New Chapters of Cybernetics. Translated from English. *Sov. radio*, 1963.
4. Berg, A. I. (ed.) Cybernetics, Thinking, Life. Collection of articles. *Mysl'*, 1964.
5. Thomson, G. *The Foreseeable Future*. 1955.
6. Il'yin, V. A. et al. (eds.) *The Philosophical Problems of Cybernetics*. Sotsekgiz, 1961.
7. Shklovskiy, I. S. The Universe, Life, Intelligence. 2nd edition. *Nauka*, 1965.
8. Engels, F. *Anti-Dühring*.
9. Encyclopaedia *Automation of Production and Industrial Electronics*. *Sovetskaya entsiklopediya*, 1962–1965.

# Index

Bold numbers indicate the page where the definition of the respective term is given.